Instrumentation
Level Four

Trainee Guide
Third Edition

PEARSON

Boston Columbus Indianapolis New York San Francisco Amsterdam
Cape Town Dubai London Madrid Milan Munich Paris Montreal Toronto Delhi
Mexico City Sao Paulo Sydney Hong Kong Seoul Singapore Taipei Tokyo

NCCER

President: Don Whyte
Vice President: Steve Greene
Chief Operations Officer: Katrina Kersch
Director of Product Development: Daniele Dixon
Instrumentation Project Managers: Jamie Carroll, Justin Street
Senior Development Manager: Mark Thomas

Senior Production Manager: Tim Davis
Quality Assurance Coordinator: Debie Hicks
Desktop Publishing Coordinator: James McKay
Permissions Specialist: Adrienne Payne
Production Specialist: Adrienne Payne
Editor: Karyn Payne

Writing and development services provided by Topaz Publications, Liverpool, NY

Lead Writer/Project Manager: Don Congdon
Desktop Publisher: Joanne Hart
Art Director: Alison Richmond

Permissions Editors: Andrea LaBarge, Briana Morgan
Writer: Don Congdon

Pearson

Head of Global Certifications/Associations: Andrew Taylor
Editorial Assistant: Collin Lamothe
Program Manager: Alexandrina B. Wolf
Digital Studio Project Managers: Heather Darby, Tanika Henderson, Jose Carchi
Director of Marketing: Leigh Ann Simms

Senior Marketing Manager: Brian Hoehl
Composition: NCCER
Printer/Binder: LSC Communications
Cover Printer: LSC Communications
Text Fonts: Palatino and Univers

Credits and acknowledgments for content borrowed from other sources and reproduced, with permission, in this textbook appear at the end of each module.

Copyright © 2016, 2003, 1992 by NCCER, Alachua, FL 32615, and published by Pearson Education, Inc., New York, NY 10013. **All rights reserved.** Printed in the United States of America. This publication is protected by Copyright and permission should be obtained from NCCER prior to any prohibited reproduction, storage in a retrieval system, or transmission in any form or by any means, electronic, mechanical, photocopying, recording, or likewise. For information regarding permission(s), write to: NCCER Product Development, 13614 Progress Blvd., Alachua, FL 32615.

PEARSON

Perfect bound ISBN-13: 978-0-13-449532-3
ISBN-10: 0-13-449532-2

Preface

To the Trainee

Instrumentation is a specialized, unique field that is critical to the success of a variety of construction ventures. Instrumentation fitters and technicians are charged with the vital tasks of assembling and installing instruments. These instruments can range from simple to complex, and can be part of control systems in enterprises such as refineries and factories.

This third edition of *Instrumentation Level Four*, revised by a diverse group of subject matter experts from across the nation, covers the fundamentals of the instrumentation trade. The four levels of the *Instrumentation* curriculum present an apprentice approach to the trade and will help you to be knowledgeable, safe, and effective on the job.

This edition of *Instrumentation* presents a training order that is tailored for instrumentation fitters and instrumentation technicians. Stackable credentials from NCCER are available as follows:

- *Instrumentation Fitter credential* – Completion of Level One, Level Two, and the first two modules of Level Three (12207-16 and 12205-16)
- *Instrumentation Technician credential* – Completion of all four levels of the third edition

Please note: Trainees must be taught by NCCER-certified instructors in order to earn credentials through NCCER.

New with *Instrumentation Level Four*

NCCER is proud to release this edition of *Instrumentation Level Four* in full color, with engaging updates to the curriculum. The design of this edition has changed to better align with the learning objectives. There are also new end-of-section review questions to complement the module review. The text, graphics, and special features have been enhanced to reflect advancements in instrumentation technology and techniques.

As part of the revision process for *Instrumentation Level Four*, the module titled *Analyzers and Monitors*, formerly in Level Three, was combined with the Level Four *Analyzers* module. The modules *Performing Loop Checks*, and *Troubleshooting and Commissioning a Loop* were combined into a new module, *Proving, Commissioning, and Troubleshooting a Loop*.

We hope you continue your training beyond this textbook to further your knowledge and skills in this field.

If you're training through an NCCER-Accredited Training Program Sponsor and successfully pass the module exams and performance tests in this course, you may be eligible for credentialing through the NCCER Registry. Check with your instructor or local program sponsor to find out. To learn more, go to **www.nccer.org** or contact us at 1.888.622.3720.

We invite you to visit the NCCER website at **www.nccer.org** for information on the latest product releases and training, as well as online versions of the *Cornerstone* magazine and Pearson's NCCER product catalog.

Your feedback is welcome. You may email your comments to **curriculum@nccer.org** or send general comments and inquiries to **info@nccer.org**.

NCCER Standardized Curricula

NCCER is a nonprofit 501(c)(3) education foundation established in 1996 by the world's largest and most progressive construction companies and national construction associations. It was founded to address the severe workforce shortage facing the industry and to develop a standardized training process and curricula. Today, NCCER is supported by hundreds of leading construction and maintenance companies, manufacturers, and national associations. The NCCER Standardized Curricula was developed by NCCER in partnership with Pearson, the world's largest educational publisher.

Some features of the NCCER Standardized Curricula are as follows:

- An industry-proven record of success
- Curricula developed by the industry for the industry
- National standardization providing portability of learned job skills and educational credits
- Compliance with the Office of Apprenticeship requirements for related classroom training (*CFR 29:29*)
- Well-illustrated, up-to-date, and practical information

NCCER also maintains a Registry that provides transcripts, certificates, and wallet cards to individuals who have successfully completed a level of training within a craft in NCCER's Curricula. *Training programs must be delivered by an NCCER Accredited Training Sponsor in order to receive these credentials.*

Special Features

In an effort to provide a comprehensive and user-friendly training resource, this curriculum showcases several informative features. Whether you are a visual or hands-on learner, these features are intended to enhance your knowledge of the construction industry as you progress in your training. Some of the features you may find in the curriculum are explained below.

Introduction

This introductory page, found at the beginning of each module, lists the module Objectives, Performance Tasks and Trade Terms. The Objectives list the skills and knowledge you will need in order to complete the module successfully. The Performance Tasks give you an opportunity to apply your knowledge to real-world tasks. The list of Trade Terms are words and phrases that will become part of your vocabulary as you study this module.

Figures and Tables

Photographs, drawings, diagrams, and tables are used throughout each module to illustrate important concepts and provide clarity for complex instructions. Text references to figures and tables are emphasized with *italic* type.

Notes, Cautions, and Warnings

Safety features are set off from the main text in highlighted boxes and categorized according to the potential danger involved. Notes simply provide additional information. Cautions flag a hazardous issue that could cause damage to materials or equipment. Warnings stress a potentially dangerous situation that could result in injury or death to workers.

Trade Features

Trade features present technical tips and professional practices from the construction industry based on real-life scenarios similar to those you might encounter on the job site.

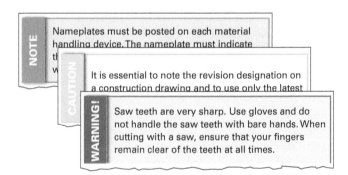

Bowline Trivia

Some people use this saying to help them remember how to tie a bowline: "The rabbit comes out of his hole, around a tree, and back into the hole."

Case History

Case History features emphasize the importance of safety by citing examples of the costly (and often devastating) consequences of ignoring best practices or OSHA regulations.

Going Green

Going Green features present steps being taken within the construction industry to protect the environment and save energy, emphasizing choices that can be made on the job to preserve the health of the planet.

Dis You Know

The *Did You Know* features introduce historical tidbits or interesting and sometimes surprising facts about the trade.

Step-by-Step Instructions

Step-by-step instructions are used throughout to guide you through technical procedures and tasks from start to finish. These steps show you not only how to perform a task but also how to do it safely and efficiently.

> Perform the following steps to erect this system area scaffold:
>
> *Step 1* Gather and inspect all scaffold equipment for the scaffold arrangement.
>
> *Step 2* Place appropriate mudsills in their approximate locations.
>
> *Step 3* Attach the screw jacks to the mudsills.

Trade Terms

Each module presents a list of Trade Terms that are discussed within the text and defined in the Glossary at the end of the module. These terms are denoted in the text with bold, blue type upon their first occurrence. To make searches for key information easier, a comprehensive Glossary of Trade Terms from all modules is located at the back of this book.

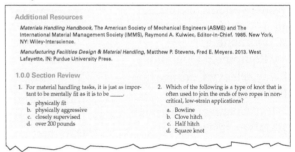

Section Review

Each section of the module wraps up with a list of Additional Resources for further study and Section Review questions designed to test your knowledge of the Objectives for that section.

Review Questions

The end-of-module Review Questions can be used to measure and reinforce your knowledge of the module's content.

NCCER Standardized Curricula

NCCER's training programs comprise more than 80 construction, maintenance, pipeline, and utility areas and include skills assessments, safety training, and management education.

Boilermaking
Cabinetmaking
Carpentry
Concrete Finishing
Construction Craft Laborer
Construction Technology
Core Curriculum: Introductory Craft Skills
Drywall
Electrical
Electronic Systems Technician
Heating, Ventilating, and Air Conditioning
Heavy Equipment Operations
Highway/Heavy Construction
Hydroblasting
Industrial Coating and Lining Application Specialist
Industrial Maintenance Electrical and Instrumentation Technician
Industrial Maintenance Mechanic
Instrumentation
Insulating
Ironworking
Masonry
Millwright
Mobile Crane Operations
Painting
Painting, Industrial
Pipefitting
Pipelayer
Plumbing
Reinforcing Ironwork
Rigging
Scaffolding
Sheet Metal
Signal Person
Site Layout
Sprinkler Fitting
Tower Crane Operator
Welding

Maritime

Maritime Industry Fundamentals
Maritime Pipefitting
Maritime Structural Fitter

Green/Sustainable Construction

Building Auditor
Fundamentals of Weatherization
Introduction to Weatherization
Sustainable Construction Supervisor
Weatherization Crew Chief
Weatherization Technician
Your Role in the Green Environment

Energy

Alternative Energy
Introduction to the Power Industry
Introduction to Solar Photovoltaics
Introduction to Wind Energy
Power Industry Fundamentals
Power Generation Maintenance Electrician
Power Generation Instrument and Controls Maintenance Technician
Power Generation Maintenance Mechanic
Power Line Worker
Power Line Worker: Distribution
Power Line Worker: Substation
Power Line Worker: Transmission
Solar Photovoltaic Systems Installer
Wind Turbine Maintenance Technician

Pipeline

Control Center Operations, Liquid
Corrosion Control
Electrical and Instrumentation
Field Operations, Liquid
Field Operations, Gas
Maintenance
Mechanical

Safety

Field Safety
Safety Orientation
Safety Technology

Supplemental Titles

Applied Construction Math
Tools for Success

Management

Fundamentals of Crew Leadership
Project Management
Project Supervision

Spanish Titles

Acabado de concreto: nivel uno (*Concrete Finishing Level One*)
Aislamiento: nivel uno (*Insulating Level One*)
Albañilería: nivel uno (*Masonry Level One*)
Andamios (*Scaffolding*)
Carpintería: Formas para carpintería, nivel tres (*Carpentry: Carpentry Forms, Level Three*)
Currículo básico: habilidades introductorias del oficio (*Core Curriculum: Introductory Craft Skills*)
Electricidad: nivel uno (*Electrical Level One*)
Herrería: nivel uno (*Ironworking Level One*)
Herrería de refuerzo: nivel uno (*Reinforcing Ironwork Level One*)
Instalación de rociadores: nivel uno (*Sprinkler Fitting Level One*)
Instalación de tuberías: nivel uno (*Pipefitting Level One*)
Instrumentación: nivel uno, nivel dos, nivel tres, nivel cuatro (*Instrumentation Levels One through Four*)
Orientación de seguridad (*Safety Orientation*)
Paneles de yeso: nivel uno (*Drywall Level One*)
Seguridad de campo (*Field Safety*)

Acknowledgments

This curriculum was revised as a result of the farsightedness and leadership of the following sponsors:

ABC Bayou Chapter
ABC Pelican Chapter
Baton Rouge Community College
Cianbro Corporation
Excel
Goodwill
ISC Constructors, L.L.C.
Jacobs Field Services

JEM Electrical Consulting Services
MMR Constructors, Inc.
Testronics
The College of Southern Maryland
The Robins and Morton Group
TIC – The Industrial Company
Total Petrochemicals & Refining USA, Inc.

This curriculum would not exist were it not for the dedication and unselfish energy of those volunteers who served on the Authoring Team. A sincere thanks is extended to the following:

Zach Boudreaux
John Clouatre
Keith Gautreau
Brenton Miller
Jim Mitchem
Janice "Dee" Morgan
Nick Musmeci

Tom Osteen
Mike Raven
Richard Sanders
Howard Smith
Jack Turner
Bruce Wall
Neal Zimmerman

NCCER Partners

American Council for Construction Education
American Fire Sprinkler Association
Associated Builders and Contractors, Inc.
Associated General Contractors of America
Association for Career and Technical Education
Association for Skilled and Technical Sciences
Construction Industry Institute
Construction Users Roundtable
Design Build Institute of America
GSSC – Gulf States Shipbuilders Consortium
ISN
Manufacturing Institute
Mason Contractors Association of America
Merit Contractors Association of Canada
NACE International
National Association of Women in Construction
National Insulation Association
National Technical Honor Society
National Utility Contractors Association
NAWIC Education Foundation
North American Crane Bureau
North American Technician Excellence

Pearson
Prov
SkillsUSA®
Steel Erectors Association of America
U.S. Army Corps of Engineers
University of Florida, M. E. Rinker Sr., School of Construction Management
Women Construction Owners & Executives, USA

Contents

Module One
Instrument Calibration and Configuration

Introduces the basic concepts of calibration, including the three- and five-point methods. Addresses pneumatic, analog, and smart instrumentation calibration methods. Also covers other process control devices that require calibration. (Module ID 12402-16; 60 Hours)

Module Two
Proving, Commissioning, and Troubleshooting a Loop

Explains the three stages in readying a loop for operation: checking, proving, and commissioning. Examines the key ideas behind each step and stresses the differences. Explores troubleshooting techniques and methodologies, with an emphasis on their use during the three stages of readying a loop. (Module ID 12410-16; 17.5 Hours)

Module Three
Tuning Loops

Introduces the techniques used in tuning loops employing PID control. Includes basic tuning theory and formulas. Examines open, closed, and visual loop tuning methods. (Module ID 12405-16; 15 Hours)

Module Four
Digital Logic Circuits

Introduces the basic ideas of digital electronics. Presents gates, combination logic, and truth tables. Addresses memory devices, counters, and arithmetic circuits as well as the numbering systems commonly used in digital systems. (Module ID 12401-16; 15 Hours)

Module Five
Programmable Logic Controllers

Introduces PLCs and their uses in industrial control. Includes hardware components, applications, communications, number systems, and programming methods. (Module ID 12406-16; 12.5 Hours)

Module Six
Distributed Control Systems

Surveys DCS technologies, including an overview of their development. Discusses key components, fieldbuses, servers, and human-machine interfaces. Also introduces maintenance and the increasingly important aspect of DCS security. (Module ID 12407-16; 15 Hours)

Module Seven

Analyzers and Monitors

Introduces the key concepts of chemistry, with an emphasis on their application in instrumentation. Explains the basic physical and chemical properties of matter. Discusses the different analytical methods used in industry to assess processes. Topics covered include pH, conductivity, ORP, gas analysis, and particulate counts. Specific instruments and techniques are described. (Module ID 12409-16; 30 Hours)

Glossary

Index

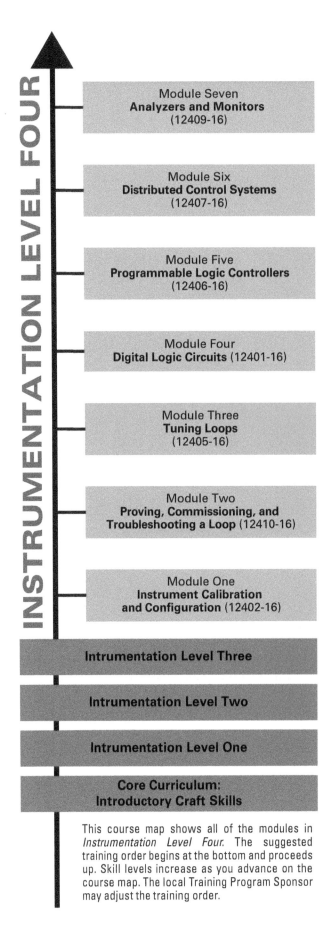

This course map shows all of the modules in *Instrumentation Level Four*. The suggested training order begins at the bottom and proceeds up. Skill levels increase as you advance on the course map. The local Training Program Sponsor may adjust the training order.

12402-16
Instrument Calibration and Configuration

Overview

This module covers calibration methods, including how to calibrate pneumatic instruments, 4–20 mA analog instruments, smart transmitters, transducers, and valve positioners. Information about how to use a HART® communicator to calibrate a smart transmitter is also included.

Module One

Trainees with successful module completions may be eligible for credentialing through the NCCER Registry. To learn more, go to **www.nccer.org** or contact us at 1.888.622.3720. Our website has information on the latest product releases and training, as well as online versions of our *Cornerstone* magazine and Pearson's product catalog.

Your feedback is welcome. You may email your comments to **curriculum@nccer.org**, send general comments and inquiries to **info@nccer.org**, or fill in the User Update form at the back of this module.

This information is general in nature and intended for training purposes only. Actual performance of activities described in this manual requires compliance with all applicable operating, service, maintenance, and safety procedures under the direction of qualified personnel. References in this manual to patented or proprietary devices do not constitute a recommendation of their use.

Copyright © 2016 by NCCER, Alachua, FL 32615, and published by Pearson Education, Inc., New York, NY 10013. All rights reserved. Printed in the United States of America. This publication is protected by Copyright, and permission should be obtained from NCCER prior to any prohibited reproduction, storage in a retrieval system, or transmission in any form or by any means, electronic, mechanical, photocopying, recording, or likewise. To obtain permission(s) to use material from this work, please submit a written request to NCCER Product Development, 13614 Progress Blvd., Alachua, FL 32615.

From *Instrumentation Level Four, Trainee Guide*, Third Edition. NCCER.
Copyright © 2016 by NCCER. Published by Pearson Education. All rights reserved.

12402-16
Instrument Calibration and Configuration

Objectives

When you have completed this module, you will be able to do the following:

1. Define the calibration process and describe the five-point method of calibration.
   a. Define the calibration process.
   b. Describe the five-point method of calibration and the related documentation requirements.
2. Describe pneumatic and analog calibration equipment and basic calibration procedures.
   a. Describe pneumatic calibration equipment and basic calibration procedures.
   b. Describe analog calibration equipment and basic calibration procedures.
3. Identify and describe smart transmitters and their calibration process.
   a. Describe various communication protocols and devices used for communication.
   b. Explain how to calibrate HART® devices.
4. Explain how to calibrate transducers and control valve positioners.
   a. Explain how to calibrate transducers.
   b. Explain how to calibrate pneumatic, electro-pneumatic, and smart control valve positioners.

Performance Tasks

Under the supervision of the instructor, you should be able to do the following:

1. Calibrate a pneumatic pressure transmitter using the proper equipment and complete the appropriate documentation.
2. Calibrate a 4–20 mA pressure transmitter using the proper calibration equipment and complete the appropriate documentation.
3. Calibrate a 4–20 mA temperature transmitter using the proper calibration equipment and complete the appropriate documentation.
4. Calibrate a smart transmitter using a HART® communicator and complete the appropriate documentation.
5. Calibrate a transducer and complete the appropriate documentation.
6. Calibrate the following valve positioners and complete the appropriate documentation:
   - Pneumatic positioner
   - Electro-pneumatic positioner
   - Smart positioner (digital valve controller)

Trade Terms

Calibration
Dry leg
Elevated zero
Factory characterization
Five-point calibration
Fulcrum
Head pressure
Inches of water
Look-up table

Multifunction loop calibrator
Range
Span
Spanning
Suppressed zero
Three-point calibration
Transfer function
Wet leg
Zeroing

Industry Recognized Credentials

If you are training through an NCCER-accredited sponsor, you may be eligible for credentials from NCCER's Registry. The ID number for this module is 12402-16. Note that this module may have been used in other NCCER curricula and may apply to other level completions. Contact NCCER's Registry at 888.622.3720 or go to **www.nccer.org** for more information.

Contents

- 1.0.0 Instrument Calibration 1
  - 1.1.0 Calibration Defined 1
    - 1.1.1 Input Signals or Energies 2
    - 1.1.2 Output Signals or Energies 2
    - 1.1.3 Calibration Concepts 2
    - 1.1.4 The Calibration Process 3
    - 1.1.5 Documentation 3
  - 1.2.0 Five-Point Calibration 3
- 2.0.0 Pneumatic and Analog Calibration 8
  - 2.1.0 Pneumatic Instrument Calibration 8
    - 2.1.1 Pneumatic Multifunction Calibrators 8
    - 2.1.2 Pneumatic Differential Pressure Transmitters 9
    - 2.1.3 Principles of Operation 9
    - 2.1.4 Pneumatic Calibration Process Overview 12
    - 2.1.5 Temperature Transmitters 12
  - 2.2.0 Analog Instrument Calibration 12
    - 2.2.1 Analog Multifunction Calibrators 13
    - 2.2.2 Analog Differential Pressure Transmitters 13
    - 2.2.3 Head Pressure 13
    - 2.2.4 Open Vessel Installations (Level Measurement) 13
    - 2.2.5 Closed, Pressurized Vessel Installations (Level Measurement) 15
    - 2.2.6 Rosemount™ 1151 Zero and Span 16
    - 2.2.7 Temperature Transmitters with Analog Outputs 18
- 3.0.0 Smart Transmitters 20
  - 3.1.0 The HART Protocol 20
  - 3.2.0 Calibrating HART Devices 21
    - 3.2.1 HART Device Internal Stage Summary 21
    - 3.2.2 The Input Stage 22
    - 3.2.3 The Middle Stage 22
    - 3.2.4 The Output Stage 22
    - 3.2.5 Smart Pressure Transmitter HART Calibration 22
    - 3.2.6 Test and Calibration Considerations 23
- 4.0.0 Transducers and Control Valve Positioners 25
  - 4.1.0 Transducers 25
  - 4.2.0 Control Valve Positioners 25
    - 4.2.1 Pneumatic and Electro-Pneumatic Positioners 25
    - 4.2.2 Smart Positioners (Digital Valve Controllers) 28

Figures

Figure 1 Multifunction calibrator 2
Figure 2 An example of a calibration documentation form 4
Figure 3 Pneumatic calibrators 9
Figure 4 Pneumatic DP transmitter... 10
Figure 5 Pneumatic DP transmitter connected for flow measurement 10
Figure 6 Foxboro® 13A pneumatic differential pressure transmitter..........11
Figure 7 Fluke® Model 725 loop calibrator 13
Figure 8 Typical DP flow transmitter installations..................................... 14
Figure 9 A level transmitter mounted below the minimum level 15
Figure 10 An open tank installation with a suppressed zero...................... 15
Figure 11 Compensating for specific gravity of liquid in wet leg 16
Figure 12 Elevated zero in a wet leg installation 16
Figure 13 Location of zero and span adjustment screws for
Rosemount™ Model 1151 transmitter.. 17
Figure 14 Handheld HART communicator 21
Figure 15 Principles of operation for a HART transmitter device................ 21
Figure 16 Fluke® 754 combined calibrator and communicator................... 22
Figure 17 Cut-away of an I/P transducer 26
Figure 18 Mechanical link of a valve actuator stem to a positioner............ 27
Figure 19 Operation schematic for a Fisher® 3582 Series
pneumatic positioner 27
Figure 20 Fisher® 3582i Series electro-pneumatic positioner 28
Figure 21 Fisher® FIELDVUE™ DVC6000 Series digital valve
controller block diagram.. 29

SECTION ONE

1.0.0 INSTRUMENT CALIBRATION

Objective

Define the calibration process and describe the five-point method of calibration.

a. Define the calibration process.
b. Describe the five-point method of calibration and the related documentation requirements.

Trade Terms

Calibration: The process of adjusting an instrument so its output signal or level of energy is representative of the level of the non-adjustable input energy at any given point. Calibration involves test instruments traceable to a national standards body.

Five-point calibration: A method of calibrating an instrument in which the output level is checked and set against a simulated input level at five percentages of span (see *span*): 0 percent, 25 percent, 50 percent, 75 percent, 100 percent, 75 percent, 50 percent, 25 percent, and 0 percent, in that order.

Inches of water: A measurement of pressure (1" H_2O = 0.03612628 psi, or 249 Pa).

Multifunction loop calibrator: A test instrument that provides an energy source and measuring capabilities in the same equipment housing. Multifunction calibrators often perform many different types of calibrations.

Range: The minimum and maximum values that an instrument can measure.

Span: The difference between an instrument's upper and lower measurement limits.

Spanning: Setting an instrument's output so that it is at its maximum value when the input process value is at the maximum value of the specified operating range.

Three-point calibration: A method of calibrating an instrument in which the output level is checked and set against a simulated input level at three different percentages of span: 0 percent, 50 percent, 100 percent, 50 percent, and 0 percent, in that order.

Zeroing: Setting an instrument's output so that it is at its minimum value when the input process value is at the minimum value of the specified operating range.

All instruments that provide a proportional output signal based on an input signal (or level of energy) must be accurately calibrated in order to function properly within the loop. Calibration is one of the primary tasks of any instrumentation technician. The methods of calibration used for one particular type of instrument can often be used with other types. However, unique methods must sometimes be applied in certain situations because of an instrument's function in the process loop.

Instruments that require calibration are generally classified under one of three signal groups: pneumatic, analog, or smart. Although smart instrumentation, programmable logic controllers (PLCs), and distributed control systems (DCSs) have become common in instrument loops, pneumatic and analog instruments are still used in some industries and require regular calibration. Some control loop instruments that require calibration include transmitters, transducers, and valve positioners.

1.1.0 Calibration Defined

All instruments that require calibration have at least two characteristics in common. First, they have some type of input signal or energy. Second, they use that input either to develop an output signal used for measurement, or they convert that energy into some other type of energy (mechanical, electrical, or pneumatic) for use by the next device in the instrument channel.

Calibration involves proportionally matching the output signal or level of energy to the input signal or level of energy. This can be accomplished by using loop calibration test equipment that provides a variable energy source to the instrument's input, while a second device reads the output signal using an appropriate meter. Calibration can also be done using a multifunction loop calibrator, which provides the energy source and measuring capabilities in a single unit (*Figure 1*). A multifunction calibrator can often perform a variety of calibrations on different types of equipment, making it a useful piece of equipment to own.

Instruments used for calibration must themselves be calibrated and referenced to accepted standards. In most cases, for a calibration to be valid, the instruments must be traceable to a national standards body such as the National Institute of Standards and Technology (NIST). Normally, instruments are sent away to a calibration lab on a scheduled basis. A calibration sticker

12402-16 Instrument Calibration and Configuration

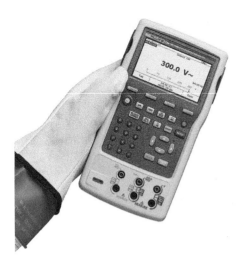

Figure 1 Multifunction calibrator.

on the instrument and a matching certificate confirm its calibration and traceability status and shows the date of calibration. Calibrations then performed using that instrument are deemed to be both valid and traceable.

Be aware that the procedures outlined in this module are general. They are designed to provide a basic understanding of the calibration process. Remember that each instrument and piece of test equipment is unique. Always consult the manufacturer's documentation before calibrating anything, and follow your company guidelines and procedures precisely.

1.1.1 Input Signals or Energies

Instruments that are in direct contact with the process are designed and manufactured to operate within a specified process input energy range. The input energy may be pressure, temperature, voltage, current, or some other form of energy. The level of this process energy typically represents the measured process variable, so the level is not generally regulated at the instrument input. For example, transmitters are usually designed and manufactured to operate within specific ranges such as 0–200 inches of water (" H_2O), 0–100 psi, or 1.3–5.3 kPa. Operating ranges typically have maximum and minimum values, usually specified on the instrument tag. It's important to note that the minimum value doesn't have to be zero—in fact, it often isn't. For example, an instrument may have a range of 60–400" H_2O.

The instrument should never be installed in a situation in which the measured variable does not normally operate within the specified range. Doing so could cause the instrument to output an inaccurate reading, or even fail.

> **WARNING!** Never expose an instrument to a greater maximum operating range limit than is specified by the manufacturer. Exceeding the maximum operating range limit may damage the instrument and could result in personnel injury if the instrument disintegrates.

1.1.2 Output Signals or Energies

Output signals or energies from an instrument are proportional to the input signal and represent it accurately. As with input signals, output signals operate within a specific range, such as 0–60 psig, 3–15 psi, 20–100 kPa, 4–20 mA, 10–50 mA, or 1–5 VDC. Remember, the minimum value isn't always zero. Unlike the input signal or energy level, the output signals on most instruments must be adjusted so that they proportionally represent the input signal at any given point across their range. Instrument calibration is the process of adjusting the output signal to be truly representative of the input signal.

> **NOTE** Some alarm instruments, pressure gauges, and other indicators may not have adjustable output signals or levels; therefore, they cannot be calibrated. On the other hand, some devices like thermocouples, which cannot be adjusted, still require a calibration process in which their behavior is characterized against a standard. This information is then used by other instruments to adjust the data that the thermocouple supplies, thereby effectively calibrating the thermocouple.

1.1.3 Calibration Concepts

In order to perform a calibration, it is crucial to understand the key terms involved as well as the calculations that must be performed. Two terms that are often misunderstood are *range* and *span*. Both of these terms apply to inputs as well as outputs.

An instrument's range consists of its lower and upper input or output limits. For example, if an instrument can measure a process value between 60" H_2O and 400" H_2O, its range is 60–400" H_2O. Similarly, if an instrument can output a pressure between 3 psi and 15 psi, its range is 3–15 psi.

Span refers to the difference between the upper and lower limits. For example, if an instrument has an input range of 60–400" H_2O, its span is 340" H_2O (400 − 60 = 340). Similarly, if an instrument has an output pressure range of 3–15 psi, its span is 12 psi (15 − 3 = 12).

As part of the calibration process, it is often necessary to calculate percentages of input or output spans. To perform percentage calculations, turn the percentage into its decimal form (divide it by 100) and multiply by the span value. For example, if you wish to calculate the 50 percent point of a span of 340" H_2O, multiply 340 by 0.50 (the decimal form of 50 percent), which gives a value of 170" H_2O.

When the minimum (zero) value for a range is a number other than zero, an additional step is required. For example, if an instrument has a range of 60–400" H_2O, the 50 percent point of its span is not 170" H_2O, because the zero value (the minimum value) of the range is 60" H_2O, not 0" H_2O. In this case, add the calculated 50 percent span point (170) to the zero value (60) to determine the true 50 percent span point of 230" H_2O (170 + 60 = 230).

As another example, if the same instrument has an output range of 3–15 psi, it has a span of 12 psi (15 − 3 = 12). But since its zero point is 3 psi, not 0 psi, a 50 percent output value will be 9 psi [(12 × 0.50) + 3 = 9].

The following formula can be used to calculate the value for any percentage of span, regardless of its zero value (a scientific calculator can be used to help make these calculations):

$$\text{value} = [(\text{max} - \text{min}) \times \text{decimal percentage}] + \text{min}$$

1.1.4 The Calibration Process

Conceptually, calibration is a very simple process: apply a specific input signal to the instrument and read the corresponding output. The output should have a definite proportional relationship to the input. Establish whether the output is providing the correct values for given inputs by measuring a series of points that represent various percentages of the instrument's span. If it is not, adjust the instrument until the output yields the correct value for a given input.

One of the simplest forms of calibration is zeroing and spanning the instrument. When you zero the instrument, you set its output value to its minimum range limit, which represents a zero input from the process. For example, consider a pneumatic transmitter used with a process whose range is 20–150" H_2O. The zero value is 20" H_2O. If the transmitter's output range is 3–15 psi, its zero value is 3 psi. A 20" H_2O input to the transmitter should produce an output of 3 psi. If it doesn't, you must adjust the output until it does.

Spanning involves setting the output's upper limit to correspond to the input's upper limit. Using the previous example, the maximum process input of 150" H_2O should produce a transmitter output of 15 psi. Again, if it doesn't, the instrument must be adjusted until it does. Once you've completed this step, the instrument is spanned, meaning that it measures across the desired range.

In some cases, setting the lower and upper limits are all that's needed. However, adjusting one value sometimes affects the other. Getting the instrument zeroed and spanned may require several adjustment cycles.

Calibration frequently involves more than zeroing and spanning. Often, additional points must be checked and adjustments made in a specific order. These additional points represent various percentages of the span. This is why it is necessary to know how to calculate percentages of input and output spans.

While there are a number of different calibration procedures, two popular methods are three-point calibration and five-point calibration. The procedure to be used depends on the desired accuracy, and the industry and/or company standards being followed.

1.1.5 Documentation

Documentation—keeping records of what you have done—is easy to forget, but it's extremely important. Most companies require that instrument calibrations be logged in some way. With older systems, logging is often done on paper, usually on a standard form (*Figure 2*). In modern smart systems, particularly those that are part of a DCS, calibrations may be logged electronically as a part of the process. In all cases, good documentation is important because it provides evidence that instruments are being maintained and operated according to industry and company procedures.

1.2.0 Five-Point Calibration

When performing a five-point calibration on an instrument, the instrument's output is checked for accuracy and proportionality against an input signal that is 0 percent, 25 percent, 50 percent, 75 percent, 100 percent, 75 percent, 50 percent, 25 percent, and 0 percent of span, in that order. Although nine (not five) points are actually collected, there are only five unique percentages of span used. Nine points are collected because it's useful to ramp up and then down through the instrument's span. This approach simulates real process behavior and also reveals whether there is a hysteresis (lagging behind) problem in the system. The procedure often requires numerous

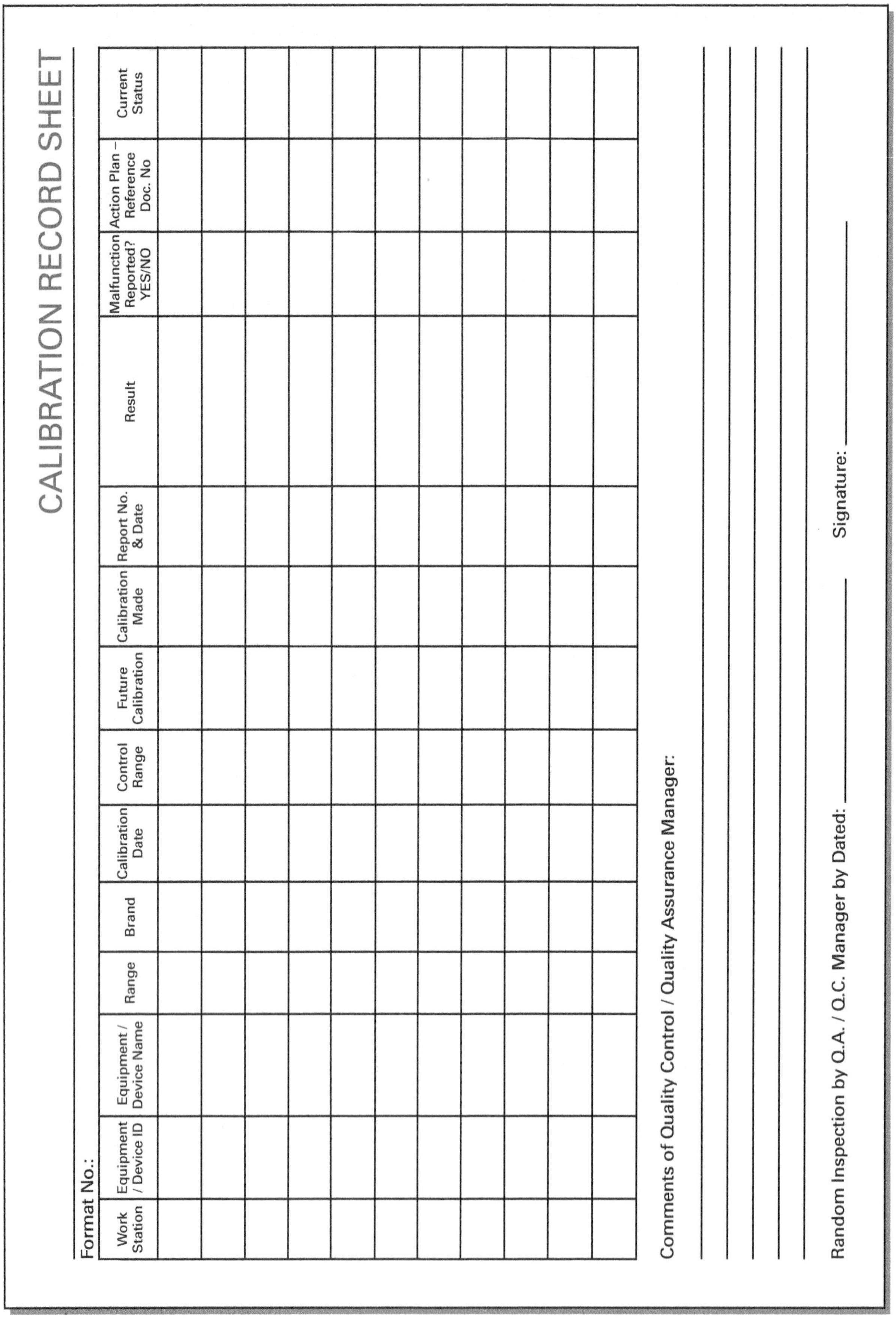

CALIBRATION RECORD SHEET

Format No.:

Work Station	Equipment / Device ID	Equipment / Device Name	Range	Brand	Calibration Date	Control Range	Future Calibration	Calibration Made	Report No. & Date	Result	Malfunction Reported? YES/NO	Action Plan – Reference Doc. No	Current Status

Comments of Quality Control / Quality Assurance Manager:

Random Inspection by Q.A. / Q.C. Manager by Dated: ____________ Signature: ____________

Figure 2 An example of a calibration documentation form.

rechecks and repeat adjustments of the previous settings.

A three-point calibration is essentially the same, except with fewer points (0 percent, 50 percent, 100 percent, 50 percent, and 0 percent).

The following sequence of steps is provided as a general guideline for calibrating instruments with the five-point method:

> **NOTE**
> When following these steps, always remember to account for a zero value that is not actually zero.

Step 1 Determine the type of input signal that must be applied to the instrument's input in place of the process signal.

Step 2 Check the process range of the instrument to be calibrated (for example, 0.14–1.4 MPa, 0–60 psi or 60–200" H₂O) to make sure that it corresponds to the process input. Also identify the means by which the instrument will be adjusted during calibration (screws, knobs, calibration screen, etc.)

Step 3 Obtain appropriate calibration equipment that will provide the correct input signal and also read the output. Confirm that the test equipment itself has been calibrated and that its calibration is current. Be sure to follow all applicable company standards.

Step 4 Determine point #1 (0 percent). In most cases, it is necessary to isolate the instrument from the process (follow company procedures to do this). Connect the calibration device to the instrument and apply a signal that corresponds to the minimum limit of the operating range (the zero point). Remember that although this value is considered the zero point, it won't necessarily be zero in value.

Step 5 Read the output and adjust the instrument so that the value corresponds to the correct zero value. For example, an instrument that outputs a 4–20 mA signal should output 4 mA. An instrument that outputs a 1–5 V signal should output 1 V.

Step 6 Determine point #2 (25 percent). Apply a signal that corresponds to 25 percent of the input span. Calculate this value using the procedure discussed earlier.

Step 7 Calculate the expected output value. For example, the 25 percent point for a 4–20 mA output is 8 mA.

$$[(20 - 4) \times 0.25] + 4 = 8$$

An instrument that outputs 1–5 V should output 2 V.

$$[(5 - 1) \times 0.25] + 1 = 2$$

Read the output and adjust the instrument so that the value corresponds to the correct 25 percent point.

Step 8 Determine point #3 (50 percent). Apply a signal that corresponds to 50 percent of the input span. Calculate this value using the procedure discussed earlier.

Step 9 Calculate the expected output value. For example, the 50 percent point for a 4–20 mA output is 12 mA.

$$[(20 - 4) \times 0.50] + 4 = 12$$

An instrument that outputs 1–5 V should output 3 V.

$$[(5 - 1) \times 0.50] + 1 = 3$$

Read the output and adjust the instrument so that the value corresponds to the correct 50 percent point.

Step 10 Determine point #4 (75 percent). Apply a signal that corresponds to 75 percent of the input span. Calculate this value using the procedure discussed earlier.

Step 11 Calculate the expected output value. For example, the 75 percent point for a 4–20 mA output is 16 mA.

$$[(20 - 4) \times 0.75] + 4 = 16$$

An instrument that outputs 1–5 V should output 4 V.

$$[(5 - 1) \times 0.75] + 1 = 4$$

Read the output and adjust the instrument so that the value corresponds to the correct 75 percent point.

Step 12 Determine point #5 (100 percent). Apply a signal that corresponds to 100 percent of the input span.

Step 13 Read the output and adjust the instrument so that the value corresponds to the correct maximum value. For example, an instrument that outputs a 4–20 mA signal should output 20 mA. An instrument that outputs a 1–5 V signal should output 5 V.

Step 14 Now, work back down through the sequence in reverse order (75 percent, 50 percent, 25 percent, and 0 percent), applying appropriate input values and checking the output values. Make adjustments as required.

Step 15 Repeat Steps 4 through 14 until you have successfully gone through the entire nine-point sequence and observed satisfactory results. It may require several cycles and multiple adjustments to achieve this condition.

Step 16 Disconnect the calibration equipment. Follow company procedures to return the instrument to service. Document what you've just done using the appropriate form or electronic logging system. Always follow company procedures meticulously.

To perform a three-point calibration, follow the same steps but omit points that are not part of the three-point sequence. Measure only the following points: 0 percent, 50 percent, 100 percent, 50 percent, and 0 percent.

> **NOTE**
>
> Do not confuse calibrating an instrument with tuning. Tuning involves controllers, and it requires many more parameters to be set. Tuning is covered in the "Turning Loops" module of this curriculum.

Automatic Calibration

Many modern multifunction calibrators can complete the entire five-point calibration process almost automatically at the touch of a button. Simply specify the number of points you want to measure and the calibrator changes the input signal, reads the output, and generates the data. The instrument technician still has to perform any required manual instrument adjustments, but if the instrument itself is smart, the calibrator may be able to make adjustments as well.

Know you calibrator and what it can do. You may find that it saves you a lot of work!

Additional Resources

Measurement and Control Basics, Thomas A. Hughes. Fifth Edition. 2014. Research Triangle Park, NC: International Society of Automation.

The following websites offer resources for products and training:

Emerson Process Management, **www.emersonprocess.com**

Fluke Corporation, **www.fluke.com**

1.0.0 Section Review

1. An instrument measures process values between 50 and 1,000 psi. What is the instrument's range?
    a. 1,000 psi
    b. 0–1,000 psi
    c. 950 psi
    d. 50–1,000 psi

2. A transmitter outputs a signal that varies between 10 and 50 mA. What value should it output to represent the 75 percent point of its span?
    a. 30 mA
    b. 35 mA
    c. 40 mA
    d. 45 mA

3. What is the correct sequence of measurements for a five-point calibration?
    a. 0 percent, 25 percent, 50 percent, 75 percent, 100 percent
    b. 0 percent, 25 percent, 50 percent, 75 percent, 100 percent, 75 percent, 50 percent, 25 percent, 0 percent
    c. 0 percent, 100 percent, 50 percent, 75 percent, 25 percent
    d. 0 percent, 100 percent, 25 percent, 50 percent, 75 percent, 100 percent, 0 percent, 25 percent, 50 percent, 75 percent

Section Two

2.0.0 Pneumatic and Analog Calibration

Objective

Describe pneumatic and analog calibration equipment and basic calibration procedures.
a. Describe pneumatic calibration equipment and basic calibration procedures.
b. Describe analog calibration equipment and basic calibration procedures.

Performance Tasks

1. Calibrate a pneumatic pressure transmitter using the proper equipment and complete the appropriate documentation.
2. Calibrate a 4–20 mA pressure transmitter using the proper calibration equipment and complete the appropriate documentation.
3. Calibrate a 4–20 mA temperature transmitter using the proper calibration equipment and complete the appropriate documentation.

Trade Terms

Dry leg: A connection from the process that contains a dry, non-condensing gas.

Elevated zero: A zero value that's above the true zero point.

Fulcrum: The pivot point around which a lever turns.

Head pressure: The pressure that results from a liquid column of a certain height.

Suppressed zero: A zero value that's below the true zero value.

Wet leg: A connection from the process that's intentionally filled with a reference fluid to prevent gas condensation from affecting the measurement.

Legacy instrumentation systems are often a mixture of pneumatic and analog electronic technology. Calibrating this type of equipment is relatively straightforward, but it requires appropriate tools. In this section, you'll learn the basic principles of calibrating both pneumatic and electronic instruments.

2.1.0 Pneumatic Instrument Calibration

For many decades, pneumatic instrumentation has been widely used by a number of industries. Today, electronic instrumentation—particularly smart technology—is replacing it. Smart devices offer several advantages, particularly their ability to work with PLCs and DCSs. Yet many industrial facilities still utilize pneumatic control loops that have performed for many years. They may not be high on the priority list for upgrades because of their use in a non-critical application, or because they perform well so there is little incentive to replace them. Consequently, instrument technicians will likely encounter pneumatic instruments and must know how to calibrate them.

2.1.1 Pneumatic Multifunction Calibrators

Pneumatic calibration equipment must be able to simulate the process input to the instrument being calibrated and measure the pneumatic output signal from the instrument. This can be done in a number of ways. It may involve several separate pieces of equipment, such as a meter of some kind and a pressure source, such as a hand pump. Generally, however, a multifunction calibrator is a popular device to use for pneumatic calibration.

Multifunction pneumatic calibrators are extremely versatile devices that can perform other types of calibrations as well. Many are modular and come with a variety of adapters that allow different signals in various ranges to be measured. In most cases, the unit will also include a pressure source, such as a built-in hand pump or electric pump; some calibrators rely on an external hand pump. Good multifunction calibrators are rugged and liquid-proof. Many are also small enough to be handheld, much like a multimeter.

Figure 3 shows a number of popular pneumatic calibrators. Typical models include the Fluke® 719, GE Druck® DPI 620-IS, and Crystal® HPC40. A classic unit that may be found in older shops is the Wallace and Tiernan® (now WIKA® Instrument, LP) Wally Box®. For many years, it was a mainstay of the industry. Today, newer designs have replaced it in many situations.

Since every calibrator is a bit unique in use and configuration, the information in this section is somewhat general. Always consult the product documentation that comes with the calibrator you are using. Remember, it's important to verify that the test equipment has been calibrated professionally and has a certificate of traceability to a national standards body. The calibration date must also be current, as specified by the manufacturer's guidelines or your company's calibration standards.

(A) Fluke® 719

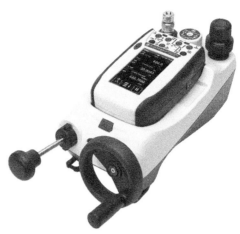
(B) GE Druck® DPI 620-IS

(C) Crystal® HPC 40

(D) WIKA® Wally Box®

Figure 3 Pneumatic calibrators.

2.1.2 Pneumatic Differential Pressure Transmitters

There are several types of pneumatic transmitters, which vary according to the type of process, state of process, variable being monitored, ambient environment, and type of control system. The most common type of pneumatic transmitter is the differential pressure (DP) transmitter, shown in *Figure 4*. It can be used in flow, pressure, and level control applications.

Pneumatic differential pressure transmitters can be installed in processes to measure pressure in a line or vessel, flow in a line by measuring the pressure drop across an orifice restriction (differential pressure), or level in tanks or vessels by measuring the difference in pressure between the top and bottom elevations of the tank or vessel.

2.1.3 Principles of Operation

A pneumatic DP transmitter has a two-sided differential pressure capsule that receives two different levels of pressure and converts the difference between the pressures into a mechanical movement via a force bar. The movement of the force bar operates a balance beam by pivoting it on a fulcrum. A flat plate, called a flapper, is attached to the balance beam. The flapper moves against the opening of a nozzle that is continuously venting air.

As the flapper partially closes off the nozzle, backpressure is created in the tubing connected to the nozzle, which is connected on the other end to the top side of a diaphragm in a pneumatic relay. As the backpressure increases and decreases on the diaphragm in response to the position of the flapper against the nozzle, the pneumatic relay proportionally regulates a pneumatic output by means of a spring-loaded valve.

The principles of operation of a pneumatic DP transmitter installed as a flow transmitter are illustrated in *Figure 4* and *Figure 5*.

The most common pneumatic output signal range is 3–15 psi (20–100 kPa), which is usually sent to a controller, recorder, gauge, indicator, or pneumatically operated alarm. *Figure 6* shows a classic pneumatic differential pressure transmitter, a Foxboro® 13A.

> **NOTE:** The pneumatic DP transmitter requires a regulated instrument air supply to operate the pneumatic relay and provide air to the nozzle assembly. Calibrators usually do not supply this air, so the normal supply must be left on. Calibration should always be performed with the instrument control loop isolated from the process.

DP transmitters can be used in pressure, flow, or level applications. Another classic line of pneumatic transmitters, the Siemens Series 50, was available in models having the following input range and span limits:

- 0–50" H_2O (range), 5–50" H_2O (span)
- 0–225" H_2O (range), 20–225" H_2O (span)
- 0–850" H_2O (range), 150–850" H_2O (span)

Notice that the spans of all three models have minimum span limits that are higher than the true zero point. A minimum span limit above the zero minimum range is referred to as elevated zero.

Elevated zeros are necessary in order to make the instrument accurate at the low end of the range. The span limits designate both the 0 and 100 percent input levels used during calibration.

Differential pressure transmitter capsules have two sides, with one designated as the high side and the other designated as the low side. The difference in pressure between the two sides is considered the differential pressure and is the process measurement that determines the output signal from the transmitter. When installed as a flow transmitter, the high side is always connected to the upstream side of the orifice restriction; the low side is always connected to the downstream side of the restriction. If the transmitter is being used to measure pressure in a line or vessel, the low side is vented to the atmosphere. In these applications, the transmitter still measures differential pressure between the high and low sides, but the low side is always zero pressure.

To calibrate the transmitter when it is isolated from the process, you must simulate these values of differential pressure within the span limits of the instrument. For instance, with the three models of the Siemens Series 50 transmitters, you would set the output pneumatic signal of the transmitter at an equivalent zero output value of 3 psi whenever you supply a differential pressure between the two sides of 5" H_2O, 20" H_2O, or 150" H_2O (the 0 percent point), depending on which one of the models you are calibrating. For the 100 percent point, you would supply a differential pressure of 50" H_2O, 225" H_2O, or 850" H_2O.

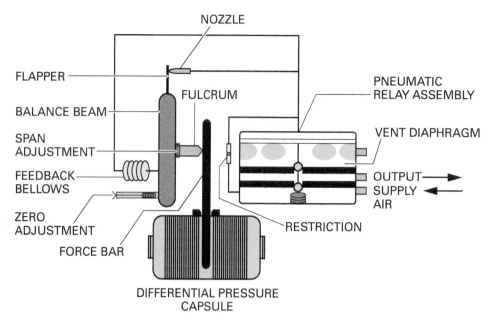

Figure 4 Pneumatic DP transmitter.

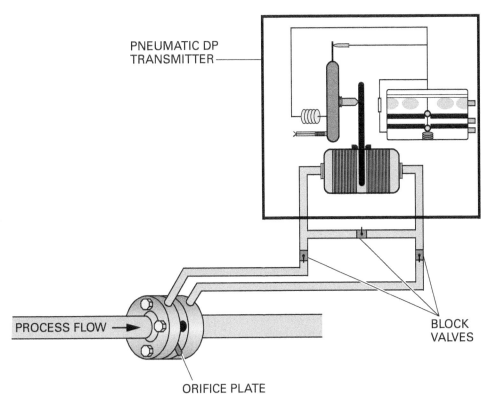

Figure 5 Pneumatic DP transmitter connected for flow measurement.

For the other calibration points, determine the actual range of the span and figure the differential pressures at each of those points. The range of the span for 5–50" H_2O is 45" H_2O (50 – 5 = 45); for the 20–225" H_2O, it is 205 (225 – 20 = 205); and for the 150–850" H_2O, it is 700" H_2O (850 – 150 = 700).

It is not necessary to supply pressure both to the high and low sides of the transmitter. All that is required is a difference between the sides, and applying pressure to the high side can accomplish this. Suppose you open (vent) both sides of the capsule to the atmosphere. What is the differential pressure between the two sides at this point? Naturally, it's zero, but in the case of the Siemens Series 50, you cannot set the transmitter's output signal at 3 psi (zero equivalent) with an input differential pressure of zero because these instrument spans have elevated zeros. However, you can leave the low side vented to the atmosphere and supply a pressure to the high side of the capsule that is equal to the minimum pressure of the span shown on the instrument. For the three Siemens Series 50 transmitters, it will be 5" H_2O, 20" H_2O, or 150" H_2O, depending on which of the transmitters you are calibrating.

Figure 6 Foxboro ® 13A pneumatic differential pressure transmitter.

2.1.4 Pneumatic Calibration Process Overview

Since each multifunction calibrator is different, the following steps for calibrating a pneumatic transmitter are fairly general, but they provide a basic outline of what's involved. It's essential to consult the equipment's documentation before performing a calibration in order to ensure good results.

Step 1 Isolate the transmitter from the process by coordinating the isolation process with the proper operations personnel. (This usually involves operating a loop controller in the manual position.)

Step 2 Follow drain and blowdown procedures at the process connections to the transmitter. Verify that the transmitter is isolated from the process. Equalize the static pressure on both sides of the transmitter's capsule by opening the equalizer valve between the high and low sides.

Step 3 Remove the calibration pipe plugs (usually ⅛" [6 mm] or ¼" [8 mm] NPT) from each side of the transmitter. Install the proper tubing connector to accommodate the test tubing supplied with the calibrator to the high side's calibration pipe plug hole. Connect the calibrator's pressure source to this side. Close the equalizer valve between the high and low sides.

Step 4 Disconnect the transmitter's pneumatic output signal line. Cover the open end to protect it. Connect the transmitter's pneumatic output to the appropriate input on the calibrator.

Step 5 Verify again that the two block valves to the process taps are securely closed (high- and low-side manifold valves). Verify that the instrument air supply to the transmitter remains on.

Step 6 Using the procedure outlined in the calibrator's documentation, apply an appropriate input signal (pressure) to the instrument and read the output result (also pressure). Make any transmitter adjustments necessary to bring the output into line with the input signal. Repeat the process for each point of the calibration cycle. Repeat as many times as necessary to achieve a good calibration.

Step 7 When finished, disconnect all calibration equipment. Reconnect the transmitter and replace all lines and plugs. Return the instrument to normal operation using company standard procedures. Observe all safety rules. Document the calibration process according to company standards.

2.1.5 Temperature Transmitters

Some older types of temperature control loops that require a pneumatic signal to operate use a temperature transmitter that operates on the principle of differential pressure, with the low side of the capsule open to the atmosphere. The high-side pressure is generated by a gas-filled tube-and-bulb system (capillary) in which the tube is connected and sealed to a special differential pressure transmitter capsule. The tube and bulb are filled with a gas, such as helium, that expands and contracts with changes in temperature. The bulb acts as the primary detecting element and is inserted directly into the process. As the temperature of the process changes, the gas within the bulb expands or contracts in proportion to the temperature change. The changing pressure of the gas is applied to the high side of the capsule, causing the force-balance mechanism in the transmitter to respond accordingly.

The only field calibration that can be performed on an electro-mechanical temperature transmitter is a range change by elevating or suppressing the zero, using a screwdriver to make the adjustments. A bench calibration check can be performed on a bulb-type instrument by providing a calibrated temperature-controlled bath in which the bulb is inserted to simulate the process, while monitoring the pneumatic output of the signal using a multifunction calibrator. Typically, the transmitter will have to be supplied with dry, regulated air since the calibrator doesn't furnish this.

2.2.0 Analog Instrument Calibration

An analog signal is a voltage, current, or other signal that is continuously variable with time. In instrumentation, the most common form of analog signal that is applied in control loops is the current loop signal, which is typically 4–20 mA, although other values are used in some settings. A 4 mA current represents the minimum (0 percent) end of the range or span, while a 20 mA current represents the maximum (100 percent) end of the range or span.

For example, if the process span of a transmitter is 20–200" H$_2$O, 20" H$_2$O (0 percent) of the process should produce a 4 mA output when the transmitter is properly calibrated. Similarly, 200" H$_2$O (100 percent) should provide a 20 mA output under the same calibration settings.

2.2.1 Analog Multifunction Calibrators

Many of the same calibrators used for pneumatic devices also have calibration features for analog electronic instrumentation. The Fluke® Model 725 loop calibrator shown in *Figure 7* is one example. This unit is capable of measuring an analog output signal while providing a variety of input signals, including voltage (V and mV), current (mA), frequency (Hz), and resistance (Ω). When coupled with any one of 28 compatible pressure modules, it can also measure pressure and differential pressure and provide a pneumatic pressure source. In any of the calibration modes, and when used with the proper accessory modules, the Model 725 is capable of providing the source and measuring the signal at the same time.

This calibrator can also provide an auto-step or auto-ramp source in which the source function can be automatically set to provide a complete range or span in incremental steps, such as 4 mA, 8 mA, 12 mA, 16 mA, and 20 mA. It may also be programmed to provide automatic steps of source pressure based on any range or span. This feature allows the technician to set the calibrator for auto-step or auto-ramp as an input at one location and check or test other parameters in a different location as the calibrator steps through the various levels of its source.

Figure 7 Fluke® Model 725 loop calibrator.

2.2.2 Analog Differential Pressure Transmitters

Like the pneumatic differential pressure transmitter, the analog DP transmitter can be used to measure flow, pressure, and level. It functions in the same manner except for its output, which is an analog electronic signal rather than a pneumatic one.

Many analog signal DP transmitters can be calibrated for elevated or suppressed zero. Some offer up to 500 percent above suppression and 600 percent above elevation, depending on the brand and model. The Rosemount™ Model 1151DP differential pressure transmitter is an example of an older transmitter with this feature.

Figure 8 shows some common differential pressure installations using the Model 1151DP to measure flows of various processes, including steam, liquid, and gas. Note the difference between the piping or tubing installations and the location of the taps. These differences in installation are designed to eliminate error caused by condensation buildup in the piping or tubing of gas or steam process, and air or gas bubbles in the liquid flow process.

2.2.3 Head Pressure

DP transmitters used for liquid level applications measure static pressure, commonly called head pressure. Factors that must be determined when calibrating DP transmitters in liquid level applications are the liquid level and the specific gravity of the liquid. Head pressure is equal to the liquid's height above the tap multiplied by the specific gravity of the liquid. Head pressure is independent of volume or vessel shape.

2.2.4 Open Vessel Installations (Level Measurement)

A differential pressure transmitter can be used to measure level in an open tank. The high-pressure side of the transmitter is tubed to the tank and the low-pressure side is vented to atmosphere. The transmitter will measure the head pressure above the tank connection point, thereby indirectly measuring the tank level.

If the transmitter is mounted below the physical minimum point of the desired level range of the tank, a zero suppression calibration must be performed. In the rare situation where the transmitter is mounted above the minimum level, the transmitter will have to be calibrated for the additional elevation above the minimum level.

Figure 9 shows a differential pressure level transmitter (LT) mounted on the side of an open tank or vessel. If the transmitter were mounted at the desired minimum level point (the dashed line), the range of the transmitter would be 0–300". However, in this installation, the transmitter is mounted 75" below the desired minimum level point. This additional head pressure must be suppressed by calculating a new range. The following formula can be used to calculate the range:

$$\text{range} = (e) \text{ to } (e + h)$$

Where

- e = the head pressure produced by Y (in inches of water)
- h = the maximum head pressure to be measured (in inches of water)

To calculate the range using this formula, you must first determine the values for *e* and *h* in the formula. The following equation can be used to calculate *e*:

$$e = (Y)(sg)$$

Where

- e = the head pressure produced by Y (in inches of water)
- Y = the vertical distance between the transmitter's high-side inlet and the minimum measurable level
- sg = the specific gravity of the liquid in the tank

In this example, Y is 75" (as shown in *Figure 9*), and the specific gravity is 0.9. Using this information and the equation, you can determine the value of *e*:

$$e = (Y)(sg)$$
$$e = (75")(0.9)$$
$$e = 67.5" \; H_2O$$

Now you must determine the value of *h*. The following equation can be used to calculate its value:

$$h = (X)(sg)$$

Where

- h = the maximum head pressure to be measured (in inches of water)
- X = the vertical distance between the minimum and maximum measurable levels
- sg = the specific gravity of the liquid in the tank

In this example, X is 300" (as shown in *Figure 9*). Using the necessary information with the equation, you can now determine the value of *h* (the maximum head pressure to be measured):

$$h = (X)(sg)$$
$$h = (300")(0.9)$$
$$h = 270" \; H_2O$$

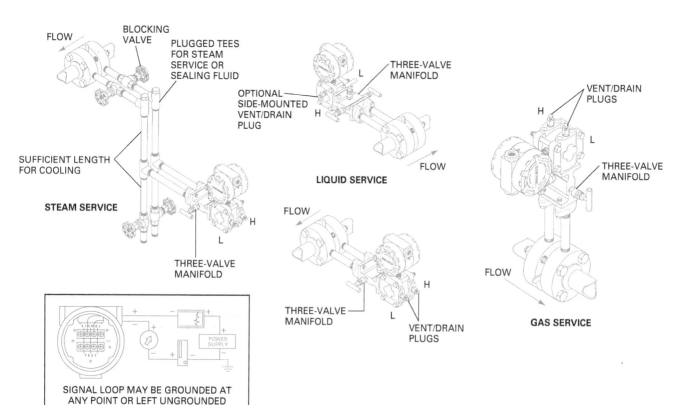

Figure 8 Typical DP flow transmitter installations.

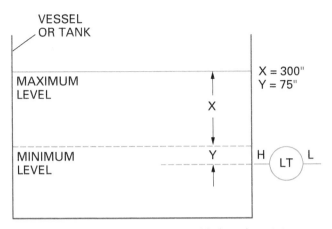

Figure 9 A level transmitter mounted below the minimum level.

Now that you know the values of *e* and *h*, you can use them with the range formula to determine the range, as follows:

range = (e) to (e + h)
range = (67.5") to (67.5" + 270")
range = 67.5" H$_2$O to 337.5" H$_2$O

This calculation is shown graphically in *Figure 10*.

2.2.5 Closed, Pressurized Vessel Installations (Level Measurement)

The system pressure above the upper level of a liquid in a closed tank or vessel affects the pressure measured at the bottom of the vessel. The pressure at the bottom of the vessel is equal to the system pressure added to the product of the liquid's specific gravity and height.

Because the differential pressure transmitter may not be capable of operating with such a significant pressure difference across it, the vessel pressure must be subtracted from the pressure at the bottom of the vessel. To accomplish this, both sides of the DP transmitter capsule must be connected to the vessel. The low side of the transmitter is connected to a tap at the top of the vessel, and the high side is connected to a tap at the bottom of the vessel. In this installation, the vessel's pressure is applied equally to both the high and low sides of the transmitter. The resulting differential pressure is proportional to the liquid's height multiplied by its specific gravity. The effects of the system pressure are nulled out, and the transmitter operates as before.

Low-side transmitter piping in a closed, pressurized vessel installation will remain empty as long as the gas above the liquid does not condense. This is referred to as a dry leg installation. The procedures applied in determining the range calculations for this type of application are the same as those applied for bottom-mounted transmitters in open vessels or tanks.

However, if the gas in a closed, pressurized vessel condenses, the condensation will eventually fill the low-side piping or tubing. To address this issue, the low-side tubing or piping is purposely filled with a convenient reference fluid. This filled pipe is known as a wet leg. The reference fluid will exert a head pressure on the low side of the transmitter capsule, and a zero elevation of the range must be made to compensate for the added head pressure. Refer to *Figure 11* as you work through the calculations for zero elevation in a closed, pressurized vessel in a wet leg condition.

The following formula can be used to find the range:

range = (e − s) to (e + h − s)

Where

e = the head pressure produced by Y (in inches of water)
s = the head pressure produced by Z (in inches of water)
h = the maximum head pressure to be measured (in inches of water)

To calculate the range using this formula, you must first find the values of *e*, *s*, and *h* in the formula. The value of *e* can be found using the following equation:

e = (Y)(sg1)

Where

e = the head pressure produced by Y (in inches of water)
Y = the vertical distance between the transmitter's high-side inlet and the minimum measurable level
sg1 = the specific gravity of the liquid in the tank

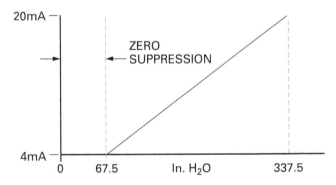

Figure 10 An open tank installation with a suppressed zero.

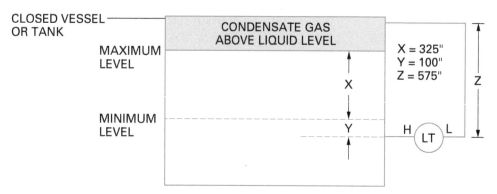

Figure 11 Compensating for specific gravity of liquid in wet leg.

In this example, Y is 100" (as shown in *Figure 11*), and the specific gravity of the liquid in the tank (sg1) is 1.0. Using this information and the equation, you can determine the value of e:

$$e = (Y)(sg1)$$
$$e = (100")(1.0)$$
$$e = 100" \, H_2O$$

Now you must determine the value of s. The following equation can be used to calculate its value:

$$s = (Z)(sg2)$$

Where

- s = the head pressure produced by Z (in inches of water)
- Z = the vertical distance between the top of the liquid in the wet leg and the transmitter's bottom tap
- $sg2$ = the specific gravity of the liquid in the wet leg

In this example, Z is 575", and the specific gravity of the liquid in the wet leg is 1.1. Using this information and the equation, you can calculate the value of s:

$$s = (Z)(sg2)$$
$$s = (575")(1.1)$$
$$s = 632.5" \, H_2O$$

The last value you must determine in order to calculate the range is h, which can be found using the following equation:

$$h = (X)(sg1)$$

Where

- h = the maximum head pressure to be measured (in inches of water)
- X = the vertical distance between the minimum and maximum measurable levels
- $sg1$ = the specific gravity of the liquid in the tank

In this example, X is 325" (*Figure 11*). Using the necessary information and the equation, you can calculate the value of h:

$$h = (X)(sg1)$$
$$h = (325")(1.0)$$
$$h = 325" \, H_2O$$

Now that you know the values of e, s, and h, you can use them with the range formula to calculate the range, as follows:

Range = $(e - s)$ to $(e + h - s)$
Range = $(100" - 632.5")$ to $(100" + 325" - 6325.5")$
Range = $-532.5" \, H_2O$ to $-207.5" \, H_2O$

This calculation is shown graphically in *Figure 12*.

2.2.6 Rosemount™ 1151 Zero and Span

The span on all Rosemount™ Model 1151 transmitters is continuously adjustable to allow calibration anywhere between maximum span and one-sixth of the maximum span. For example, the span on a Rosemount™ Model 1151 Range 4 transmitter can be adjusted between 25" H_2O and 150" H_2O.

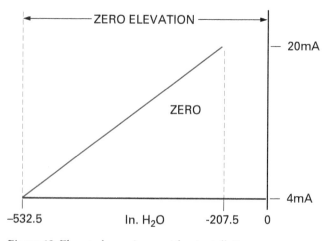

Figure 12 Elevated zero in a wet leg installation.

The zero on a Rosemount™ Model 1151 with the E or G output options can be adjusted for up to 500 percent suppression or 600 percent elevation. The zero may be elevated or suppressed to these extremes as long as no applied pressure within the calibrated range exceeds the full-range pressure limit. For example, a Range 4 transmitter cannot be calibrated for 100–200" H_2O because 200" H_2O exceeds the full-range limit of a Range 4, which is 150" H_2O.

The zero and span calibration screws are externally accessible behind the nameplate on the terminal side of the electronics housing, as shown in *Figure 13*. To increase the output of the transmitter, turn the screws in a clockwise direction. Adjusting the zero adjustment screw does not affect the span setting; however, changing the span adjustment does affect the zero setting. This effect is minimized with those spans that have zero as their base or minimum setting.

When calibrating for elevated or suppressed zeros, it is much easier on the Model 1151 to make a zero-based calibration first, and then compensate for the elevated or suppressed zero by adjusting the zero adjustment screw, because changing it does not affect the span.

The following calibration procedure steps are for a Rosemount™ 1151DP Range 4, but many other types of analog DP transmitters may be calibrated in a similar manner. While all of the steps are important and must be performed in order, pay close attention to Step 6 to reduce the number of readjustments required due to the span affecting the zero setting. In this example, the desired calibration is 0–100" H_2O.

Step 1 Connect the multifunction calibrator to the transmitter output. Set the unit for a current-based calibration.

Step 2 With zero differential pressure applied to the transmitter (pressure source vent open), adjust the zero adjustment screw until the output of the transmitter reads 4 mA on the calibrator.

Step 3 Apply a differential air pressure of 100" H_2O to the high side of the transmitter, with the low side vented to atmosphere.

Step 4 Adjust the span adjustment screw until the transmitter's output reads approximately 20 mA on the calibrator. An exact value is not needed at this stage.

Step 5 Release the input differential pressure to zero and readjust the zero output of the transmitter to again read 4 mA (±0.032 mA).

Step 6 Reapply 100" H_2O to the high side of the transmitter capsule. If the output reads greater than 20 mA, multiply the difference between the reading and 20 mA by 0.25, and subtract the result from 20 mA. If the output reads less than 20 mA, multiply the difference by 0.25, and add the result to 20 mA. Adjust the 100 percent output to this value.

> **NOTE**
> For example, if at 100" H_2O the transmitter output reads 20.100 mA, multiplying 0.100 (20.100 – 20) by 0.25 results in 0.025 mA. Subtracting 0.025 from 20 mA results in 19.975 mA. Adjust the transmitter output to 19.975 mA with 100" H_2O input differential pressure.

Step 7 Release the pressure to the high-side capsule and readjust the zero setting.

Step 8 Reapply 100" H_2O to the high-side capsule, and repeat Steps 5 through 7 until the full-scale output reads 20 mA (±0.032 mA).

Step 9 Document what you've just done using the appropriate form or electronic logging system. Follow company procedures meticulously.

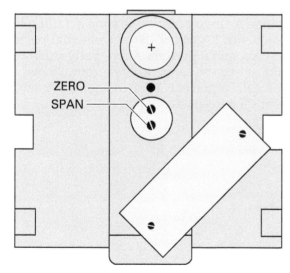

Figure 13 Location of zero and span adjustment screws for Rosemount™ Model 1151 transmitter.

2.2.7 Temperature Transmitters with Analog Outputs

Measuring or monitoring temperature plays a key role in many industrial and commercial processes. Examples include measuring cooking temperature in food processing, monitoring the temperature of molten steel in a mill, verifying the temperature in a cold storage warehouse, or regulating temperatures in the drying rooms of a paper manufacturer.

In an analog temperature control loop, a temperature transmitter uses some type of measuring device to sense the temperature and then regulates a 4–20 mA feedback loop to a temperature control element. The final control element might consist of a temperature control valve (TCV) that opens or closes to allow more steam into a heating process or more fuel to a burner. Conversely, it could be a relay or valve that controls a cooling system.

The two most common types of temperature sensing devices connected to a temperature transmitter are thermocouples (TCs) and resistance temperature detectors (RTDs). There are a broad range of temperature calibration tools to help you quickly and reliably calibrate temperature instrumentation. Many multifunction calibrators have this capability. Dedicated temperature calibrators, such as the Fluke® 724, also perform this task.

A good calibrator offers several temperature-related functions, including thermocouple identification and simulation. This second feature is especially handy when calibrating temperature transmitters.

The following is a general temperature transmitter calibration procedure that identifies the major steps in this process. Consult your calibrator documentation for specific details.

Step 1 Connect the calibrator's measuring leads to the temperature transmitter's output terminals and set the calibrator to provide loop power.

Step 2 Connect the calibrator's thermocouple jack to the transmitter's thermocouple input. Set the calibrator to simulate the correct type of thermocouple.

Step 3 Select the correct temperature scale (C or F).

Step 4 Set the thermocouple simulator to output the temperature that corresponds to the zero point (0 percent).

Step 5 Read the transmitter output current. It should be 4.00 mA. If it's not, adjust the transmitter's zero control to bring the reading to 4.00 mA.

Step 6 Set the thermocouple simulator to output the temperature that corresponds to the maximum point (100 percent).

Step 7 Read the transmitter output current. It should be 20.00 mA. If it's not, adjust the transmitter span control to bring the reading to 20.00 mA.

Step 8 Repeat Steps 4 through 7 until both values remain correct.

Step 9 Disconnect all equipment from the transmitter and restore it to normal operation, following company procedures meticulously. Document the calibration using company procedures and forms.

If you wish to perform a three- or five-point calibration instead of the simple zero and span calibration described above, follow the same procedure but substitute the appropriate sequence of points. A three-point calibration should follow the sequence 0 percent, 50 percent, 100 percent, 50 percent, and 0 percent. A five-point calibration should follow the sequence 0 percent, 25 percent, 50 percent, 75 percent, 100 percent, 75 percent, 50 percent, 25 percent, and 0 percent.

Additional Resources

Measurement and Control Basics, Thomas A. Hughes. Fifth Edition. 2014. Research Triangle Park, NC: International Society of Automation.

The following websites offer resources for products and training:

Crystal Engineering Corporation, **www.crystalengineering.net**

Emerson Process Management, **www.emersonprocess.com**

Fluke Corporation, **www.fluke.com**

GE, Measurement & Control, **www.gemeasurement.com**

2.0.0 Section Review

1. A pneumatic pressure transmitter has an input range of 0–100" H_2O and an input span of 50–100" H_2O. This transmitter has zero _____.
   a. elevation
   b. suppression
   c. nulling
   d. spanning

2. When calibrating a pneumatic transmitter, what is the first step?
   a. Equalize the pressure across it.
   b. Open the calibration ports.
   c. Isolate it from the process.
   d. Disconnect the pneumatic output.

3. Filling a process connection with a reference fluid to deal with a process gas that can condense is an example of creating a(n) _____.
   a. suppressed leg
   b. elevated leg
   c. dry leg
   d. wet leg

SECTION THREE

3.0.0 SMART TRANSMITTERS

Objective

Identify and describe smart transmitters and their calibration process.
  a. Describe various communication protocols and devices used for communication.
  b. Explain how to calibrate HART® devices.

Performance Task

4. Calibrate a smart transmitter using a HART® communicator and complete the appropriate documentation.

Trade Terms

Factory characterization: Information describing a sensor's characteristics that has been generated by the manufacturer and stored within the device.

Look-up table: A table of input and output values stored within an instrument. The device compares its input signal with the input side of the look-up table until it finds a match. It then outputs the corresponding output value.

Transfer function: A mathematical equation that describes an input-to-output relationship.

Smart transmitters operate differently from ordinary analog transmitters. A smart transmitter contains an embedded processor that manipulates information digitally and transmits it in a number of different forms. It also stores data about the attached sensor's specific characteristics and compensates for these sensor variations. The factory process for generating the sensor performance profile is called factory characterization. Today's smart transmitters communicate using a number of different digital protocols. The most common one is the Highway Addressable Remote Transducer (HART®). It can be used to communicate device configuration, status, measurement, and diagnostic information and is invaluable for calibrating instruments.

3.1.0 The HART Protocol

The HART protocol is a globally used open protocol for automation communication among smart instruments. It was developed by Rosemount, Inc., but all rights now belong to the independent HART Communication Foundation, an international organization that supports the protocol and oversees its development. The HART protocol can be used with two- and four-wire HART-capable devices. Such devices typically include transmitters, valve positioners and actuators, magnetic flowmeters, level devices, liquid and gas analyzers, and control devices.

HART smart devices are designed to provide a wide range of data that is accessible using standard commands. Typically, there are about 35 accessible data items in every HART device. These include device status, diagnostic alerts, process variables and units, loop current and percent range, basic configuration parameters, and manufacturer and device-specific information.

The HART protocol provides two-way digital communications compatible with the industry standard 4–20 mA analog signaling used by traditional instrumentation equipment. It allows for simultaneous communication of the continuous 4–20 mA analog signal with a second digital communications signal superimposed at a low level on top of the analog signal. Data for up to four measurements can be transmitted in a single message, and HART protocol multi-output devices have been developed to take advantage of this. In addition, if only digital communication is needed, several devices can be connected in parallel on a single pair of wires. Each device has its own address, so a host device can communicate with each one in turn.

Communication with HART devices can be done with a computer that has a HART interface. PLCs and DCSs also can include HART interfaces. One of the most common tools for working with this protocol is a HART communicator (*Figure 14*). These handheld devices are convenient for use in the field and offer a wide variety of services. Many can communicate using other digital protocols, such as FOUNDATION™ Fieldbus, as well. Handheld communicators should be kept updated so they can interact with the most recent devices. Consult the manufacturer's documentation for this procedure.

Figure 14 Handheld HART communicator.

3.2.0 Calibrating HART Devices

As with all instruments, HART devices require periodic calibration. Unlike many conventional instruments, the specific calibration requirements for any particular HART device depend on the application. Calibration of a HART device involves the following three functions performed by the processor embedded within the device:

- Calibrating the process variable
- Scaling the process variable
- Producing the process signal

3.2.1 HART Device Internal Stage Summary

Figure 15 shows the three stages of the HART device that are associated with the calibration process. As shown, the input to the first stage is a raw digital count that represents the measured value of the analog process signal. Depending on the device, this measured value can be in volts, current, frequency, capacitance, or some other property. Conversion of the analog process signal to a corresponding digital equivalent prior to being sent to the first stage is done by an analog-to-digital converter (ADC) in the processor.

In the input stage, a built-in equation or look-up table is used to correlate the process signal raw count to the actual process variable (PV), such as temperature, pressure, or flow. The output of this stage is a digital representation of the measured process signal. This PV value can be accessed and read using a HART communicator. The look-up table used in this stage is normally provided by the manufacturer. Most devices also have the capability for making field adjustments; this is called *sensor trim*.

The second stage converts the digital PV input signal into a corresponding digital milliamp (mA) representation, with a 4 mA value being the lower end of the range and a 20 mA value being the upper end. Percent range is also determined in this stage. These conversions are done mathematically, using the range values of the device as they relate to the device zero and span values, in conjunction with the device's transfer function. The transfer function is typically linear, but some devices, such as pressure transmitters, have a square root transfer function instead. The output of this stage is a value that can be read with a HART communicator when reading loop current.

Note that many HART devices can be commanded to enter a test mode in which they output a fixed signal. This mode causes a specific current (mA) value to be substituted for the normal output from the second stage.

The third stage converts the digital milliamp value from the second stage into a digital form that controls a digital-to-analog converter (DAC). This device produces an actual 4–20 mA analog

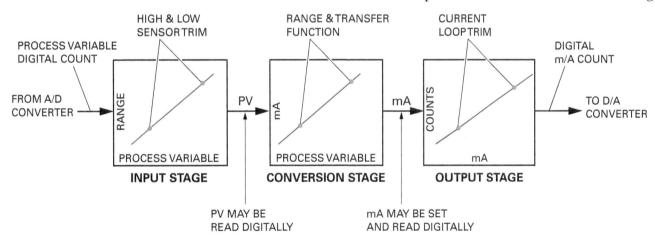

Figure 15 Principles of operation for a HART transmitter device.

signal. In this stage, internal calibration factors are used to obtain the correct output signal values (4 mA corresponds to 0 percent and 20 mA corresponds to 100 percent). Adjustment of these factors is commonly called *current loop trim*.

3.2.2 The Input Stage

Calibration of the input stage is required for all device applications. Note that if the HART device is used as a digital device only (the analog current loop output is not used), calibration of the input stage is all that is required. When testing the input stage, a multifunction calibrator is used to measure and possibly also supply a known input signal. A HART communicator is used to measure the resultant output (PV) signal. Because there is always a linear relationship between the input and output signals, and both are recorded in the same measurement units, error calculations are more easily made. The test results should meet the accuracy specified by the device manufacturer.

If the test fails, the stage should be adjusted by following the manufacturer's procedure for trimming the input stage. The procedure usually involves one or two trim points. Some devices, such as pressure transmitters, may also have a zero trim, where the stage is adjusted to read exactly zero, instead of low range.

> **NOTE**
> Do not confuse a trim with any form of re-ranging or any procedure that involves using zero and span buttons.

3.2.3 The Middle Stage

When both the input and output stages of a HART device are calibrated, it should operate correctly. No calibration is required for the second stage because it only performs calculations. This feature allows the range, units, and transfer functions to be changed without affecting the calibration of the input or output stages.

3.2.4 The Output Stage

Calibrating the output stage involves using a HART communicator to put the device into a fixed current output mode. When you specify the desired output current, the device outputs this value. A multifunction calibrator is then used to measure the resulting output current. Because there is a linear relationship between the input and output signals and both are recorded in milliamps, error calculations are more easily made. The test result should meet the accuracy specified by the device manufacturer.

If the test fails, the output stage should be adjusted following the manufacturer's procedure for trimming the stage. The manufacturer may call this procedure *current loop trim*, *D/A trim*, or *4–20 mA trim* in its service literature. Adjustment involves two trim points close to or just outside of 4 and 20 mA.

3.2.5 Smart Pressure Transmitter HART Calibration

HART devices may be calibrated using a multifunction calibrator and a HART communicator. However, a number of manufacturers produce units that combine all functions in a single unit. The Fluke® 754 is one such example (*Figure 16*). When using this device to calibrate a pressure transmitter, you'll also need a suitable pressure module and a hand pump pressure source.

The following is a generic procedure for calibrating a smart pressure transmitter with a combined calibrator/communicator. As always, consult your equipment's documentation for specific details.

Step 1 Connect the calibrator/communicator to the transmitter outputs and set it to energize the loop. Some communicators require a suitable resistor to be connected in series to simulate the correct loop conditions. If this is required, connect it as well.

Step 2 Connect the calibrator to the transmitter inputs using the correct pressure module and tubing. Connect the pressure source as well.

Figure 16 Fluke® 754 combined calibrator and communicator.

Step 3 Open the pressure source vent so the transmitter input sees zero pressure. Push the pressure-zero button on the calibrator.

Step 4 If the process is based on true zero, set the sensor lower value. If it is not zero-based, apply the pressure equivalent to the sensor lower value. Close the hand pump vent and pressurize close to the desired value. Use the pump vernier knob to set the pressure exactly. Some calibrators can detect the correct value and alert you when it has been reached.

Step 5 Follow the same procedure to set the sensor upper value.

Step 6 Adjust the pressure and repeat the procedure for each additional calibration point.

3.2.6 Test and Calibration Considerations

Some other factors that must be taken into consideration when calibrating HART devices are identified as follows:

- *Damping* – Some HART devices support a damping parameter that causes a delay between a change in the device input and the detection of that change for output from the device. This delay can exceed the settling time used for the test and calibration process. Settling time is the time the test or calibration process waits between sending the input to the device and reading the resulting output. For this reason, the device damping value should always be set to zero before performing any test or adjustment. After the test or calibration is done, the damping value should be reset to its required value.

- *Digital range change* – Many technicians think that using a HART communicator to change the range of a HART device calibrates the device. This is a common misconception. Changing the range causes a configuration change only. This is because a change in range affects only the conversion process performed by the second stage (refer to *Figure 15*) of the HART calibration process. It has no effect on the digital process value as read by a communicator. Calibration always requires a traceable external device, such as a multifunction calibrator, to supply and monitor an input signal to the device under calibration.

- *Zero and span adjustment* – Employing only the zero and span procedures used with conventional analog devices to calibrate a HART device can yield a 4 mA output indicating correct adjustment. However, the internal digital readings are often corrupted. This can be determined by reading the internal digital values using a HART communicator. Corruption of the internal digital values occurs because the zero and span buttons change the range. This affects only the second stage conversion process of the HART calibration process. A zero trim procedure should be performed to make the appropriate internal adjustments to correct for a zero shift condition in a HART device. This adjusts the input stage so that the digital PV value corresponds to a calibration standard.

  The following example illustrates this type of problem. Assume that a technician installs and tests a differential pressure device that was set at the factory for a range of 0–100" H_2O. When the device is tested, it is found that it now has a zero shift of +1" H_2O. Therefore, with both sides of the capsule vented (zero), its output is 4.16 mA instead of 4.00 mA. When 100" H_2O is applied, its output is 20.16 mA instead of 20.00 mA. To remedy this, the technician vents both sides and presses the zero button on the device, causing the output to go to 4.00 mA. Based on getting the desired 4 mA output, the technician assumes that the adjustment was successful. However, when the device is checked using a HART communicator, the range is 1–101" H_2O at the test points, and the PV reads 1" H_2O instead of 0. In other words, the problem has been hidden, not fixed.

- *Loop current adjustment* – Technicians often use a handheld HART communicator to adjust a device's loop current so that an accurate input to the device agrees with some display device on the loop. On some communicators, this parameter is called *current loop trim*. Unfortunately, this adjustment strategy can result in corrupted internal digital readings because it masks a calibration problem in the input stage by compensating for it in the output stage.

  To illustrate this type of problem, use the zero shift example described earlier. Assume there is a digital indicator in the loop that displays 0.0 at 4 mA and 100.0 at 20 mA. During testing, the display reads 1.0 with both sides of the capsule vented (zero), and 101.0 with 100" H_2O applied.

Using the communicator, the technician adjusts the loop current so that the display device reads 0 and 100. While this appears to solve the problem, there is a fundamental calibration issue. The HART communicator will show that the PV still reads 1 and 101" H$_2$O at the test points. The digital reading of the mA output still reads 4.16 mA and 20.16 mA, even though the actual output from the device is 4 mA and 20 mA. In other words, the problem has been hidden but not fixed. Instead of using loop current trim, the technician should perform a complete calibration of the input stage to eliminate the problem at its source. Only when the PV values are both internally and externally correct can the device be said to be properly calibrated. The HART communicator is extremely useful here since it allows the values of each stage to be examined.

Additional Resources

Measurement and Control Basics, Thomas A. Hughes. Fifth Edition. 2014. Research Triangle Park, NC: International Society of Automation.

Maintenance and Calibration of HART® Field Instrumentation (PDF), R. Pirret. Fluke Corporation, Everett, WA, USA. Accessed at **www.plantservices.com**.

The following websites offer resources for products and training:

Emerson Process Management, **www.emersonprocess.com**

Fluke Corporation, **www.fluke.com**

3.0.0 Section Review

1. The HART protocol is carried over _____.
    a. pneumatic signal lines
    b. 1–5 V analog voltage lines
    c. 4–20 mA analog current loops
    d. FOUNDATION Fieldbus

2. The third stage of a HART device may be adjusted to modify the _____.
    a. current loop trim
    b. input trim
    c. transfer function
    d. look-up table

3. Calibrating a HART device always requires two tools, a HART communicator and a _____.
    a. current loop
    b. multifunction calibrator
    c. voltage loop
    d. power meter

Section Four

4.0.0 Transducers and Control Valve Positioners

Objective

Explain how to calibrate transducers and control valve positioners.
a. Explain how to calibrate transducers.
b. Explain how to calibrate pneumatic, electro-pneumatic, and smart control valve positioners.

Performance Tasks

5. Calibrate a transducer and complete the appropriate documentation.
6. Calibrate the following valve positioners and complete the appropriate documentation:
   - Pneumatic positioner
   - Electro-pneumatic positioner
   - Smart positioner (digital valve controller)

Although calibration is considered an activity associated with measuring instruments, it's equally important for other types of instruments as well. Certain kinds of actuators, such as valve positioners, require calibration. Transducers often require calibration as well. In both cases, as with measuring instruments, failure to calibrate can lead to problems throughout the instrument channel.

4.1.0 Transducers

The function of the transducer is to receive a linear signal in one energy form and convert it into a linear signal in another energy form. The incoming signal may be pneumatic, voltage, current, vibration, force, or sound. The outgoing signal may be in almost any form, although pneumatic and electrical signals are the most common.

Many control valves operate on pneumatic energy. But instrumentation control loops are often 4–20 mA analog. Consequently, I/P transducers are needed to convert the current signal into a linear and proportional pneumatic signal that the control valve can use. Transducers that convert a voltage signal to a pneumatic signal are also available and are referred to as E/P transducers.

Calibrating any transducer is fairly simple, as long as the correct calibration equipment is used. Generally, the process involves supplying the correct form of input signal within the operating range of the transducer and then measuring and adjusting the output signal so that it is linear and proportional to the incoming signal. Most transducers can be bench-calibrated and taken to the field for installation.

Figure 17 shows the internal components of an I/P transducer. It is an electro-pneumatic device that accepts a 4–20 mA current signal and converts it to a pneumatic signal. This conversion is accomplished by applying the current to a coil that creates a magnetic field, which causes a flexible arm to move.

Like the pneumatic transmitter, the transducer's pneumatic section operates on a force-balance principle. The arm manipulates a flapper that floats over a nozzle that is continuously exhausting air. The varying backpressure controls the diaphragm of the pneumatic relay directly below. This device, in turn, provides the pneumatic output signal. Since the magnetic field is directly proportional to the input current in the coil, it is the current that ultimately controls the pneumatic output through a number of intermediate stages.

4.2.0 Control Valve Positioners

The function of a control valve positioner is to compare the valve's stem position to the input signal and manipulate the actuator so that its position matches the input signal. The positioner is usually connected to the actuator's stem by a mechanical link, as shown in *Figure 18*. The position of the mechanical link indicates the stem's position, which is compared to the incoming signal. If the two don't match, the positioner either increases or decreases the control signal to the valve diaphragm in order to bring the stem to the correct position.

> **NOTE**
> Positioners can be as sensitive as any other instrument. Use care when adjusting them, and use the proper tools.

4.2.1 Pneumatic and Electro-Pneumatic Positioners

A pneumatic valve positioner is generally used with a diaphragm-actuated, sliding-stem control valve assembly. The pneumatic valve positioner receives a pneumatic input signal from a control device and modulates the supply pressure to

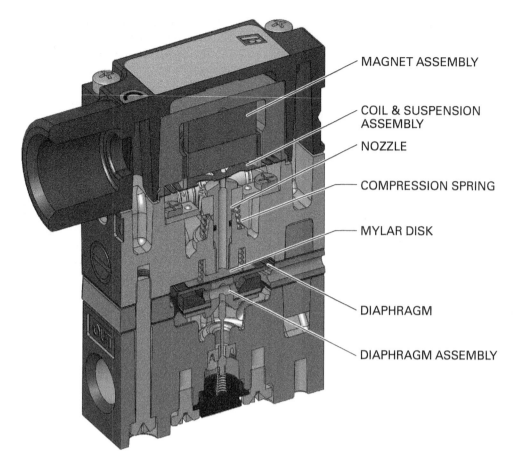

Figure 17 Cut-away of an I/P transducer.

the control valve actuator, providing an accurate valve stem position that is proportional to the pneumatic input signal.

Figure 19 is an operational schematic of a direct-acting Fisher® 3582 Series pneumatic positioner. The pneumatic positioner receives a pneumatic input signal from a control device. A dry, regulated instrument air supply is connected to the pneumatic relay. A fixed restriction in the relay limits flow to the nozzle so that when the flapper is not restricting the nozzle, air can bleed out faster than it is being supplied.

The input signal from the control device is connected to the bellows. When the input signal increases, the bellows expands and moves the balance beam. The beam pivots about the input axis, moving the flapper closer to the nozzle. As the nozzle backpressure increases, it increases the pressure on the valve actuator diaphragm. This increased output pressure to the actuator causes the actuator stem to move downward.

Stem movement is fed back to the beam through the mechanical linkage connected to a cam in the positioner. As the cam rotates, the beam pivots about the feedback axis to move the flapper slightly away from the nozzle. The nozzle pressure decreases and reduces the output pressure to the actuator. Stem movement continues, which causes the flapper to back away from the nozzle until equilibrium is reached.

When the input signal decreases, the bellows contracts (aided by an internal range spring), and the beam pivots about the input axis to move the flapper away from the nozzle. Nozzle backpressure decreases, and the pneumatic relay vents the diaphragm casing pressure to the atmosphere. The actuator stem moves upward due to spring pressure. Through the mechanical linkage and cam movement, the stem's movement is fed back to the beam, which repositions the flapper closer to the nozzle. When equilibrium conditions are obtained, stem movement stops, and the flapper is positioned to prevent any further decrease in diaphragm case pressure.

The principle of operation for reverse-acting pneumatic valve positioners is similar except that as the input signal increases, the diaphragm casing pressure is decreased. Conversely, a decreasing input signal causes an increase in the pressure to the diaphragm casing. The action of many pneumatic positioners, whether direct or reverse, can be easily changed by a simple field adjustment.

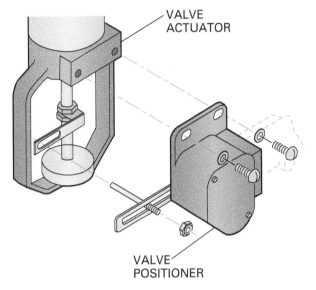

Figure 18 Mechanical link of a valve actuator stem to a positioner.

Electro-pneumatic valve positioners are very similar to purely pneumatic positioners except that a DC current signal is received by an electro-pneumatic signal converter (transducer) attached to the positioner. This is shown in *Figure 20*, an illustration of a Fisher® 3582i Series electro-pneumatic positioner.

Calibrating positioners is nearly as simple as calibrating transducers in that the output of the pneumatic relay must be linear and proportional to the input signal. However, in the process of calibrating the positioner's output to the input, the valve's actuator stem position must also be set so that it responds correctly with the positioner output signal.

In the case of direct-acting positioners, the valve's actuator stem should just reach its fully downward position as the positioner's output signal reaches 15 psi (100 kPa). Likewise, at 3 psi (20 kPa) output, the actuator stem position should

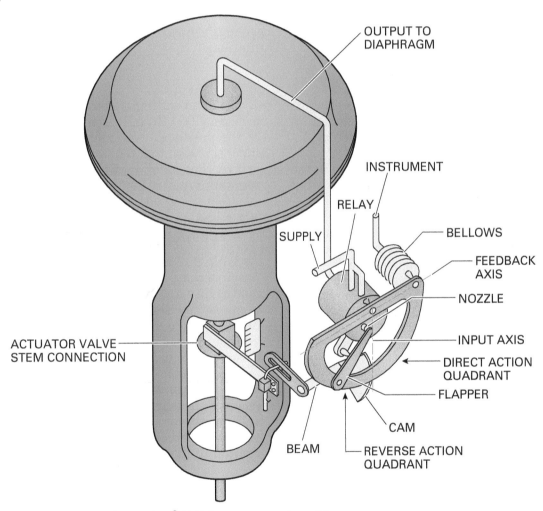

Figure 19 Operation schematic for a Fisher® 3582 Series pneumatic positioner.

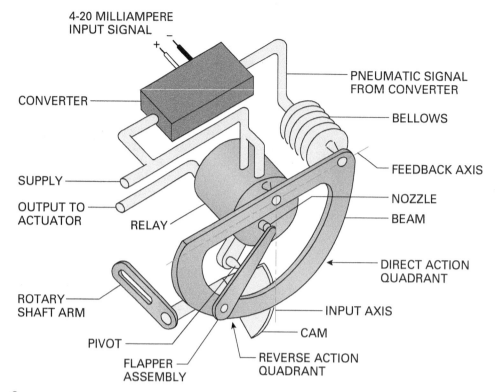

Figure 20 Fisher ® 3582i Series electro-pneumatic positioner.

just reach its fully upward position. It is important that even the slightest pressure change in the output signal of the positioner should begin a movement in the actuator stem.

Just because a valve stem is all the way in the downward position at 15 psi (100 kPa) does not mean that the positioner and valve stem movement are calibrated to one another. For example, the positioner and stem are not properly calibrated unless at a value just barely below 15 psi (100 kPa) the valve's stem begins a slight movement up. The same goes for the opposite end of the range. The valve stem should show movement anywhere within the range of 3 to 15 psi (20–100 kPa).

With the cover removed from the positioner and a 3 psi (20 kPa) or 4 mA signal applied to the input of the positioner (depending on type), check the 0 mark on the cam and the valve seating. At the 0 mark, it is typically desired to have the valve plug just on the seat for direct-acting valves, and the cam at the 0 mark with the positioner still dynamic (not dead-ended). Adjust the zero adjustment so that these parameters are met.

The span setting is accomplished by applying a 15 psi (100 kPa) or a 20 mA signal to the input of the positioner and adjusting the span adjustment screw so that the upward travel of the stem just reaches its upper limits while the positioner remains dynamic (not dead-ended).

4.2.2 Smart Positioners (Digital Valve Controllers)

As with most instrumentation, control valves and positioners are also being replaced with smart technology instruments. One example of a smart positioner is the Fisher® FIELDVUE™ DVC6000 Series digital valve controller, illustrated by the block diagram in *Figure 21*. It is a communicating, microprocessor-based current-to-pneumatic instrument.

In addition to the traditional electro-pneumatic function of converting a current signal to a valve-position pressure signal, DVC6000 Series digital valve controllers give easy access to information critical to process operation, using the HART communications protocol. This can be done using

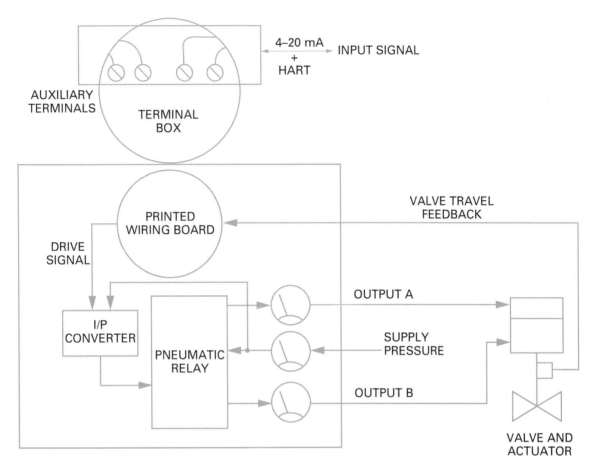

Figure 21 Fisher ® FIELDVUE™ DVC6000 Series digital valve controller block diagram.

a HART communicator at the valve or at a field junction box, or by using a personal computer or a system console within the control room. Using the HART communication protocol, information can be integrated into a control system or received on a single loop basis.

The digital valve controller uses two-wire, 4–20 mA loop power and receives feedback of the valve's travel position as well as supply and actuator pneumatic pressure. This allows the instrument to diagnose not only itself but also the valve and actuator to which it is mounted. These digital valve controllers can directly replace older analog instruments because they operate with a traditional 4–20 mA control signal. Calibration can be performed in the field using the HART handheld communicator, or it may be performed from the control room if the system is integrated with DCS technology.

Additional Resources

Measurement and Control Basics, Thomas A. Hughes. Fifth Edition. 2014. Research Triangle Park, NC: International Society of Automation.

Maintenance and Calibration of HART® Field Instrumentation (PDF), R. Pirret. Fluke Corporation, Everett, WA, USA. Accessed at **www.plantservices.com**.

The following websites offer resources for products and training:

Emerson Process Management, **www.emersonprocess.com**

Fluke Corporation, **www.fluke.com**

4.0.0 Section Review

1. An I/P transducer converts an electrical signal to a pneumatic one by applying the current to a _____.
   a. float
   b. limiter
   c. coil
   d. sequencer

2. A valve actuator's stem is moved to a precise position by a _____.
   a. strain gauge
   b. spring
   c. sequencer
   d. positioner

3. A smart digital valve controller may be calibrated from the control room if the controller is part of a(n) _____.
   a. DCS
   b. IANA
   c. TCP/IP
   d. UDP

SUMMARY

Only those instruments that incorporate both an input and output signal require calibration. The primary instrument that requires calibration is the field transmitter because it is the instrument that receives the signal that represents what is being monitored and controlled in the process control loop.

This module covered the transmitters that are common in many installations throughout the industry, including pneumatic, analog, and smart transmitters. Manufacturers often specify calibration procedures for their instruments, and these specific procedures should be followed when calibrating the instrument in order to achieve accuracy and repeatability. This module presented some general procedures that can be applied to most instruments during the calibration process.

Smart instruments must always be calibrated by following the specific steps found in the user's manual for the instrument, as these instruments are generally matched to loop parameters and cannot be calibrated using standard analog calibration procedures. Always use the proper equipment, such as a HART communicator, when calibrating or testing these instruments.

This module introduced the basic principles of the HART protocol and provided an introduction to calibrating HART transmitters. It also described digital valve controllers, which are rapidly replacing the older analog positioner technology.

Review Questions

1. All instruments that provide a proportional output signal based on an input signal or level of energy must be accurately _____.
   a. aligned
   b. balanced
   c. calibrated
   d. converted

2. For a device calibration to be valid, the instruments used to calibrate the device must have a _____.
   a. manufacturer's receipt
   b. certificate of calibration
   c. dated warranty document
   d. certificate of origin

3. Instruments that are in direct contact with the process are designed and manufactured to operate within a specified process _____.
   a. liquid level
   b. input energy range
   c. operating range limit
   d. minimum operating limit

4. In an operating range of 40–200" H$_2$O, 50 percent of the span would be _____.
   a. 80" H$_2$O
   b. 100" H$_2$O
   c. 120" H$_2$O
   d. 140" H$_2$O

5. Setting the minimum output signal of a transmitter to proportionally represent the minimum input value is referred to as _____.
   a. setting the span
   b. nulling the instrument
   c. zeroing the instrument
   d. adjusting the deadband

6. A three-point calibration checks an instrument at the following percentages of span: 0 percent, 50 percent, _____.
   a. 100 percent, 0 percent, and 50 percent
   b. 25 percent, 75 percent, and 100 percent
   c. 75 percent, 25 percent, and 100 percent
   d. 100 percent, 50 percent, and 0 percent

7. What does a pneumatic DP transmitter do with its high- and low-pressure sides?
   a. It adds them.
   b. It subtracts them.
   c. It multiplies them.
   d. It divides them.

8. Some multifunction calibrators can sequence through input signal values and take output readings at key calibration points. This feature is called _____.
   a. incrementing
   b. auto-ramp
   c. cycling
   d. decrementing

9. When used for liquid level applications, DP transmitters measure _____.
   a. head pressure
   b. pneumatic differential pressure
   c. specific gravity
   d. density reduction

10. A differential pressure transmitter installation used for level measurement in which the low side of the transmitter piping remains empty is called _____.
    a. dry leg
    b. force balance
    c. inches of water
    d. control loop

11. Many multifunction calibrators used for temperature transmitter calibration include a feature in which a specific thermocouple is _____.
    a. zeroed
    b. characterized
    c. spanned
    d. simulated

12. The factory process of generating a sensor's performance profile for a smart instrument is called factory _____.
    a. three-point calibration
    b. five-point calibration
    c. characterization
    d. zeroing

12402-16 Instrument Calibration and Configuration

13. Which stages in a HART device can be adjusted during the calibration process?
    a. Stage 1 only
    b. Stages 1 and 2 only
    c. Stages 1 and 3 only
    d. Stages 1, 2, and 3

14. A parameter that causes a delay between a change in the device input and a change in the output is called _____.
    a. zeroing
    b. damping
    c. converting
    d. ranging

15. A valve positioner compares a valve stem's position against the actuator's _____.
    a. input signal
    b. output signal
    c. valve damping
    d. valve cycling

Trade Terms Introduced in This Module

Calibration: The process of adjusting an instrument so its output signal or level of energy is representative of the level of the non-adjustable input energy at any given point. Calibration involves test instruments traceable to a national standards body.

Dry leg: A connection from the process that contains a dry, non-condensing gas.

Elevated zero: A zero value that's above the true zero point.

Factory characterization: Information describing a sensor's characteristics that has been generated by the manufacturer and stored within the device.

Five-point calibration: A method of calibrating an instrument in which the output level is checked and set against a simulated input level at five percentages of span: 0 percent, 25 percent, 50 percent, 75 percent, 100 percent, 75 percent, 50 percent, 25 percent, and 0 percent, in that order.

Fulcrum: The pivot point around which a lever turns.

Head pressure: The pressure that results from a liquid column of a certain height.

Inches of water: A measurement of pressure (1" H_2O = 0.03612628 psi, or 249 Pa).

Look-up table: A table of input and output values stored within an instrument. The device compares its input signal with the input side of the look-up table until it finds a match. It then outputs the corresponding output value.

Multifunction loop calibrator: A test instrument that provides an energy source and measuring capabilities in the same equipment housing. Multifunction calibrators often perform many different types of calibrations.

Range: The minimum and maximum values that an instrument can measure.

Span: The difference between an instrument's upper and lower measurement limits.

Spanning: Setting an instrument's output so that it is at its maximum value when the input process value is at the maximum value of the specified operating range.

Suppressed zero: A zero value that's below the true zero value.

Three-point calibration: A method of calibrating an instrument in which the output level is checked and set against a simulated input level at three different percentages of span: 0 percent, 50 percent, 100 percent, 50 percent, and 0 percent, in that order.

Transfer function: A mathematical equation that describes an input-to-output relationship.

Wet leg: A connection from the process that's intentionally filled with a reference fluid to prevent gas condensation from affecting the measurement.

Zeroing: Setting an instrument's output so that it is at its minimum value when the input process value is at the minimum value of the specified operating range.

Additional Resources

This module presents thorough resource for task training. The following resource material is recommended for further study.

Measurement and Control Basics, Thomas A. Hughes. Fifth Edition. 2014. Research Triangle Park, NC: International Society of Automation.

Maintenance and Calibration of HART®Field Instrumentation (PDF), R. Pirret. Fluke Corporation, Everett, WA, USA. Accessed at **www.plantservices.com**.

The following websites offer resources for products and training:
- Crystal Engineering Corporation, **www.crystalengineering.com**
- Emerson Process Management, **www.emersonprocess.com**
- Fluke Corporation, **www.fluke.com**
- GE, Measurement & Control, **www.gemeasurement.com**

HART®, *Wireless*HART®, and HART-IP™ are registered trademarks of the FieldComm Group™, **www.fieldcommgroup.org**

Figure Credits

©Curraheeshutter/Dreamstime.com, Module Opener

Fluke Corporation, reproduced with permission, Figures 1, 3A, 7, 16

GE, Measurement & Control, Figure 3B

Crystal Engineering Corporation, Figure 3C

WIKA Instrument, LP, Figure 3D

Topaz Publications, Inc., Figure 6

Emerson Process Management, Figures 8, 14, 19, 20, 21

Fairchild Industrial Products Company, Figure 17

Section Review Answer Key

Answer	Section Reference	Objective
Section One		
1. d	1.1.3	1a
2. c	1.1.3	1a
3. b	1.2.0	1b
Section Two		
1. a	2.1.3	2a
2. c	2.1.4	2a
3. d	2.2.5	2b
Section Three		
1. c	3.1.0	3a
2. a	3.2.1, 3.2.4	3b
3. b	3.2.5	3b
Section Four		
1. c	4.1.0	4a
2. d	4.2.0	4b
3. a	4.2.2	4b

NCCER CURRICULA — USER UPDATE

NCCER makes every effort to keep its textbooks up-to-date and free of technical errors. We appreciate your help in this process. If you find an error, a typographical mistake, or an inaccuracy in NCCER's curricula, please fill out this form (or a photocopy), or complete the online form at **www.nccer.org/olf**. Be sure to include the exact module ID number, page number, a detailed description, and your recommended correction. Your input will be brought to the attention of the Authoring Team. Thank you for your assistance.

Instructors – If you have an idea for improving this textbook, or have found that additional materials were necessary to teach this module effectively, please let us know so that we may present your suggestions to the Authoring Team.

NCCER Product Development and Revision
13614 Progress Blvd., Alachua, FL 32615

Email: curriculum@nccer.org
Online: www.nccer.org/olf

❏ Trainee Guide ❏ Lesson Plans ❏ Exam ❏ PowerPoints Other _____

Craft / Level: _____ Copyright Date: _____

Module ID Number / Title: _____

Section Number(s): _____

Description: _____

_____

_____

_____

_____

Recommended Correction: _____

_____

_____

_____

_____

Your Name: _____

Address: _____

_____

Email: _____ Phone: _____

12410-16

Proving, Commissioning, and Troubleshooting a Loop

Overview

Once an instrument loop has been assembled, it must go through a series of steps so that it can be brought on line with live process. It is very likely that problems will arise along the way. These will require troubleshooting, followed by corrective action. This module describes the sequence of events that should culminate in a working loop, ready to perform its duties in the overall process.

Module Two

Trainees with successful module completions may be eligible for credentialing through the NCCER Registry. To learn more, go to **www.nccer.org** or contact us at 1.888.622.3720. Our website has information on the latest product releases and training, as well as online versions of our *Cornerstone* magazine and Pearson's product catalog.

Your feedback is welcome. You may email your comments to **curriculum@nccer.org**, send general comments and inquiries to **info@nccer.org**, or fill in the User Update form at the back of this module.

This information is general in nature and intended for training purposes only. Actual performance of activities described in this manual requires compliance with all applicable operating, service, maintenance, and safety procedures under the direction of qualified personnel. References in this manual to patented or proprietary devices do not constitute a recommendation of their use.

Copyright © 2016 by NCCER, Alachua, FL 32615, and published by Pearson Education, Inc., New York, NY 10013. All rights reserved. Printed in the United States of America. This publication is protected by Copyright, and permission should be obtained from NCCER prior to any prohibited reproduction, storage in a retrieval system, or transmission in any form or by any means, electronic, mechanical, photocopying, recording, or likewise. To obtain permission(s) to use material from this work, please submit a written request to NCCER Product Development, 13614 Progress Blvd., Alachua, FL 32615.

From *Instrumentation Level Four, Trainee Guide*, Third Edition. NCCER.
Copyright © 2016 by NCCER. Published by Pearson Education. All rights reserved.

12410-16
PROVING, COMMISSIONING, AND TROUBLESHOOTING A LOOP

Objectives

When you have completed this module, you will be able to do the following:

1. Describe how to inspect loop components and perform continuity checks prior to proving the loop.
   a. Describe how to visually inspect various loop components.
   b. Describe how to conduct loop continuity tests on electrical and pneumatic devices.
2. Describe how to prove and calibrate a loop.
   a. Describe how to prove a loop.
   b. Describe how to calibrate a loop.
3. Describe how to commission a new loop.
   a. Describe the documents associated with commissioning.
   b. Describe the commissioning process.
4. Identify the fundamental steps in loop troubleshooting and describe the troubleshooting process.
   a. Identify the fundamental steps in loop troubleshooting.
   b. Describe the loop troubleshooting process for oscillating loops.

Performance Tasks

Under the supervision of the instructor, you should be able to do the following:

1. Perform a continuity check on an electrical system and document the findings.
2. Perform a continuity check on a pneumatic system and document the findings.
3. Prove a loop and document its completion.
4. Commission a loop.
5. Troubleshoot a newly installed control loop.
6. Troubleshoot an oscillating process.

Trade Terms

Cavitation
Commissioning
Continuity tester
Current loop trim
Ground loop

Instrument asset management system (IAMS)
Interlock
Loop checking
Oscillating process

Proving
Selling
Sensor trim
Stroke
Zero and span calibration

Industry Recognized Credentials

If you are training through an NCCER-accredited sponsor, you may be eligible for credentials from NCCER's Registry. The ID number for this module is 12410-16. Note that this module may have been used in other NCCER curricula and may apply to other level completions. Contact NCCER's Registry at 888.622.3720 or go to **www.nccer.org** for more information.

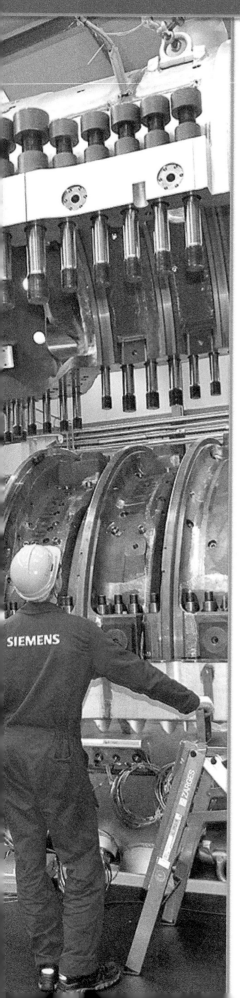

Contents

1.0.0 Loop Checking 1
   1.1.0 Loop Inspection 1
      1.1.1 Mechanical Inspection 1
      1.1.2 Primary Element 2
      1.1.3 Field Transmitter 2
      1.1.4 Field Wiring, Conduit, and Tubing 3
      1.1.5 Control Room Components 3
      1.1.6 Tag Numbers and Loop Sheets 4
   1.2.0 Loop Continuity 4
      1.2.1 Electrical Continuity Test 5
      1.2.2 Simplified Methods 5
      1.2.3 Pneumatic Testing 6

2.0.0 Proving and Calibrating a Loop 9
   2.1.0 Proving a Loop 9
      2.1.1 Simulation 9
      2.1.2 Required Test Equipment 9
      2.1.3 Testing the Programming 11
      2.1.4 Testing Interlocks and Alarms 11
   2.2.0 Calibrating a Loop 11
      2.2.1 Calibrating Analog Instrument Loops 12
      2.2.2 Calibrating Smart (HART) Instrument Loops 12
      2.2.3 Connecting a HART Communicator 13
      2.2.4 Asset Management 14

3.0.0 Commissioning a Loop 16
   3.1.0 Gathering the Documentation 16
   3.2.0 The Commissioning Process 16

4.0.0 Troubleshooting a Loop 19
   4.1.0 Fundamentals of Troubleshooting a Loop 19
      4.1.1 Identifying the Problem 19
      4.1.2 Identifying the Loop Components 20
      4.1.3 Understanding the Loop 20
      4.1.4 Troubleshooting the Loop 21
      4.1.5 Repairing the Loop 24
   4.2.0 Troubleshooting an Oscillating Process 24
      4.2.1 Verifying That a Problem Exists 24
      4.2.2 Gathering Information 24
      4.2.3 Identifying Possible Causes of the Problem 25
      4.2.4 Checking Instruments in the Field 25
      4.2.5 Checking the Transmitter 26
      4.2.6 Locating the Problem 26
      4.2.7 Using a Troubleshooting Flowchart 26

Figures

Figure 1	An example of a P&ID	2
Figure 2	Proper method of installing tubing to a DP transmitter on a steam process line	3
Figure 3	Loop sheet	4
Figure 4	Fluke® Model 789 ProcessMeter	5
Figure 5	Connecting a loop power source meter to a field instrument	6
Figure 6	Tone tester pairs	6
Figure 7	Hand pump for pneumatic testing	6
Figure 8	Bubbler	7
Figure 9	Fluke® Model 725 multifunction process calibrator	10
Figure 10	Pressure modules for the Fluke® Model 725	10
Figure 11	Differential pressure connection using a Fluke® pressure module with a Fluke® Model 725 multifunction process calibrator	11
Figure 12	HART communicator	12
Figure 13	Rosemount™ Model 3051 Smart Transmitter	13
Figure 14	Connecting a HART communicator to transmitter terminals	13
Figure 15	Example loop diagram	17
Figure 16	Loop sheet	21
Figure 17	Example panel graphic	24
Figure 18	Loop diagram	25
Figure 19	Diagnostic flowchart	27

Section One

1.0.0 Loop Checking

Objective

Describe how to inspect loop components and perform continuity checks prior to proving the loop.

a. Describe how to visually inspect various loop components.
b. Describe how to conduct loop continuity tests on electrical and pneumatic devices.

Performance Tasks

1. Perform a continuity check on an electrical system and document the findings.
2. Perform a continuity check on a pneumatic system and document the findings.

Trade Terms

Commissioning: This step involves documenting and testing an instrument channel using live process. Checks are usually incremental, beginning with manual control and progressing to full automatic.

Continuity tester: A basic piece of test equipment that verifies an electrical current can flow through a circuit.

Ground loop: An undesirable current flow through a cable's shield caused by grounding it at each end. The difference of ground potentials causes the current. Ground loops result in noise and signal distortion.

Loop checking: The process of performing a series of checks on a completed instrument chain and verifying that each device is installed correctly and has continuity to other devices.

Proving: A step involving the testing of an instrument channel using a simulated process. Control behavior is assessed and devices are calibrated.

Selling: The final step of loop checking, which involves turning the commissioned loop over to the client or to another contractor.

Before a completed instrument control loop is ready to handle live process, it goes through a series of checks that verify its fitness for the task. The terminology associated with these steps can cause confusion about what they involve. For this reason, it is essential to understand what these terms mean and what they involve before going into detail about the steps themselves.

The first term, and the topic of this section, is loop checking. A loop check verifies that the recently completed work has been done correctly. Each link in the instrument chain is examined to confirm that it was installed and connected properly. The second step is proving. This task involves simulated testing and calibration of the various components in the chain. A live process is not used at this stage. The next step is called commissioning. This incremental step involves extensive documentation and using live process to continue the testing. If this step is completed successfully, the loop is deemed ready for actual work. The final step is selling the loop, which involves officially turning it over to the client or to the next contractor in the process.

1.1.0 Loop Inspection

Once an instrumentation loop has been installed, it must be thoroughly inspected for proper mechanical installation and for verification that the components installed belong in the loop. This can be accomplished by matching the tag numbers on the individual components with the tag numbers found on the corresponding loop sheet. The next step is to check the loop for continuity from the primary element to the final element. This includes continuity through the transmitter, any installed transducer, through the controller or PLC (programmable logic controller), and finally, to the control valve in the field. Regardless of whether a loop is electrical or pneumatic, all loops must be intrinsically sound before activation.

1.1.1 Mechanical Inspection

Construction and installation of instrumentation loops, like most phases of construction, are usually subject to time constraints that require rapid and sometimes less-than-perfect installation methods and techniques. The materials and components installed may also have been damaged during installation, or they may have arrived from the supplier in an inoperable or damaged condition. For these reasons, among others, it is necessary to verify the mechanical installation of all instrumentation loops before assuming that they are ready for calibration or normal operation.

The first stage of verifying mechanical installation is a visual inspection of the loop, starting with the primary element and ending with the final element, which usually is a control valve. In order to accurately perform a visual inspection you must be familiar with the loop. Always consult the piping and instrumentation diagram, or P&ID (*Figure 1*), as well as the loop sheet for each particular loop you are going to inspect.

In addition to checking for damage and improper installation, the mechanical check also includes confirming that components are of the correct size and rating. This check is particularly important for wiring and tubing. The components themselves should always be verified to be the correct model or size as well. Mistakes in these areas are common and very easy to make.

1.1.2 Primary Element

The primary element is the first component in a control loop that is in contact with the process being controlled. Depending on the process variable, the primary element could be an orifice plate, a thermocouple, an RTD, a float, a displacer, or a pressure sensor. Regardless of its type, it functions to make either direct or indirect contact with the process being controlled.

Mechanical inspection of these elements goes beyond that of the installation of the element itself. It must also include the installation orientation of the device as well as any tubing, wiring, conduit, fittings, connections, terminals, and methods used in conjunction with its installation. Common faults to look for include the following:

- Orifice plates that are installed backwards, as indicated by the flow indicator on the plate handle
- Broken conduit or conduit fittings that may have been damaged after installation
- Damaged thermocouple or RTD heads
- Damaged and/or leaking pneumatic tubing
- Loose or shorted wiring terminals
- Any obvious primary element installation flaw that might interrupt proper operation of the loop

1.1.3 Field Transmitter

The field transmitter receives the signal from the primary element and typically sends that signal to either a controller or a PLC. Like the primary element, the transmitter is usually located in the field and is therefore subjected to a greater degree of potential damage than components located in the more secure environment of a control room.

When inspecting a field transmitter, verify that it is mounted in a location that is as close as possible to the primary element. This prevents the need for excessively long runs of conduit or tubing to interconnect the two devices. At the same time, make sure that the location provides some physical protection for the transmitter. If the surrounding environment experiences temperature extremes or contains corrosive substances, the transmitter must be rated for that environment or

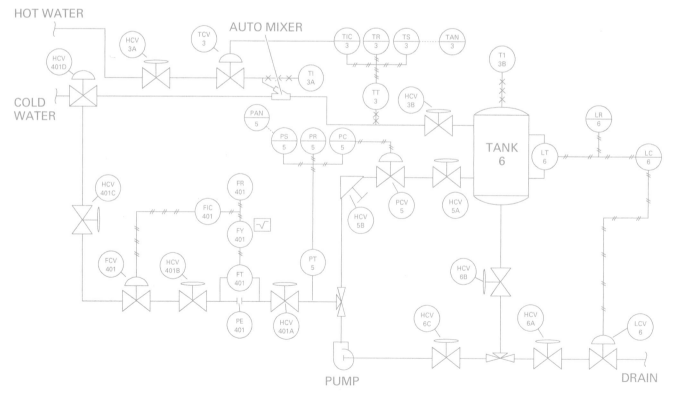

Figure 1 An example of a P&ID.

placed in a suitable enclosure. Newer smart transmitters must be accessible in order to allow handheld communicators to be connected to them for programming and calibration of the instrument.

Differential pressure (DP) transmitters should be checked for proper tubing and piping. Proper slope, cooling lengths in hot processes, and drain (blow-out) capabilities on both the high- and low-side process connections are needed in order to keep the process lines free of contaminants and condensates. This type of installation is shown in *Figure 2*.

All electrical connections should be checked for terminal tightness and sound conduit and fittings. Instrument tags should be readily accessible and easy to read. Calibration terminals should be located to allow the technician ease of access without having to be exposed to potential injury caused by climbing or harsh environments. Visual field readouts on smart transmitters should be oriented to permit easy reading with minimal glare and obstructions. Instrument stands should be made of materials compatible with the environment and fabricated to sufficiently support the instrument, while also remaining aesthetically pleasing.

1.1.4 Field Wiring, Conduit, and Tubing

It is common to find damaged wiring, conduit, and tubing in the field environment due to their exposure during construction as well as to daily maintenance activities within an industrial facility. Slightly damaged signal wiring or broken conduit can cause intermittent problems in a control system. Signal distortion is caused by electrical noise and/or changes in resistance in the wiring or cable. These small degrees of damage often cause difficulty in controlling a system or loop due to the inconsistency of their effects on the instrument signal.

While in the field, be on constant lookout for minor damage, and make repairs immediately if possible. At a minimum, note them so that once enough of them are listed, repairs can be made to all of them during the same time period. No physical damage should be neglected, even if it seems minor. It will only grow in size, affecting the accuracy and response of the system. Network cables and sophisticated cabling systems, such as fiber optic, should be left to IT professionals unless a properly certified instrument technician is available.

1.1.5 Control Room Components

Most loop components installed in the control room, including single-loop controllers and PLCs, are protected from physical damage. However, these components must always be inspected after installation to ensure that they were not damaged prior to or during installation. Wiring and tubing connections should be checked for tightness and damage. Make sure to check all terminations under, behind, and overhead in cabinets. It is in these areas that many cables, connections, and terminals may not be adequately connected or terminated because of poor accessibility. Check for proper routing of conductors, cables, and tubing, looking for sharp bends. Restrictions can cause poor signal transmission and can cause a loop to be slow in response or inaccurate. Always repair any and all of these types of discrepancies immediately; this can save time and trouble later on if the conditions worsen.

Always be aware of signal noise and how it develops, both in field and control room wiring. Make sure signal cable shields or drain wires are properly grounded, usually at one end only, and try to provide adequate spacing between power conductors and signal conductors. Never ground a signal cable shield in the control room unless you have physically located the other end and confirmed that the shield is not grounded at that end. Double-grounding can cause ground loops. (A ground loop is a current-carrying circuit that causes noise and signal distortion.)

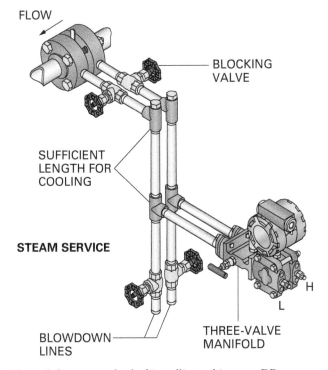

Figure 2 Proper method of installing tubing to a DP transmitter on a steam process line.

1.1.6 Tag Numbers and Loop Sheets

Tag numbers are unique numbers assigned to each component in a loop, all sharing the common loop number as part of their tag number. A loop sheet, like the example shown in *Figure 3*, contains all of the components that make up a single loop, including their tag numbers and specifications. Every component installed in every loop should be inspected and verified against the information in the loop sheet, including tag number and calibration range. If any component is found to have a tag number on it other than the one found on the loop sheet, it should be investigated immediately. The next step in the process (proving) cannot begin until the discrepancy is resolved.

1.2.0 Loop Continuity

Loop continuity tests involve applying a signal generated by test equipment to the wiring or tubing of a loop to determine if the signal passes through the complete loop without leakage or excessive resistance.

WARNING! Before performing any continuity test on any loop, always properly lock out and tag out the loop energy source according to your company's guidelines and policies. Never assume that the circuitry or tubing in the loop will only carry a safe level of voltage, current, or air. Remember that at this phase of testing, the loop has not yet been tested for proper installation. Continuity tests must never be performed with the loop in an energized state, whether the energy is electrical or pneumatic.

PLCs and DCSs (distributed control systems) can be damaged by some continuity test equipment. You must isolate these components from the instrument loops before testing the loop's wiring or tubing to avoid damage to the digital equipment. You can simulate the instrument's being in the loop in order to create continuity by connecting the output leads of a loop calibrator or tester directly into the loop.

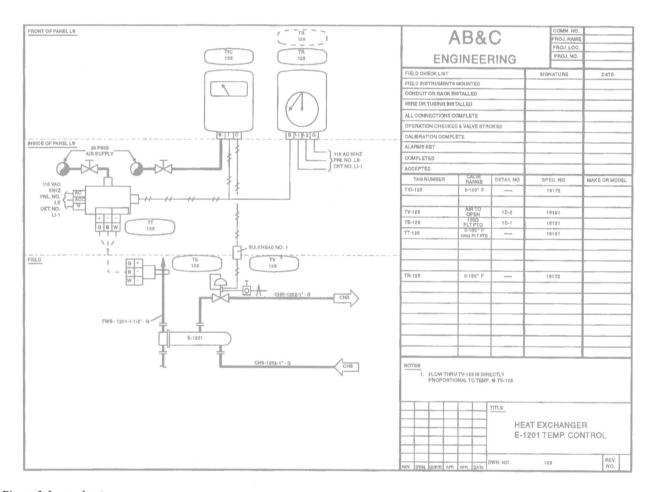

Figure 3 Loop sheet.

1.2.1 Electrical Continuity Test

Once the circuitry has been locked out and tagged out according to your plant's safety policies, and sensitive components in the loop have been isolated, you may begin your electrical continuity test on the wiring in the loop. This test will ensure accurate signal transmission once the loop is put into operation.

There are many test instruments that simulate the output signal of a typical electronic field transmitter. Fluke® offers a very complete line of process control test and calibration meters. Other companies that sell similar equipment include Extech®, GE®/Druck®, Omega®, and Crystal Engineering®. Ideally, you should select a device that can energize the loop using its own power supply, allowing the loop's normal supply to be locked out during testing.

An example of a Fluke® meter that may be used for electrical loop continuity checks is the Fluke® Model 789 ProcessMeter™ shown in *Figure 4*. The 789 is a handheld, battery-operated tool for measuring electrical parameters, supplying steady or ramping current to test process instruments, and providing a loop power supply. It has all the functions of a digital multimeter, as well as current output capability.

Because the loop's operating power source should be locked and tagged out during continuity check, you can use the Model 789 or similar test equipment to simulate the loop's operation by placing the meter in the Loop Power mode and connecting the output leads of the meter to the field transmitter. While in Loop Power mode, the meter acts like a battery and supplies voltage at a nominal 24 VDC, while the process instrument regulates the current in the loop. At the same time, the metering section on the Model 789 measures the current running through the loop. This type of test will provide a verifiable electrical continuity check of the loop, while at the same time indicating the proper operation of the loop. The meter's source leads must be connected in series with the instrument current loop, as shown in *Figure 5*.

With the newer smart instruments that now make up many of the loops controlled by PLCs and DCSs, it is necessary to connect a precision (±1%) resistor of 250 Ω into the circuitry in order to simulate the loop load. The Model 789 contains an internal series resistor that can be switched in for communication with HART® and other smart devices.

1.2.2 Simplified Methods

Because continuity testing an electrical loop can be defined as verifying that the wiring and terminations in a loop are connected, there are many more simple and less technical methods that may be used to test the wiring from one end to the other. Two people, each equipped with two-way radios or cell phones and a simple battery-powered continuity tester, can check wiring from one end to the other as one person intentionally grounds one end of a wire while the person on the other end checks the continuity between the wire and ground.

Another simple method for testing continuity and identifying wires at both ends is to use battery-powered telephone receivers. This method is

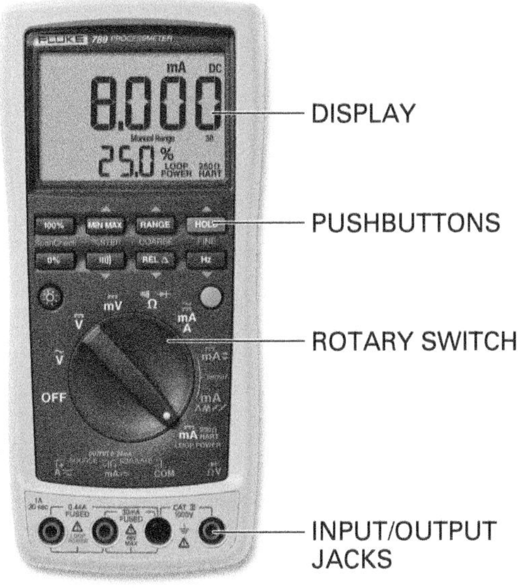

Figure 4 Fluke® Model 789 ProcessMeter.

Figure 5 Connecting a loop power source meter to a field instrument.

still used in some facilities but for the most part has been replaced by more sophisticated and reliable multi-purpose loop testing equipment.

A tone tester pair (*Figure 6*) is an excellent way of confirming basic continuity and requires just one person. The transmitter is attached to the loop at one end and switched on. The technician then uses the probe to detect the signal at the other end. Tone testers are especially handy in tracing unlabeled or mislabeled signal cables since the probe only needs to get reasonably close to the correct wire to detect the transmitter's signal.

Network cables used in DCSs and PLCs may be continuity tested for basic functionality, but more sophisticated tests are needed to confirm that they will actually carry the data signal reliably and at full speed. Network certification testing verifies the cable at its specified rating. Many larger IT departments have a tester that performs this task. Instrumentation personnel that regularly run network cabling should consider obtaining one. Fiber optic cabling requires special skills and tools to install, test, and repair. Most companies contract this task out to specialist companies or have a few persons specifically trained and certified to do these tasks.

1.2.3 Pneumatic Testing

Even though many of the older instrumentation technologies in the field have been replaced by electronic or digital loops, there is still a substantial amount of pneumatic instrumentation that requires maintenance and calibration. Because it is unlikely that new facilities or expansions of existing ones will involve pneumatic technology, this module does not include detailed information on testing and calibrating pneumatic instrumentation. The information presented serves mainly to provide familiarity with the equipment available for testing and calibration of pneumatic control loops.

Pneumatic testing is usually done with a multi-function calibrator that supports pressure testing.

Fluke®, GE Druck®, WIKA®, and Crystal Engineering® all make suitable units. Some calibrators require specific pressure modules to be added to the basic unit. Many include built-in pressure sources in the form of a hand or electric pump. Others require an external hand pump (*Figure 7*), such as the type supplied with a pump-up kit.

The line under test for continuity is plugged or capped on the other end and pressurized. Once the air supply is isolated from the pressurized line under test, the gauge or readout can be monitored to see if the pressure drops. If it does drop, a leak or open is present in the loop and must be repaired before the loop is proved. However, if the pressure remains the same after a period of time (the length of which may be dictated in the loop specifications), the loop's continuity and integrity are sound and the loop is ready to be proved.

Continuity and integrity of pneumatic tubing and piping can also be checked using a bubbler like the one shown in *Figure 8*. The pneumatic line must be capped or plugged at one end and the bubbler connected in series with the line at a suitable point. The bubbler provides a visible indication if air is flowing through it, even at a very small flow rate. If the bubbler is installed in a leak-free line that is capped or plugged and is then pressurized, no air will flow through the system. However, if a tubing fitting or valve is slightly loosened or opened to the atmosphere,

Figure 6 Tone tester pairs.

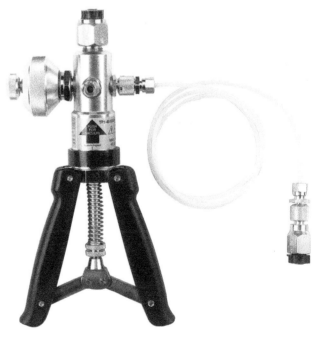

Figure 7 Hand pump for pneumatic testing.

air will flow through the bubbler, showing itself as bubbles in the water-filled glass bowl.

Continuity of a lengthy run of tubing or piping can be checked by plugging one end, installing the water-filled bubbler on the other end, and then pressurizing the line. No bubbles should appear. If they do, there is a leak somewhere in the system that is allowing the air to flow. If no bubbles appear after the bubbler is installed, the plug or cap is slightly loosened on the opposite end of the bubbler to allow a small flow of air to escape into the atmosphere. If bubbles appear in the bowl, the line is both tight and continuous from one end to the other. This is a simple and inexpensive method of testing both for leaks and continuity in a pneumatic system. Once the test is complete, the fittings may be retightened and the loop proved.

Figure 8 Bubbler.

Additional Resources

Instrumentation for Process Measurement and Control, Norman A. Anderson. Third Edition. 1997. Boca Raton, FL: CRC Press.

Maintenance and Calibration of HART® Field Instrumentation (PDF), Richard Pirret. Fluke Corporation, Everett, WA, USA. Accessed at **www.plantservices.com**.

The following websites offer resources for products and training:

Amtek Calibration, **www.crystalengineering.net**

Emerson Process Management, **www.emersonprocess.com**

Extech Instruments, **www.extech.com**

Fluke Corporation, **www.fluke.com**

GE Digital Solutions, **www.gemeasurement.com**

Omega Engineering, **www.omega.com**

WIKA Instrument, LP, **www.wika.us**

1.0.0 Section Review

1. Verifying a loop using an incremental sequence of tests involving a live process is known as _____.
    a. loop checking
    b. proving
    c. commissioning
    d. selling

2. What part of a loop check involves confirming that the components are correct and that nothing is damaged?
    a. A continuity test
    b. A mechanical check
    c. A pneumatic check
    d. An electrical check

3. If a calibrator is used to perform an electrical continuity check, what must it include?
    a. A loop power source
    b. A pressure module
    c. A pressure source
    d. A hand pump

Section Two

2.0.0 PROVING AND CALIBRATING A LOOP

Objective

Describe how to prove and calibrate a loop.
a. Describe how to prove a loop.
b. Describe how to calibrate a loop.

Performance Task

3. Prove a loop and document its completion.

Trade Terms

Current loop trim: An adjustment made at the third stage of a smart instrument, which adjusts the relationship between the calculated output signal and the actual analog output signal.

Instrument asset management system (IAMS): A software application that combines a database and communications system to keep detailed records about instrumentation. An IAMS guides calibrations and maintenance, keeps inventory, and provides alerts to instrument personnel.

Interlock: A mechanical, electrical, or software safety feature that prevents certain events from occurring unless other necessary conditions are first met.

Sensor trim: An adjustment made at the first stage of a smart instrument, which establishes the relationship between the input signal and the lower and upper limits of the calculated process variable.

Zero and span calibration: A basic calibration procedure in which an instrument's zero and span points are adjusted to produce the proper relationship between the input and output minimum and maximum points.

Once a loop has been checked, the next step is proving it. Some instrument technicians consider proving as just another aspect of checking the loop, and in many ways, that is true. Proving continues the checking process but takes it a step further. Proving begins the process of checking actual loop functionality by operating it under controlled conditions.

2.1.0 Proving a Loop

In the previous section, you learned that performing a loop continuity check means verifying that the loop components and their associated wiring or tubing are connected from one end to the other. It also includes confirming that no shorts or loose wires are present in electrical loops and that no leaks are present in pneumatic ones.

Proving is the next step toward getting a control loop to an operational state. This step operates the devices in the instrument chain to confirm that they work properly and communicate with each other. Zeroing and calibration are a part of this step as well, since these tasks are necessary for the loop's operation. However, proving does not involve live (or wet) process. Instead, it substitutes simulated processes. This is done for both convenience and safety, since the loop may still harbor hidden problems at this stage.

As an example, consider a simple flow-control loop. Proving would include confirming that the primary element (an orifice plate) creates a differential pressure that is read by the transmitter, which in turn sends its output signal to the next instrument in the loop and finally on to the controller or PLC. The last step in this phase is making sure that the final control element receives the output signal from the controller or PLC and responds to that output. Calibration and zeroing come after these steps have been successfully completed.

2.1.1 Simulation

There are only two ways to verify that instruments in a loop are interacting with one another as intended and designed: putting the loop into actual operation with live process, or simulating the operation of the loop. Because the first option could have serious consequences, you must simulate the loop's operation in order to prove the loop.

As with the other tests previously performed on the loop, you must have the P&ID available, as well as loop sheets for the loops that you are going to prove, since they list each instrument and its function in the loop. In addition to the proper documentation, you must have test equipment that has the ability to simulate the process signal.

2.1.2 Required Test Equipment

Proving a loop requires test equipment that can supply the loop's power or energy, whether it is air or electrical. The device must also be able to read the output signals of each of the instruments in the loop, regardless of their signal forms (volts, milliamperes, psi, etc.). If the loop includes signal conditioners or transducers, it may be necessary to have more than one piece of test equipment on hand in order to read the various intermediate signal forms. However, test and calibration equipment that provides all of these functions in one unit is available.

> **NOTE**
> The use of a calibration device to perform this phase of testing does not mean that you are actually calibrating the instruments yet.

In order to simulate a loop, your equipment must be able to provide the form of signal required. It must also supply the signal at the level at which the loop would normally operate. For example, most electronic loops are operated by a permanent 24 VDC power supply that is capable of supplying many instruments at one time without experiencing a drop in voltage. Because you are going to simulate the loop, the power source on your test equipment must also be able to handle all of the loads that are interconnected in the loop. Most process control loop calibrators are equipped to do so.

One multi-channel loop calibrator that provides output in various forms and is capable of receiving input signals in volts, milliamperes, psi, and even differential pressure, is the Fluke® Model 725 multifunction process calibrator (*Figure 9*). It has capabilities to measure and source current, voltage, temperature (RTDs and thermocouples), frequency, resistance, and pressure. Add-on modules are required for the pressure-measuring capability (*Figure 10*).

The Model 725 has a split display that lets you view input and output values simultaneously. This allows you to power up a field transmitter from the meter and read its output on the meter at the same time. For valve and I/P transducer tests, you can source electrical current while measuring pressure. The Model 725 is also equipped with auto-stepping and auto-ramping for remote testing. This function lets you connect the meter into the loop and set its output to step through the range so that you may take measurements or readings at various points in the loop to determine if all the instruments in the loop are functioning properly. This is a very handy method to use when proving a loop.

Other manufacturers provide instruments with similar capabilities. Likewise, calibration equipment that is not multifunctional can be used in conjunction with other tools, where one provides the power source and the other reads the loop. The choice of test and calibration equipment depends on factors such as cost, familiarity, and preference. *Figure 11* shows a typical differential flow calibration setup using a differential pressure module in conjunction with the Fluke® Model 725.

> **WARNING!**
> Multifunction calibrators are not designed to be connected directly to process lines. NEVER make a connection of this type unless you have verified that the device in question may be connected in this way. Improper use of calibration and test equipment may result in destruction of equipment and possible injury to personnel.

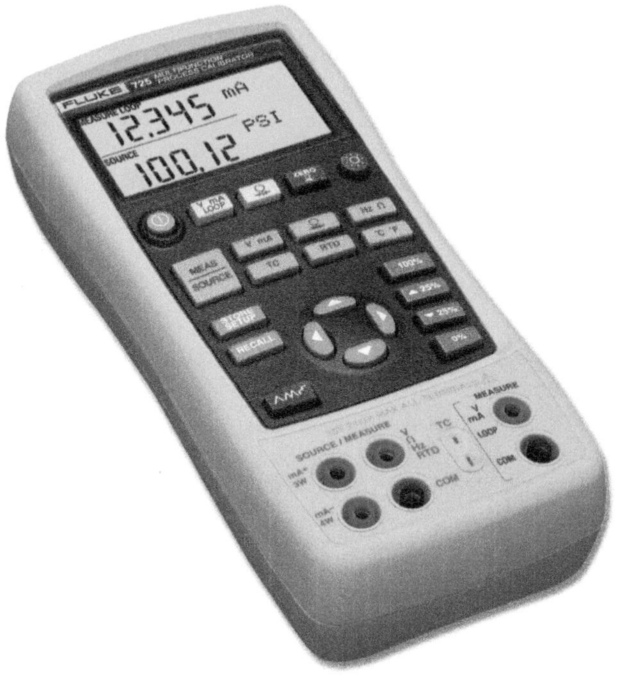

Figure 9 Fluke® Model 725 multifunction process calibrator.

Figure 10 Pressure modules for the Fluke® Model 725.

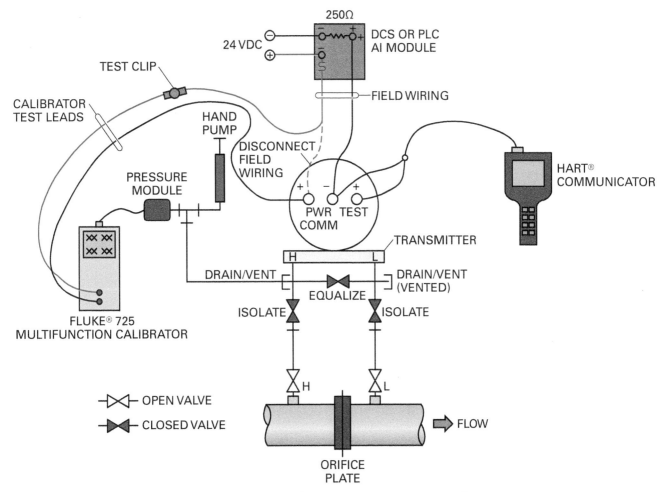

Figure 11 Differential pressure connection using a Fluke® pressure module with a Fluke® Model 725 multifunction process calibrator.

2.1.3 Testing the Programming

This step may or may not be necessary, depending on the details of the particular control system. If there are no programmable components, then it may be skipped. But if a PLC or programmable controller is involved, the software component must also be tested as part of the proving step.

Control software should respond to inputs with specific output responses. These may be either discrete (on/off) or analog (voltage or current) signals. If there is an HMI/OWS, it should also work correctly and respond to changes in the system condition. Outputs under its control should respond appropriately. Regardless of the specific details, you'll need to provide sufficient simulated signals to ensure that the control software is working correctly. Depending upon how your workplace handles programming, you may need to coordinate with a programmer for this stage of testing.

2.1.4 Testing Interlocks and Alarms

Another possible step involves checking interlocks in the control system. (An interlock is a feature that prevents certain events from occurring unless necessary conditions are first met.) These may or may not be present, depending on the system's complexity. If they are, they should be checked and their proper operation confirmed. Similarly, if there are alarm devices or screens, they should be checked to ensure that they go off / appear in response to appropriate signal values.

2.2.0 Calibrating a Loop

The next stage in the proving process is calibrating the devices in the loop. Calibration should not be confused with loop tuning; they are completely separate steps. The loop cannot be tuned until it has been calibrated. Tuning is covered in a separate module of this curriculum.

Calibrating a loop is the process of adjusting an instrument to set the correct relationship between its input signal and its output signal. Some instrument components (such as recorders and indicators) and control elements (such as control valves) receive an input, but have no output. In this case, calibration involves setting the instrument to indicate an output that has an accepted or desired relationship to the input signal.

The "Instrument Calibration and Configuration" module of this curriculum presents topics on calibrating instruments in existing and functional instrument loops. Because most of the principles involved are identical, calibration as a step in the proving process will be addressed only briefly in this module. Refer to the "Instrument Calibration and Configuration" module for more detailed information about calibration.

2.2.1 Calibrating Analog Instrument Loops

To verify the overall accuracy of a conventional 4–20 mA analog instrument, a multiple-point test using a multifunction calibrator that simulates the input and measures the output can be used. The normal calibration adjustment involves setting only the zero value and the span value because there is effectively only one adjustable operation between the input and output. This procedure is often referred to as a zero and span calibration. If the relationship between the input and output range of the instrument is not linear, the transfer function must be known before expected outputs can be calculated for each input value. Without knowing the expected output values, you cannot calculate the performance errors.

As with all calibrations, it is important to confirm that your test equipment itself is properly calibrated and traceable to a national standards body, such as the NIST. It is also important that the test equipment's calibration be within date, otherwise any calibrations that you perform with it are suspect.

2.2.2 Calibrating Smart (HART) Instrument Loops

Technology in process control has made great advances into digital signaling. Instruments that incorporate digital technology are designed with embedded processors so that they can communicate with other digital instruments in the loop, including PLCs or DCSs. In order for a technician to communicate with these instruments to verify or calibrate them, digital communicators must be connected into the loop. One such communicator is the HART communicator (*Figure 12*).

Figure 12 HART communicator.

HART is an acronym for "Highway Addressable Remote Transducer". It is a protocol that defines communications rules between smart devices. It has the additional advantage of being able to travel over conventional 4–20 mA analog signal lines and co-exist with the analog signal. Consequently, it is a popular digital protocol in the instrumentation industry.

For a HART instrument, like the Rosemount™ Model 3051 smart transmitter shown in *Figure 13*, a multiple-point test between input and output (as performed on a traditional analog 4–20 mA loop) does not provide an accurate representation of the transmitter's operation. Just like a conventional transmitter, the measurement process begins with a technology that converts a physical quantity into an electrical signal. However, the resulting calibration process is much different.

Instead of a purely mechanical or electrical path between the input and the resulting 4–20 mA output signal, a HART transmitter has an embedded processor that manipulates the input data digitally. There are typically three internal stages involved in a HART device, and each of these may be individually tested and adjusted.

The instrument's microprocessor measures some electrical property that represents the process variable. The measured value may be a voltage, current, or other electrical measurement. However, before the value can be used by the microprocessor, it must be transformed to a digital count by an analog-to-digital converter (ADC).

In the first instrument stage, the microprocessor converts the digital count to an internal representation of the sensor units (temperature,

Figure 13 Rosemount™ Model 3051 Smart Transmitter.

calculate this value. Although a linear transfer function is the most common, pressure transmitters often have a square root option. The output of the second stage is a digital representation of the desired instrument output current. This value may also be read with a HART communicator. Many HART instruments support a command that puts the instrument into a fixed output test mode. This command overrides the normal output of the second instrument stage and substitutes a specified fixed output value.

The third instrument stage is the output section where the calculated current output is converted to a value that can be loaded into a digital-to-analog converter (DAC). This produces an actual analog electrical signal that exits the instrument through its outputs. Once again, the microprocessor must rely on some internal calibration factors in order to provide the correct output. Adjusting these factors is often referred to as a current loop trim or *4–20 mA trim*.

2.2.3 Connecting a HART Communicator

For the HART communicator to function properly, 250 Ω of resistance must be present in the loop. Often, an external resistor must be used to ensure that this load is present. The HART communicator does not measure loop current directly. *Figure 14* shows the correct method for connecting a HART communicator to a compatible device.

If the smart transmitter is used in a digital-only instrument loop (with a PLC or DCS, for example), only the first instrument stage requires calibration. If the transmitter must provide a conventional analog (4–20 mA) output signal, both the first and third stages must be calibrated. Note

flow, pressure, etc.) by using an equation or lookup table. The manufacturer creates this table or equation during production. However, it is possible to tweak this stage in the field. Doing so is referred to as adjusting the sensor trim. The output of the first section is a digital representation of the process variable. You can read this value using a HART communicator and see the process variable in real units.

The second instrument stage performs strictly mathematical operations on the process variable to convert it to an equivalent current value in milliamperes (mA). The range values of the instrument (related to the zero and span values) are used in conjunction with its transfer function to

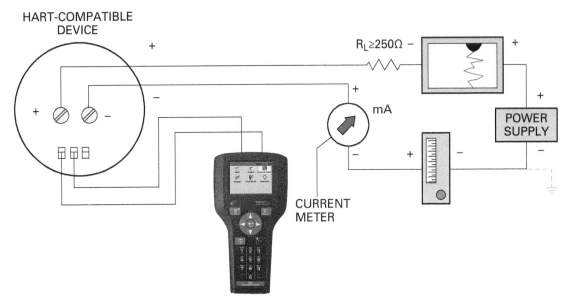

Figure 14 Connecting a HART communicator to transmitter terminals.

that the third stage calibration is independent of the first and has nothing to do with the zero and span settings. The second stage does not require calibration.

Always be aware that merely setting the device using a HART communicator is not calibration. A standardized input signal, perhaps provided by a multifunction calibrator, must be applied to the instrument's inputs. This device, in conjunction with the HART communicator, can then be used to calibrate the first and third sections properly. Some multifunction calibrators also include a HART communicator, so it's possible to perform the entire process with a single instrument.

2.2.4 Asset Management

One aspect of working with instrument loops, which is becoming increasingly important, is the instrument asset management system (IAMS). A typical plant has a huge number of assets, which include machines, control components, computers, replacement parts, tools, and countless other objects. Keeping track of everything can be a challenge, particularly in a large plant. In the past, asset management has been handled on paper or managed with a computer database but even databases require a lot of manual data entry.

Where control systems are concerned, asset management goes beyond simple inventory management. Instruments and other control components require maintenance and regular calibration. Worn equipment must be replaced. For major work, the entire process might have to be shut down, possibly at significant cost. For these and other reasons, keeping a detailed record of every part of an instrumentation system could definitely pay off, but doing so was extremely difficult until recently.

When smart instruments began to dominate industry, a major piece of the asset management puzzle fell into place. Since smart instruments can communicate through fieldbuses and other networking technologies, they can provide a wealth of data about themselves to an asset management system. They can report their settings, calibration status, maintenance history, and current operational performance. As devices start to wear out, their performance values change, which can be an early warning sign that maintenance or replacement is needed.

Instrument asset management systems are software packages that run on a networked computer or server. They communicate with smart instruments and keep detailed records automatically. Using rules established by plant personnel, they can schedule calibrations, alert technicians of impending failures, and help keep track of everything without the need to do so on paper. One example of a well-known package is Emerson's AMS Suite®. Other manufacturers have comparable products. As time goes by, the companies that specialize in process control add more features to their asset management systems, making them more powerful and flexible.

For an asset management system to be useful, however, it must be able to interact with smart devices automatically. Fully digital plants dovetail neatly with these systems, but plants with several generations of control technology may have problems. However, third-party solutions exist that can enable earlier technologies to work with an instrument asset management system.

Many of the more recent HART communicators and multifunction calibrators can communicate with an IAMS as well, making it possible for less-integrated devices to have a presence in the asset database. This means that they can store the information from a calibration run and later transfer it to an IAMS system. The Fluke® 754 is a typical example of a so-called documenting calibrator, and there are many others from most major manufacturers. If you have a documenting calibrator, you should spend some time becoming acquainted with its features and learn how it can work with your IAMS system.

Getting a loop integrated into an instrument asset management system should be done as a part of the proving process. The system's first calibration occurs at this point. For this reason, the proving stage is the logical point to start tracking the control system. Then, once commissioning begins, real process will start to move through the system.

Additional Resources

Instrumentation for Process Measurement and Control, Norman A. Anderson. Third Edition. 1997. Boca Raton, FL: CRC Press.

Maintenance and Calibration of HART® Field Instrumentation (PDF), Richard Pirret. Fluke Corporation, Everett, WA, USA. Accessed at **www.plantservices.com**.

The following websites offer resources for products and training:

Amtek Calibration, **www.crystalengineering.net**

Emerson Process Management, **www.emersonprocess.com**

Extech Instruments, **www.extech.com**

Fluke Corporation, **www.fluke.com**

GE Digital Solutions, **www.gemeasurement.com**

Omega Engineering, **www.omega.com**

WIKA Instrument, LP, **www.wika.us**

2.0.0 Section Review

1. Proving a loop involves testing its functionality with _____.
    a. limited amounts of live process
    b. simulated process information
    c. digital process management
    d. a reference liquid

2. If a HART device is used in an analog control system, which sections must be calibrated?
    a. The first and third sections
    b. The first section only
    c. The third section only
    d. The first, second, and third sections

3. An IAMS can alert instrument technicians to possible upcoming failures by monitoring _____.
    a. industry statistics
    b. network traffic
    c. HART traffic
    d. performance changes

SECTION THREE

3.0.0 COMMISSIONING A LOOP

Objective

Describe how to commission a new loop.
a. Describe the documents associated with commissioning.
b. Describe the commissioning process.

Performance Task

4. Commission a loop.

Trade Terms

Stroke: To operate a valve through its entire range to confirm that it is working correctly.

When proving is complete, it's time to perform testing with some live process. It's also time to formally verify that the system works as it was designed. Handing off the completed system, either to a client or to another contractor, is known as *selling*. Of course, the system will not be accepted without tangible and well-documented proof that everything works as specified. Warranties hinge upon good documentation as well, so this step requires as much paperwork as it does process and testing.

3.1.0 Gathering the Documentation

All project drawings and documents must be assembled before starting the commissioning process. These include the following:

- *Piping and instrumentation diagram (P&ID)* – The P&ID shows the method of control for a given process, along with approximate device locations and tag numbers.
- *Control room plan, section, and detail drawings* – These drawings show the physical location of all equipment in the control room and in any auxiliary equipment and computer rooms.
- *Control room single-line diagram drawings* – These diagrams show the interconnecting cables and pneumatic connections (but not terminations) between major devices in the control room and those between auxiliary equipment, local control panels, and field devices.
- *Instrument location and conduit plan* – This drawing shows the locations and elevations of field instruments, control panels, and junction boxes. It also shows the associated raceway system (electrical conduits and cable trays).
- *Pneumatic location and tubing plan* – This drawing shows the location of the air manifold and pneumatic devices, including the routing of pneumatic tubing.
- *Control panel layout and wiring drawings* – These drawings show the size and location of all panel devices, along with their electrical and pneumatic interconnections. They also depict the control panel termination points for field device connections, which are helpful when troubleshooting open circuit problems in loops.
- *Power distribution panel schedule drawing* – This schematic diagram shows the voltage, amperage, and locations of all circuit-interrupting devices associated with the instruments and systems. It is helpful when attempting to locate problems with equipment supply power.
- *Hardware installation detail drawings* – These drawings show the actual mounting, orientation, and connection of each instrument and control device. They can be helpful when troubleshooting instruments that require special installation procedures.
- *Loop diagrams* – These are the most referenced drawings during the proving, commissioning, and troubleshooting processes. They show the connection details of the control system components arranged into a loop (*Figure 15*).
- *Instrument index* – This lists the instrument tag numbers for all instruments or control system components, including the associated drawing numbers, materials, specification sheets, and other related information for each component.
- *Specification sheets* – These documents are written in a standard Construction Specification Institute (CSI) format. They contain the technical details for one or more tag-marked components.

3.2.0 The Commissioning Process

Many of the steps carried out in commissioning are similar to those carried out in loop checking and proving. However, in commissioning, these steps are more formal and thoroughly documented. This is the final verification that the entire loop has been constructed correctly and is working according to specifications.

Commissioning normally involves component testing, followed by functional loop testing.

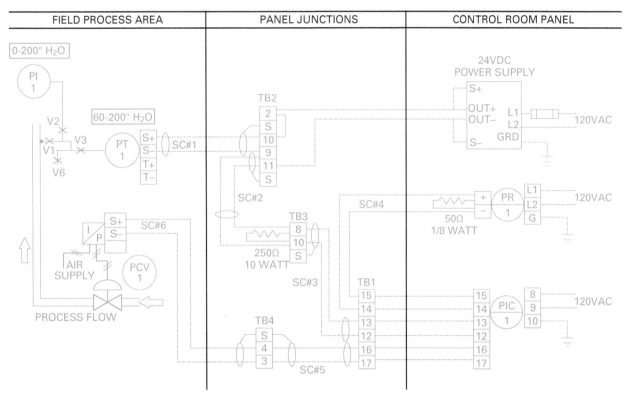

Figure 15 Example loop diagram.

Single component devices, such as switches, relays, and pressure or temperature gauges not associated with the loop, are checked as standalone devices during the commissioning procedure. The loop may be tested with simulated process first, followed by live process. Tests are typically done in stages, where the loop is first operated in manual mode, and then increasing degrees of automation are added. At the end of the process, tests are done with the loop in its normal operating configuration. Each system will include detailed commissioning checklists. The actual procedure depends on the system being tested. A typical procedure includes the following steps:

Step 1 Assemble all drawings and documents related to the project. This includes P&IDs, loop diagrams, wiring diagrams, tubing and conduit plans, section and detail drawings, instrument lists, panel schedules, and specifications.

Step 2 Perform pressure tests per the installation design values. Test parameters include duration, test medium, and pressure level.

Step 3 Verify that the instruments have been calibrated. This is normally done by the instrument manufacturer, but may be required as part of the proving procedure.

Step 4 Perform mechanical checks. This involves the following tasks:
- Make sure that each device has been securely installed in the correct location, is identified with a loop number tag, and has sufficient clearance for service and maintenance.
- Check that valves are not installed backwards and stroke all control valves (apply a variable input and verify that they open and close all the way and that there are no spots where the valve appears to stick or operate sluggishly).
- Examine instrument air supply lines for leaks, pinch points, and the presence of dirt or moisture.

Step 5 Perform electrical checks. This involves the following tasks:
- Verify the correct loop supply voltage.
- Ensure that the electrical configuration (voltage) of all instruments match the job specifications.
- Verify point-to-point wiring continuity as compared to the loop diagram.
- Check the wiring for unintentional short circuits to ground.
- Verify that specified ground connections are made and that signal cable shields (on the same cable) are not grounded at more than one point.

- Check the transmitter outputs.
- Verify the correct transmitter output display at the controller or computer control system.
- Check the operation of single component devices such as switches, relays, and pressure or temperature gauges.
- Check the controller settings and adjust if necessary.
- Confirm that the output devices go to a fail-safe mode at failure.

Step 6 Test the operation of all safety devices, interlocks, and alarms.

Step 7 Verify the operation of the functional loop either with simulated or live process.

Step 8 Correct any problems using a systematic process of troubleshooting and repair.

Step 9 When all components are functioning properly and have been checked out, assemble the completed test records. These are supplied with the as-built drawings, vendor data, and instruction manuals as part of the complete system documentation package.

Additional Resources

Instrumentation for Process Measurement and Control, Norman A. Anderson. Third Edition. 1997. Boca Raton, FL: CRC Press.

Maintenance and Calibration of HART® Field Instrumentation (PDF), Richard Pirret. Fluke Corporation, Everett, WA, USA. Accessed at **www.plantservices.com**.

The following websites offer resources for products and training:

Amtek Calibration, **www.crystalengineering.net**

Emerson Process Management, **www.emersonprocess.com**

Extech Instruments, **www.extech.com**

Fluke Corporation, **www.fluke.com**

GE Digital Solutions, **www.gemeasurement.com**

Omega Engineering, **www.omega.com**

WIKA Instrument, LP, **www.wika.us**

3.0.0 Section Review

1. The most referenced document during commissioning is the _____.
   a. control room plan
   b. pneumatic tubing plan
   c. loop diagram
   d. power distribution schedule drawing

2. What is a major difference between the proving and commissioning processes?
   a. Proving involves testing the loop devices, while commissioning does not.
   b. Commissioning involves live process, while proving does not.
   c. Commissioning involves calibrating the loop, while proving does not.
   d. Proving requires extensive documentation, while commissioning does not.

3. During commissioning, operation of the loop typically begins in _____.
   a. automatic mode
   b. manual mode
   c. tuned mode
   d. process mode

Section Four

4.0.0 TROUBLESHOOTING A LOOP

Objective

Identify the fundamental steps in loop troubleshooting and describe the troubleshooting process.
  a. Identify the fundamental steps in loop troubleshooting.
  b. Describe the loop troubleshooting process for oscillating loops.

Performance Tasks

5. Troubleshoot a newly installed control loop.
6. Troubleshoot an oscillating process.

Trade Terms

Cavitation: An undesirable pulsation in a fluid stream caused by flow changes that create pockets of low-pressure fluid mixed in with normal pressure. Cavitation can damage equipment and cause a process to oscillate.

Oscillating process: A process that continues reacting to a disturbance rather than returning to a stable condition in a reasonable period of time.

During the loop proving and commissioning processes, problems in the loop may cause it to function improperly. When these situations occur, the source of the fault must be located and repaired. This process is referred to as *troubleshooting*. Troubleshooting may be as simple as recognizing an obviously defective part that can be immediately repaired or replaced, or it may be much more complex and require a systematic approach to locate the problem and repair it. This section examines the mindset and concepts behind troubleshooting, as well as specific loop troubleshooting strategies. While the focus is on troubleshooting with the end goal of successful loop proving and commissioning, many of the principles involved are applicable to troubleshooting a loop during its normal operation in an otherwise functioning system.

4.1.0 Fundamentals of Troubleshooting a Loop

In any loop troubleshooting task, there are fundamental steps that should be followed in order to successfully remedy the problem; these include the following:

1. Identifying the problem
2. Identifying the loop components
3. Understanding the loop
4. Troubleshooting the loop
5. Repairing or replacing the instruments in the loop

4.1.1 Identifying the Problem

Troubleshooting a loop starts with identifying the problem in the loop—a problem recognized by an obvious symptom or through an operator's observation over a period of time. As the technician responsible for troubleshooting, you must listen carefully to the operator's complaint. Too often, inexperienced technicians will analyze a loop problem based on their own perception of the loop without giving full consideration to those who operate or observe the loop on a daily basis. Valuable troubleshooting input can be lost due to poor listening skills. Operators may not always use words or terms that technically describe the actual operation of the loop or process, but their descriptive explanations of what's going on can be instrumental in helping you to locate the problem and determine the cause.

Questions that you might ask an operator to help understand and diagnose the problem include the following:

- *"Did this loop just begin to malfunction or is it a chronic (recurring) problem?"*
- *"Does the problem go away when the controller is in the manual position?"*
- *"Is the problem worse or better if the weather is warm, cold, dry, or damp?"*
- *"Has anyone tried to fix the problem before?"*
- *"Who was the technician who worked on this loop before?"*

If you do not ask the operator about the problem, many tasks may have to be repeated in order to get to the point already reached by the operator. This doesn't mean that you should not see for yourself what transpires when the problem occurs, but asking questions can provide valuable and time-saving information that can be used as a starting point in the troubleshooting process.

Another troubleshooting tool is the documented history or maintenance record of the loop. Many facilities keep maintenance records that list problems and remedies of previous faults with the loop, if any have occurred. If available, always check the maintenance records for previous problems with the loop and determine if the current problem matches any of the previous ones. If it does, note what was done to remedy the problem in the past.

Recurring problems in the same loop may indicate an underlying cause that has not yet been located or resolved correctly. It may also suggest an unsatisfactory repair or replacement on a previous maintenance call. Likewise, documented notes from previous technicians can serve as a beneficial tool in locating or determining the cause of the problem. If the records indicate that there have been no previous loop problems, reviewing the history of abnormalities on similar loops can often direct you to a possible cause for the current problem.

Once the operator's complaint has been heard and the history of the loop has been checked, the next step in identifying the problem should include visually inspecting the loop for obvious abnormalities.

4.1.2 Identifying the Loop Components

In order to troubleshoot any control loop that passes through multiple zones of interconnected wiring (or tubing, in the case of pneumatic loops) and that contains several components in various locations, you must reference the loop sheet for that loop and be able to interpret it before you can take a logical approach to locating a problem in the loop.

Examine *Figure 16*, which is a drawing of a loop sheet / wiring diagram that shows all of the components and interconnected wiring in a pressure control loop. There are three zones depicted on the drawing: the field process area, the panel junction area, and the control room panel. As would be expected, the pressure transmitter (PT1) and the pressure control valve (PCV1) are located in the field process area. Likewise, because the transmitted signal in this loop is a 4–20 mA analog signal and PCV1 is a pneumatic control valve, there is also an I/P transducer located near the control valve that converts the analog signal received from the controller into a proportional pneumatic signal that can be used by the valve. In this particular loop, there is also a pressure gauge (PI1) located near PT1.

All of the field wiring interconnects with the control room wiring in the panel junction zone. This zone is typically located somewhere in the control room, usually behind the panel or in a junction box enclosure mounted on a wall in the control room. In any event, it should be in an easily accessible location. Finally, the control room panel holds the loop's power supply and a PLC with some kind of display (YIC1). You should be able to see why it may be very difficult to troubleshoot the newly installed loop without this drawing.

4.1.3 Understanding the Loop

If you do not know how the loop works, it is impossible to troubleshoot it. This does not mean that you must learn how to operate every loop in every process that you are called upon to troubleshoot, but it does mean that you must have a high level of confidence in your ability to understand what the loop is supposed to do as a part of the process. You must know the loop power source and the types of signals transmitted through the loop. If the operation of the loop or process is beyond your immediate knowledge, you must familiarize yourself with its operation by reviewing P&IDs, loop sheets, operation and maintenance (O&M) instructions, and other drawings and documents. You should also be prepared to question operators and other technicians or supervisors as well as observe the process.

> **WARNING!**
> Never attempt to troubleshoot a loop in a process without knowing what the loop's function is in the overall process picture. Not only could you possibly have an unsuccessful troubleshooting outcome—you could also subject yourself and others in the plant to potential injury or hazards by inadvertently causing undesirable and dangerous changes in the process. You must understand the instruments and their individual functions in the loop so your testing does not cause an instrument to change the process and create a dangerous situation. If you do not feel confident in your level of knowledge, you should not attempt to troubleshoot the problem.

While restoring the loop to an operational state is the desired outcome, troubleshooting involves more than just that. In order to troubleshoot a loop or process, you must compare its normal condition to its abnormal condition. To

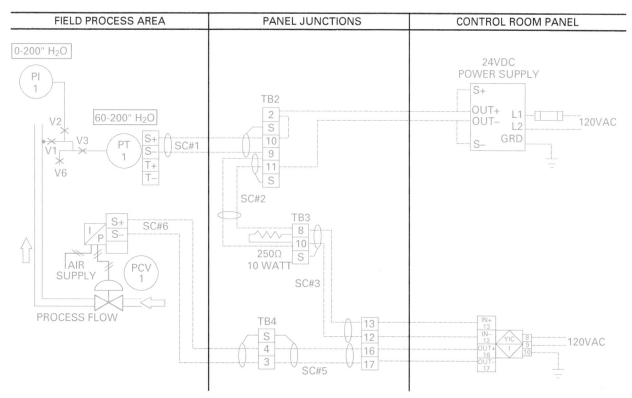

Figure 16 Loop sheet.

accomplish this step, you must have an understanding of the normal and abnormal conditions of the loop, or at least have someone present who can describe the normal condition while you witness the abnormal state. This allows you to make a comparison between the two conditions. The comparison acts as a tool in the troubleshooting process and is your first step toward achieving a solution. Once you understand the apparent difference between the normal and abnormal states of the loop, you can begin the process of figuring out the possible causes for the abnormality.

4.1.4 Troubleshooting the Loop

> **WARNING!**
> Although lockout and tagout (LOTO) procedures should be outlined in all facilities, they are not always followed by all personnel. At no time during troubleshooting a problem in any loop should changes be made to the present position of any valve, switch, circuit breaker, interrupting device, or termination without first following established lockout/tagout procedures. An existing problem or additional work may be going on that warrants the present position. That is why it is very important never to troubleshoot a loop without first isolating the loop from the process.

Every loop troubleshooting experience is unique. But a certain way of thinking tends to pervade all troubleshooting activities. This section provides an example based upon a problem in the loop illustrated in *Figure 16*; refer to this diagram as you read through this example troubleshooting process.

Assume that all you know at the beginning of the troubleshooting process is that during the loop proving, PCV1 did not respond to the simulated primary element signal from the test equipment and acts as if it is not even part of the loop. With a problem like this, any component or wiring connection in the loop could cause that symptom. So where do you begin?

You begin by going back to the first fundamental step of troubleshooting: identifying the problem. Because you are proving the loop for the first time, there's a good chance that you installed the loop, and you should therefore be able to recognize the problem as no response from the control valve.

The second and third basic troubleshooting fundamentals are identifying the loop components and understanding the loop and how it is supposed to operate. If you installed the loop, you probably have a fairly good understanding of these matters. If you did not install this loop, you must use the loop sheet / wiring diagram to get acquainted with the loop, its components, and its function.

The next focus should be on the control valve, since its unresponsiveness is the only symptom showing up that indicates a problem in the loop. This leads into the next step: testing the instruments in the loop. In testing the loop, you could begin at the power supply and verify that you have 24 VDC. However, because the power supply in this loop is in the control room, and the symptom is showing up in the field at the control valve, start at the location where the symptom of the problem appears—the control valve.

> **CAUTION**
> Troubleshooting must be performed using the proper test equipment. If you do not have loop simulating test equipment available, troubleshooting will take much longer and can only be accomplished by trial-and-error methods, which are not recommended. In situations where the proper test equipment is not available, loop troubleshooting should be left up to instrumentation personnel equipped for that type of work, or the proper test equipment should be acquired.

The best place to begin is to make sure that an adequate air supply is connected to and providing instrument air to the valve actuator, valve positioner (if installed), and I/P transducer. Without adequate air supply to all three of these components, the valve will not operate properly. This should be checked directly at the air connections to the devices either with a precision air test gauge (0–60 psi, or 0–400 kPa, works well) or with a pneumatic calibration device / multifunction calibrator. If the proper air supply is not present, trace the piping or tubing, and locate the source of the interruption.

> **CAUTION**
> Never open a closed instrument air supply valve until you have determined why the valve is in a closed position.

If the proper air supply is present at all three components, but the valve has no response to the loop, the next step is to verify loop power to the transducer by measuring between S+ and S− with a DC voltmeter or loop tester. If voltage is present, you have ruled out any problems in the power supply circuitry. You should now check the proper operation of the I/P transducer by disconnecting the loop wiring and simulating a varying 4–20 mA signal into the S+ and S− terminals with a multifunction tester or calibrator. The valve should stroke as you vary the input current to the transducer between 4 and 20 mA. If the valve responds proportionally, the transducer and valve are okay, and the problem is somewhere else in the loop, specifically in the signaling circuit that regulates the current (such as the controller). If the valve does not respond, either the transducer is defective or the valve is stuck or defective.

The transducer can be checked by disconnecting the transducer outlet air line to the valve and connecting a precision pressure gauge to the outlet. The air pressure output of the transducer should vary proportionally between 3 and 15 psi (20–100 kPa) as you vary the current input to the transducer between 4 and 20 mA. If the transducer functions properly, the problem is in the valve, and the valve should be replaced. If the transducer does not function properly, replace it.

> **NOTE**
> Pay close attention to the fact that the 24 VDC power supply is looped in series (not parallel) through the control loop. It originates at the 24 VDC S+ terminal of the power supply located in the control room, literally loops its way through the entire loop, and eventually returns to the S− of the power supply. Therefore, an absence of power at any instrument does not reveal where the problem is; it only indicates an open loop. Intentionally varying the current in a supplied with 24 VDC loop by changing the overall loop's resistance is what causes the final element to respond. Any unintentional change in resistance in a control loop, whether it is an increase or a decrease, will cause problems in the loop because the current flow will no longer be totally controlled by the loop.

For this example, assume that after performing the above tests, you discover that there is no electrical power present at the I/P transducer's inputs. If electrical power is not present at the I/P transducer, return all wiring and tubing in the field to its normal condition, proceed to the loop's power supply in the control room, and verify that you have 24 VDC across its terminals (OUT+ / OUT− or S+ / S−). If 24 VDC is not present at the power supply's terminals, you must verify that you have 120 VAC at the power supply's input terminals. If not, determine the reason and restore AC power. However, if there is AC power at the input but no DC voltage on the output, the power supply should be replaced, following all safety procedures in the process.

For this example, assume that you have a good power supply but still no voltage at the I/P transducer. At this point, you have determined the following:

- *Problem*: The valve (PCV1) does not respond.
- *Test*: You have proper air supply to all pneumatic instruments.
- *Test*: You do not have electrical power at the I/P transducer.
- *Test*: The control valve is okay.
- *Test*: The I/P is okay.
- *Test*: You have 24 VDC at the output of the power supply.
- *Conclusion*: You have an open circuit somewhere in the loop's wiring, and you must begin a systematic test at various terminal points in order to locate it.

Based on these conclusions, you can continue to perform a logical test as follows, using the loop sheet / wiring diagram (*Figure 16*):

Step 1 Test for 24 VDC between terminals 2 and 11 on TB2. This step tests the wiring connecting the 24 VDC power supply to TB2.

Step 2 Test for 24 VDC between terminals 10 and 11 on TB2. This test ensures that the jumper is installed between terminals 2 and 10 on TB2.

Step 3 Test for voltage between terminals 9 and 10 on TB2. This test confirms that shielded cable SC#1 from TB2 to the pressure transmitter (PT1) and PT1's internal circuitry don't have an open loop.

Step 4 Test for voltage between terminals 8 and 10 on TB3. This test confirms the integrity of shielded cable SC#2 from TB2 to TB3.

Step 5 Check for the installation of a 250 Ω, 10 W resistor between terminals 8 and 10 on TB3. This resistor must be in place in order for the loop to calibrate and operate properly.

Step 6 Test for voltage between terminals 12 and 13 on TB1. This test confirms the integrity of shielded cable SC#3 from TB3 to TB1.

Based on what you determine in these tests and checks, repair or replace any defective wiring and terminations. If no problems are located, and voltage is present at all of these locations, you have determined the following:

- *Problem*: The valve (PCV1) does not respond.
- *Test*: You have proper air supply to all pneumatic instruments.
- *Test*: The control valve is okay.
- *Test*: The I/P is okay.
- *Test*: You have 24 VDC at the output of the power supply.
- *Test*: Terminations and voltage levels on TB1 (except terminals 16 and 17), TB2, and TB3 are okay.
- *Test*: Shielded cables SC#1, SC#2, and SC#3 are okay.
- *Test*: A 250 Ω, 10 W resistor is installed between terminals 8 and 10 on TB3.
- *Test*: A jumper is in place between terminals 2 and 10 on TB2.
- *Conclusion*: The problem is confined to one or more of the following components:
  - YIC1 controller
  - Wiring from TB1 to YIC1
  - Wiring from TB1 to TB4 (SC#5)
  - Wiring from TB4 to I/P terminals in the field (SC#6)

Based on these conclusions, you can continue to perform a logical test as follows on the remaining terminal points and wiring, referencing the loop sheet / wiring diagram:

Step 1 Test for voltage between terminals 12 and 13 on the YIC1 terminal strip. This tests the integrity of the wiring between TB1 and YIC1.

Step 2 Test for 120 VAC between terminals 8 and 9 on the YIC1 power terminal strip. This test confirms AC controller operating power.

Step 3 Test for 24 VDC between terminals 16 and 17 on the YIC1 terminal strip. This test confirms that the output circuitry of YIC1 works.

Step 4 Test for 24 VDC between terminals 16 and 17 on TB1. This test confirms that the wiring from terminals 16 and 17 on YIC1 to terminals 16 and 17 on TB1 is intact.

Step 5 Test for 24 VDC between terminals 3 and 4 on TB4. This test confirms that shielded cable SC#5 from TB1 to TB4 is working.

For this example, let's say you conclude based on these findings that the problem is shielded cable SC#6. Your earlier test confirmed that voltage was not present at the I/P transducer S+ and S– inputs. Yet voltage is present at terminals 3 and 4 on TB4. Clearly, the cable between these two points is damaged. This damage must be located and repaired, or the entire cable should be replaced.

 12410-16 Proving, Commissioning, and Troubleshooting a Loop   Module Two   23

The question may be reasonably asked, *"With the problem located near the I/P, why didn't you start from the I/P's wiring and work back through the loop in testing the wiring?"* Keep in mind that only because of the fact that you have now completed the troubleshooting steps and located the problem close to the valve does it seem logical that you should have started at the final element end.

Where a technician chooses to start the troubleshooting procedures is up to his or her discretion. In this case, the problem could have just as easily been defective wiring at the power supply end. It only appears that you started at the wrong end of the loop because of where the problem was eventually found. Remember, however, that it was only found because you stuck to the logical approach and followed through the steps. Start wherever your instinct tells you to start, but always be systematic and logical.

4.1.5 Repairing the Loop

Once you've located the problem, the troubleshooting process switches to something much more straightforward—repair. Of course, what you must do for this step depends heavily on the kind of failure involved. If an instrument is defective, it must be repaired or replaced. Wiring or tubing problems call for repair. Components that are stuck must be freed. Whatever the case, be sure to follow your company's procedure.

An equally important step, though often neglected, is documenting what you've done both in diagnosing the problem and in correcting it. You should always leave a complete record behind so future technicians will have the benefit of your experiences. Your company probably also expects you to keep a record of the time that you spent on the process as well as the components that you had to repair or replace. If your company uses an IAMS, this process may be greatly streamlined.

4.2.0 Troubleshooting an Oscillating Process

A very common problem in some control systems is an oscillating process. This is a process that continues to react to a disturbance rather than returning to stability after a reasonable period of time. Troubleshooting a dynamic or oscillating process takes a good working knowledge of the system as well as patience. Because the process may be constantly changing, it is difficult to compare static readings. The basic procedure for troubleshooting this type of process is described in the following sections.

4.2.1 Verifying That a Problem Exists

In many cases, trouble in a system is discovered by an operator. The operator may observe symptoms of a problem, or an alarm may indicate that something is wrong. An oscillating process may first be noticed on a recorder controller. When this type of trouble is discovered, the first step is to verify that there is a genuine problem in the system, and then to learn something about the problem, such as the following:

- How was the problem discovered?
- When did the problem begin?
- Is the problem constant or intermittent?

Observe the system in operation. This allows you to see the symptoms firsthand and to verify how the system works.

4.2.2 Gathering Information

In addition to finding out about the problem, you must gather information about the system and how it is designed to operate. There are three important sources of information: the operator, the panel graphic, and the loop diagram. The operator can provide information about system operation. The panel graphic (*Figure 17*) is another source of information. While its primary function is to serve as a control center for the operator, the panel graphic also provides some information about the process.

The loop diagram provides detailed information about the components and design of the loop. The loop diagram in this example (*Figure 18*) provides the following information:

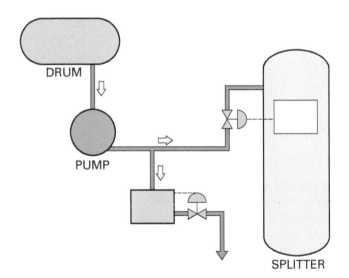

Figure 17 Example panel graphic.

- The flow element (FE 112) is an orifice.
- The flow transmitter (FT 112) senses the differential pressure across the orifice and sends a pneumatic signal to the flow recording controller.
- The recording controller (FRC 112) sends a pneumatic signal to the flow control valve.
- The flow control valve (FCV 112) responds to the controller signal to control the flow rate of the process fluid.

4.2.3 Identifying Possible Causes of the Problem

Once the symptoms of the problem have been determined and you are acquainted with how the system should work, you can begin to identify possible causes of the problem. For example, the symptom is an oscillating process in a flow control loop. There are several possible causes for this type of problem, including the following:

- The transmitter may be malfunctioning and sending improper signals.
- A faulty controller could be causing the oscillations; this can occur if the controller is not tuned properly.
- The control valve could be sticking or the actuator may not be operating properly.
- The process itself may be oscillating due to a change in process fluid characteristics.

Controller operation is the easiest and most likely initial check to make. A controller can often be ruled out as the cause of the problem simply by shifting it to manual and observing the controller and process response. The manual mode bypasses the controller's automatic components to produce an output that is independent of these components. When in manual, the oscillation should stop if the controller is the problem. If it continues, the controller is probably not the cause of the problem. One way to check a controller is as follows:

Step 1 Move the controller transfer switch to the center or seal position.

Step 2 Match the output to the setpoint. This is done to reduce upset during the transition from automatic to manual.

Step 3 Move the switch to manual.

If the output stabilizes and the input continues to oscillate, the controller is receiving an oscillating signal. This is an indication that the controller is probably not the cause of the problem.

If the controller is not the cause of the problem, the process of identifying the cause is continued by checking out the other possible causes. In this example, the valve, transmitter, and process remain to be checked.

4.2.4 Checking Instruments in the Field

Before going to the field, it is good practice to check the loop again. This information can help determine the most efficient way to approach the problem. Making visual checks and observing gauge readings can frequently identify or rule out possible causes of a problem. Use the following steps to check instruments in the field:

Step 1 Visually check the output gauge on the transmitter. If the gauge is oscillating it indicates that the transmitter output is oscillating. Oscillation in the transmitter output could be due to a problem in the transmitter, the valve, or the process.

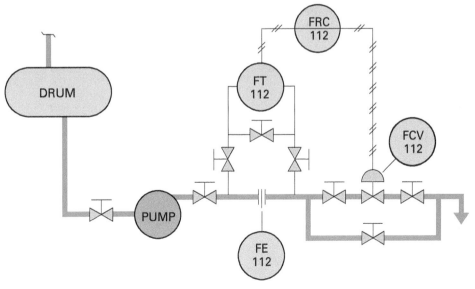

Figure 18 Loop diagram.

Step 2 Check the instrument air supply gauge to the transmitter. If the transmitter is receiving a pulsating supply, it could cause oscillation in the transmitter output.

Step 3 Visually check the valve. By looking at the valve stem, you can see if it is moving or not. It should be steady because the output from the controller in the manual mode is steady. If the stem is moving, there may be a problem with the valve actuator or positioner. Look for air supply problems as well as erratic positioner outputs.

4.2.5 Checking the Transmitter

Transmitter operation can be checked by first performing a zero check, which is done by equalizing the pressure on both sides of the differential pressure transmitter. With a zero differential input pressure, the transmitter output should have a signal equivalent to zero. If the output drops but continues to oscillate, the transmitter is probably the cause of the problem. One method for performing a zero check on a differential pressure transmitter is as follows:

Step 1 Open the bypass or equalizer valve. This should cause the output reading to go to zero and stop oscillating, indicating that the transmitter is responding to a change in differential pressure.

Step 2 Close the valve on the high-pressure side of the transmitter.

Step 3 Close the valve on the low-pressure side of the transmitter.

Step 4 With the transmitter on bypass, a stable zero output indicates that the transmitter is at least responding to changes in differential pressure.

Step 5 Once its position has been verified, the transmitter should be put back into service.

Step 6 Open the low-pressure valve.

Step 7 Open the high-pressure valve.

Step 8 Close the bypass valve.

If the oscillation re-appears when differential pressure is restored, the cause of the problem is probably in the process itself. Other possible instrument problems include a loose control valve plug or a transmitter that does not malfunction until a differential pressure is sensed. However, checking these problems is a time-consuming job, so it is most efficient to make a quick check of the process first.

4.2.6 Locating the Problem

When the cause of a problem appears to be in the process, the loop diagram or the P&ID can provide information about the equipment and its function in the process loop. Referring back to *Figure 18,* you can see that there is a drum with a pipe going from the bottom through a suction valve to a pump. From the pump, the pipe goes to a discharge valve and to the orifice plate or primary flow element. The cause of the problem could be in the process piping, one of the valves, or the pump. Checking these possibilities first may avoid the costly time involved in removing and disassembling instruments.

Assume the technician found that the suction valve between the tank and the pump was partially closed. A partially blocked valve can cause abnormally low pressures to occur in the pump suction as it attempts to draw liquid in. These low pressures can cause vapor bubbles to form that collapse at the pump discharge. This process is called cavitation. Cavitation leads to pulsating flow and pump pressure. When the pump cavitated, it caused oscillation in the pump discharge. The oscillation was accurately measured by the instrument system, but the controller could not correct this problem.

When the valve problem is corrected, the problem should disappear. However, the technician should verify that the problem truly has been corrected. Verification that the problem has been repaired usually involves repeating the same check that identified the problem. In this example, this check might involve the following steps:

Step 1 Move the controller transfer switch to the seal position.

Step 2 Match the output to the setpoint.

Step 3 Move the switch to the automatic position.

Step 4 Observe the input and the output. Both should be tracking without oscillation. If oscillation occurs, the problem has not been corrected.

4.2.7 Using a Troubleshooting Flowchart

Certain problems are common enough that they are documented through a troubleshooting tool called a flowchart. Flowcharts are diagrams that ask a series of logical questions designed to eliminate certain possibilities. As you advance through the chart, you will eventually run into the probable cause of the problem. *Figure 19* shows a typical troubleshooting flowchart. Some key characteristics of flowcharts are as follows:

- Diamonds represent decision-making points.
- Squares (or rectangles) represent facts or solutions.
- Circles (or ovals) represent a reference to another source.

When trying to troubleshoot a problem, find out if a diagnostic flowchart exists. If it does, use it as your starting point since it represents the accumulated understanding of others who have dealt with similar problems before you. However, keep in mind that flowcharts do not guarantee a solution. The problem may be unusual or even unique. Always be ready to depart from the established paths and be innovative as you attempt to solve the problem. Regardless of how you attempt to diagnose the problem, always be systematic, logical, and use good judgement.

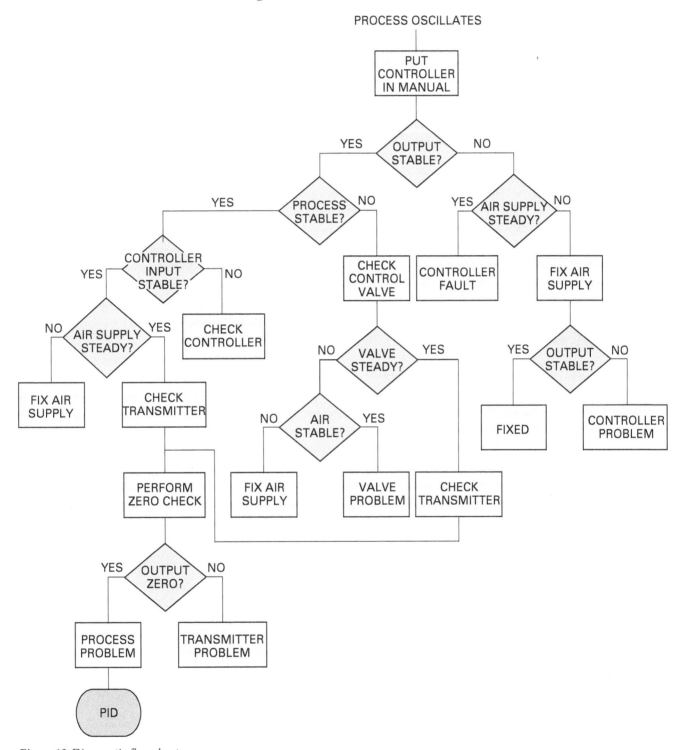

Figure 19 Diagnostic flowchart.

Additional Resources

Instrumentation for Process Measurement and Control, Norman A. Anderson. Third Edition. 1997. Boca Raton, FL: CRC Press.

Maintenance and Calibration of HART® Field Instrumentation (PDF), Richard Pirret. Fluke Corporation, Everett, WA, USA. Accessed at **www.plantservices.com**.

The following websites offer resources for products and training:

Amtek Calibration, **www.crystalengineering.net**

Emerson Process Management, **www.emersonprocess.com**

Extech Instruments, **www.extech.com**

Fluke Corporation, **www.fluke.com**

GE Digital Solutions, **www.gemeasurement.com**

Omega Engineering, **www.omega.com**

WIKA Instrument, LP, **www.wika.us**

4.0.0 Section Review

1. When trying to identify a loop problem, the best place to start is with the _____.
   a. operator of the loop
   b. diagnostic flowchart
   c. power supply
   d. O&M manual

2. When troubleshooting a loop, a key to successful problem solving is to _____.
   a. use the diagnostic flowchart
   b. start at the final element
   c. be systematic and logical
   d. have a multifunction calibrator

3. To explore a controller's possible role in an oscillating process, one should _____.
   a. disconnect its air supply
   b. switch it to manual mode
   c. change it to or from PID mode
   d. recalibrate it

SUMMARY

Once a control loop's installation is finished, it must pass a series of milestones on its way to being turned over to the client (a process known as *selling*). These steps begin with checking the loop by visually inspecting it, followed by a continuity test. Proving and calibration come next; *proving* is the testing of a loop using simulated process. Finally, the loop is completely documented and tested with live process, a step known as *commissioning*. Once it passes this point, the loop is ready for handoff.

Since problems are likely to surface at each stage of this chain of events, the instrument technician must be skilled at troubleshooting problems so they can be corrected. This module also includes guidelines designed to establish an effective troubleshooting mindset.

Review Questions

1. The process of documenting a loop occurs during _____.
   a. checking
   b. proving
   c. commissioning
   d. selling

2. Mechanical inspection of a loop includes looking for _____.
   a. damaged components
   b. oscillating processes
   c. uncalibrated instruments
   d. untuned controllers

3. When performing tests on electrical loops, it may be necessary to simulate the load with a resistor of _____.
   a. 100 Ω
   b. 150 Ω
   c. 200 Ω
   d. 250 Ω

4. What do some pneumatic calibrators require in addition to the device itself in order to perform a pneumatic integrity check?
   a. A loop excitation source
   b. A power supply
   c. A pressure module
   d. A blow-down valve

5. The main goal of proving a loop is to confirm _____.
   a. mechanical integrity
   b. component functionality
   c. proper documentation
   d. electrical continuity

6. Loop proving usually simulates the process using a _____.
   a. multifunction calibrator
   b. thermocouple
   c. DP cell
   d. loopback tester

7. Calibration should not be confused with _____.
   a. zeroing and spanning
   b. loop tuning
   c. traceability
   d. sensor trim

8. If a HART device is used in digital-only mode, it is necessary to calibrate the device's _____.
   a. first internal section
   b. second internal section
   c. third internal section
   d. first and third internal sections

9. A good instrument asset management system (IAMS) can keep track of and schedule _____.
   a. loop checks
   b. network traffic
   c. loop oscillations
   d. calibrations

10. When troubleshooting instruments that have special installation procedures, you should consult the _____.
    a. loop diagrams
    b. hardware installation detail drawings
    c. P&ID
    d. specification sheets

11. Loop commissioning generally involves testing the loop with varying degrees of _____.
    a. tuning
    b. calibration
    c. automation
    d. faults

12. A loop problem that appears repeatedly in the maintenance records suggests that _____.
    a. the problem has never truly been fixed
    b. the problem is due to an uncalibrated component
    c. the problem is the fault of the process operator
    d. a stuck valve is almost certainly the problem

13. To get a feel for how a loop operates, you should consult the _____.
    a. hardware specification sheet
    b. operator of the loop
    c. maintenance records
    d. indicating transmitter

 12410-16 Proving, Commissioning, and Troubleshooting a Loop Module Two 29

14. A good basic check of an analog instrument loop begins by testing each instrument input for _____.
    a. oscillation
    b. frequency
    c. resistance
    d. voltage

15. If a pneumatic transmitter is pulsing, a good place to begin troubleshooting is to check the _____.
    a. air supply
    b. calibration date
    c. tuning parameters
    d. bypass valves

Trade Terms Introduced in This Module

Cavitation: An undesirable pulsation in a fluid stream caused by flow changes that create pockets of low-pressure fluid mixed in with normal pressure. Cavitation can damage equipment and cause a process to oscillate.

Commissioning: This step involves documenting and testing an instrument channel using live process. Checks are usually incremental, beginning with manual control and progressing to full automatic.

Continuity tester: A basic piece of test equipment that verifies an electrical current can flow through a circuit.

Current loop trim: An adjustment made at the third stage of a smart instrument, which adjusts the relationship between the calculated output signal and the actual analog output signal.

Ground loop: An undesirable current flow through a cable's shield caused by grounding it at each end. The difference of ground potentials causes the current. Ground loops result in noise and signal distortion.

Instrument asset management system (IAMS): A software application that combines a database and communications system to keep detailed records about instrumentation. An IAMS guides calibrations and maintenance, keeps inventory, and provides alerts to instrument personnel.

Interlock: A mechanical, electrical, or software safety feature that prevents certain events from occurring unless other necessary conditions are first met.

Loop checking: The process of performing a series of checks on a completed instrument chain and verifying that each device is installed correctly and has continuity to other devices.

Oscillating process: A process that continues reacting to a disturbance rather than returning to a stable condition in a reasonable period of time.

Proving: A step involving the testing of an instrument channel using a simulated process. Control behavior is assessed and devices are calibrated.

Selling: The final step of loop checking, which involves turning the commissioned loop over to the client or to another contractor.

Sensor trim: An adjustment made at the first stage of a smart instrument, which establishes the relationship between the input signal and the lower and upper limits of the calculated process variable.

Stroke: To operate a valve through its entire range to confirm that it is working correctly.

Zero and span calibration: A basic calibration procedure in which an instrument's zero and span points are adjusted to produce the proper relationship between the input and output minimum and maximum points.

Additional Resources

This module presents thorough resources for task training. The following reference material is recommended for further study.

Instrumentation for Process Measurement and Control, Norman A. Anderson. Third Edition. 1997. Boca Raton, FL: CRC Press.

Maintenance and Calibration of HART®Field Instrumentation (PDF), Richard Pirret. Fluke Corporation, Everett, WA, USA. Accessed at **www.plantservices.com**.

The following websites offer resources for products and training:

Amtek Calibration, **www.crystalengineering.net**

Emerson Process Management, **www.emersonprocess.com**

Extech Instruments, **www.extech.com**

Fluke Corporation, **www.fluke.com**

GE Digital Solutions, **www.gemeasurement.com**

Omega Engineering, **www.omega.com**

WIKA Instrument, LP, **www.wika.us**

HART®, *Wireless*HART®, and HART-IP™ are registered trademarks of the FieldComm Group™, **www.fieldcommgroup.org**

Figure Credits

©Rawi Rochanavipart/Dreamstime.com, Module Opener

Emerson Process Management, Figures 2, 12–14

Fluke Corporation, reproduced with permission, Figures 4, 5, 9, 10

Greenlee/A Textron Company, Figure 6

Dwyer Instruments, Inc., Figure 7

Topaz Publications, Inc., Figure 8

Section Review Answer Key

Answer	Section Reference	Objective
Section One		
1. c	1.0.0	1
2. b	1.1.1	1a
3. a	1.2.1	1b
Section Two		
1. b	2.1.0, 2.1.1	2a
2. a	2.2.3	2b
3. d	2.2.4	2b
Section Three		
1. c	3.1.0	3a
2. b	3.2.0	3b
3. b	3.2.0	3b
Section Four		
1. a	4.1.1	4a
2. c	4.1.4	4a
3. b	4.2.3	4b

NCCER CURRICULA — USER UPDATE

NCCER makes every effort to keep its textbooks up-to-date and free of technical errors. We appreciate your help in this process. If you find an error, a typographical mistake, or an inaccuracy in NCCER's curricula, please fill out this form (or a photocopy), or complete the online form at **www.nccer.org/olf**. Be sure to include the exact module ID number, page number, a detailed description, and your recommended correction. Your input will be brought to the attention of the Authoring Team. Thank you for your assistance.

Instructors – If you have an idea for improving this textbook, or have found that additional materials were necessary to teach this module effectively, please let us know so that we may present your suggestions to the Authoring Team.

NCCER Product Development and Revision
13614 Progress Blvd., Alachua, FL 32615

Email: curriculum@nccer.org
Online: www.nccer.org/olf

❏ Trainee Guide ❏ Lesson Plans ❏ Exam ❏ PowerPoints Other _____

Craft / Level: _____ Copyright Date: _____

Module ID Number / Title: _____

Section Number(s): _____

Description: _____

_____

_____

_____

_____

Recommended Correction: _____

_____

_____

_____

_____

Your Name: _____

Address: _____

_____

Email: _____ Phone: _____

12405-16
Tuning Loops

Overview

A control loop that has been tested and proved is fully functional. However, before it can perform well under real process conditions, it must be tuned. If it is a PID-type controller, the P, I, and D settings must be tweaked to optimal values. Controllers with fewer modes still require some adjustment. This process is known as "tuning", and it requires a certain understanding of process behavior as well as some basic math skills. This module provides the information necessary to learn to tune loops.

Module Three

Trainees with successful module completions may be eligible for credentialing through the NCCER Registry. To learn more, go to **www.nccer.org** or contact us at 1.888.622.3720. Our website has information on the latest product releases and training, as well as online versions of our *Cornerstone* magazine and Pearson's product catalog.

Your feedback is welcome. You may email your comments to **curriculum@nccer.org**, send general comments and inquiries to **info@nccer.org**, or fill in the User Update form at the back of this module.

This information is general in nature and intended for training purposes only. Actual performance of activities described in this manual requires compliance with all applicable operating, service, maintenance, and safety procedures under the direction of qualified personnel. References in this manual to patented or proprietary devices do not constitute a recommendation of their use.

Copyright © 2016 by NCCER, Alachua, FL 32615, and published by Pearson Education, Inc., New York, NY 10013. All rights reserved. Printed in the United States of America. This publication is protected by Copyright, and permission should be obtained from NCCER prior to any prohibited reproduction, storage in a retrieval system, or transmission in any form or by any means, electronic, mechanical, photocopying, recording, or likewise. To obtain permission(s) to use material from this work, please submit a written request to NCCER Product Development, 13614 Progress Blvd., Alachua, FL 32615.

From *Instrumentation Level Four, Trainee Guide*, Third Edition. NCCER.
Copyright © 2016 by NCCER. Published by Pearson Education. All rights reserved.

12405-16
Tuning Loops

Objectives

When you have completed this module, you will be able to do the following:

1. Describe the function of tuning and basic proportional control concepts.
   a. Describe the importance and function of loop tuning.
   b. Describe basic proportional control and define terms relevant to tuning.
2. State the basic equations needed for loop tuning and describe various loop tuning processes.
   a. State the basic equations needed for loop tuning.
   b. Describe open loop tuning processes.
   c. Describe closed loop tuning processes.
   d. Describe a visual loop tuning process.

Performance Tasks

Under the supervision of the instructor, you should be able to do the following:

1. Perform open loop tuning.
2. Perform closed loop tuning.
3. Perform visual loop tuning.

Trade Terms

Amplitude
Bump test
Critical frequency
Dampened oscillation method
Deadtime
Derivative
Differentiator
Integral
Integrating process
Integrator
Lag
Non-integrating process
Oscillations
Overshoot
Over-the-hump time
Reaction rate method
Reciprocal
Repeats per minute (rpm)
Saturate
Stability
Steady state
Stiction
Time constant
Time constant method
Tuning
Ultimate period method

Industry Recognized Credentials

If you are training through an NCCER-accredited sponsor, you may be eligible for credentials from NCCER's Registry. The ID number for this module is 12405-16. Note that this module may have been used in other NCCER curricula and may apply to other level completions. Contact NCCER's Registry at 888.622.3720 or go to **www.nccer.org** for more information.

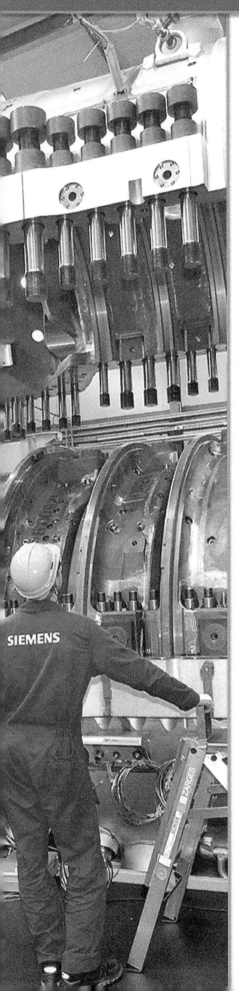

Contents

- 1.0.0 Tuning Concepts 1
  - 1.1.0 The Reason for Loop Tuning 1
    - 1.1.1 Factors that Influence Loop Stability 1
  - 1.2.0 Loop Tuning Concepts 2
    - 1.2.1 Proportional (P) Control Review 2
    - 1.2.2 Proportional + Integral (PI) Control 2
    - 1.2.3 Proportional + Integral + Derivative (PID) Control 3
    - 1.2.4 Terms and Definitions Associated with Loop Tuning 3
    - 1.2.5 Process Type Identification 4
- 2.0.0 Tuning Methods 7
  - 2.1.0 Tuning Equations 7
    - 2.1.1 Energy Balance 7
    - 2.1.2 Time Constant 8
    - 2.1.3 Complete Response 8
    - 2.1.4 Process Gain 8
    - 2.1.5 Proportional Band 9
    - 2.1.6 Integral Time 10
    - 2.1.7 Derivative Time 10
    - 2.1.8 PID Loop 11
  - 2.2.0 Open Loop Tuning 11
    - 2.2.1 Procedure Overview 11
    - 2.2.2 Time Constant Method 13
    - 2.2.3 Reaction Rate Method 15
  - 2.3.0 Closed Loop Tuning 16
    - 2.3.1 Ultimate Period Method 16
    - 2.3.2 Dampened Oscillation Method 18
  - 2.4.0 Visual Loop Tuning 19
    - 2.4.1 Incremental Changes 19
    - 2.4.2 Apparent Instability 19
    - 2.4.3 Sluggish Response 20

Figures

Figure 1	Graph of a control system's behavior	4
Figure 2	Process response curve of a non-integrating process	5
Figure 3	Process response curve of an integrating process	6
Figure 4	Surge tank and its response to a disturbance	9
Figure 5	Proportional plus derivative output	11
Figure 6	PID output	12
Figure 7	Block diagram of a process being open loop tuned	12
Figure 8	Process reaction curve for an open loop tested process	13
Figure 9	Time constant method for open loop tuning	13
Figure 10	Test data for open loop time constant method	14
Figure 11	Reaction rate method for open loop tuning	15
Figure 12	Test data for open loop reaction rate method	16
Figure 13	Control system at the ultimate gain or proportional band	17
Figure 14	Responses of the measured variable to a supply disturbance	17
Figure 15	Test data for the ultimate period tuning method	18
Figure 16	Test data for the dampened oscillation method	19
Figure 17	Waveform of a sticking control valve (stiction)	21

SECTION ONE

1.0.0 TUNING CONCEPTS

Objective

Describe the function of tuning and basic proportional control concepts.
a. Describe the importance and function of loop tuning.
b. Describe basic proportional control and define terms relevant to tuning.

Trade Terms

Amplitude: The strength of a signal, often represented by the height of the plot in a graph.

Bump test: A method of testing the response of a control system by suddenly changing the controller setpoint to produce a process disturbance.

Critical frequency: The frequency at which a control system can become unstable if its gain is greater than 1.

Deadtime: The delay interval between the point when the controlled process variable changes and the point when the controller begins to respond to the change.

Derivative: A control method in which the controller's output depends on the rate of change of the process variable or error.

Integral: A control method in which the controller's output depends on the amount and duration of the error signal.

Integrating process: A process that changes without limit as long as the controller is outputting a value above the process equilibrium level.

Lag: The delay between an input change and a corresponding output response.

Non-integrating process: A process that reaches a new equilibrium level in response to a change in controller output but does not keep changing once it reaches the new level. Also known as a self-regulating process.

Oscillations: Repetitive changes in signal amplitude.

Overshoot: The tendency for the controlled variable to exceed the setpoint.

Reciprocal: The value by which you can multiply a given number in order to obtain a result of 1. The reciprocal of a number is usually written as a fraction, and is equal to is 1 over that number. For example, the reciprocal of 3 is ⅓.

Stability: The condition in which a control system responds to disturbances by returning quickly to the setpoint with oscillations that either remain constant or die out.

Steady state: The condition in which the controlled variable is essentially at the setpoint and is not changing significantly.

Time constant: A measurement of time that describes the response behavior of a particular control system. A single time constant is the time required for the controlled variable to change by 63.2 percent of the amount that it must change to reach steady state.

Tuning: Adjustments made to a loop controller to produce a desired response behavior to process disturbances.

Once a loop has been commissioned and placed into service, it generally requires tuning. Tuning involves adjusting the loop controller to produce a desired response. The term *response* describes the dynamic behavior of the entire system after a disturbance, which may occur either as a result of demand or supply variations. Response is the combined behavior of the process and the control system. The overall quality of the response is judged by the speed with which the controlled variable returns to the setpoint, by the amount of overshoot that occurs, and by the stability of the system during the transient condition.

1.1.0 The Reason for Loop Tuning

Ideally, a control system responds quickly to disturbances and returns the controlled variable to the setpoint value promptly. It should be as stable as possible so that it remains near the setpoint, rather than swinging far away from it every time there is a disturbance. The factors that influence the control system's behavior can be complex. Tuning involves adjusting those factors in order to achieve the best response possible. Since some factors cannot be adjusted, tuning is often a compromise between different factors.

1.1.1 Factors that Influence Loop Stability

The factors that affect a control system's behavior often conflict with one another. An increase in response speed usually decreases the system's stability. Attempting to achieve too rapid a response may end up creating a completely unstable system.

A stable control system is one that is undergoing oscillations that are constant or decreasing in amplitude. An unstable system's oscillations increase in amplitude and usually become uncontrolled and potentially destructive.

These undesirable oscillations normally occur at a certain frequency, which is known as the critical frequency. At this frequency, there is a 360-degree phase shift around the control loop. This 360-degree phase shift is critical because under this condition, energy or material entering the process will maintain the oscillations or even increase their amplitude.

A 360-degree phase shift alone will not sustain the oscillations, however. The control loop must have a gain greater than or equal to 1 for this to happen. If the loop's gain is equal to 1, there is just enough energy or material entering the system to sustain oscillations at a constant level. The phase shift ensures that the energy or material enters the loop at exactly the right moment to keep the oscillations going, much like pushing a child on a swing—the key is to time the pushes at just the right moment in the cycle.

If the gain is greater than 1, the system acquires more energy or material with each cycle. Its oscillations become larger and larger (increasing amplitude). Eventually, they will go out of control in a number of possible ways, none of them particularly good. Consequently, the system is completely unstable.

During controller tuning, the goal is to adjust the loop's gain so that it is less than 1 at the critical frequency. If this is done correctly, process oscillations will die out as quickly as possible, and the controlled variable will remain near the setpoint. This section explores different ways this goal can be reached.

1.2.0 Loop Tuning Concepts

The "Process Control Theory" module from *Instrumentation Level Three* covers process control, as well as various forms of proportional control (including PI, PD, and PID), in some detail. It may be advantageous to review that material before continuing with this module. The following sections highlight some key ideas related to proportional control and also introduce several terms crucial to loop tuning.

1.2.1 *Proportional (P) Control Review*

Proportional control acts to position the final control element in a loop (a valve, for example) in proportion to the degree of error. Error, as you may recall, is the difference between the setpoint and the actual measured value of the controlled variable (temperature, pressure, flow, etc.). Proportional control provides a stepless (smooth) output that can position the final control element at intermediate positions as well as fully on/off or open/closed.

With proportional control, the final control element assumes a definite position for each value of the measured variable. The relationship between the two is usually linear. The setpoint is typically in the middle of a specified proportional range or band, resulting in an offset between the active loop control point and the setpoint. The proportional band is usually expressed as a percentage of full scale or degrees. Although the terms *gain* and *proportional band* are sometimes used interchangeably, gain is actually the reciprocal of the proportional band.

The main advantage of proportional control is that it immediately produces a proportional output when the controller detects a nonzero error value; this enables the final control element to be repositioned in a relatively short amount of time. The main disadvantage is that a residual offset error exists between the measured variable and the setpoint at almost all times. The offset will only drop to zero when the final control element is positioned so that the controlled variable exactly matches the setpoint.

It is important to note that with a proportional controller, it is typically necessary to make a small adjustment manually to bring the controlled variable to the setpoint on initial startup. Manual control may also be necessary if the process alters significantly and the controller cannot handle the change.

Although proportional control is generally a good method and works well for many situations, it is far from perfect. For this reason, controllers often incorporate other control modes that act in addition to the proportional mode. These other modes act only at certain times and serve to strengthen the weak aspects of proportional control.

1.2.2 *Proportional + Integral (PI) Control*

Some control loops use a proportional + integral (PI) mode of control. This mode can be significantly better than proportional mode alone. It combines the immediate response output characteristics of the proportional mode with an integral control function. The integral control function provides for an output that is proportional to the time integral of the input. This means that it continues to change as long as an error signal exists. The integral function acts only when an error exists between the controlled variable and the reset value. The reset value is the setpoint for a set time period. As a result, the integral function eliminates any offset that exists between the measured variable and the setpoint, which can occur with proportional control alone. The addition of the integral action to proportional control automatically performs a gain resetting. For this reason, PI controllers are sometimes called *proportional plus automatic reset*, or *proportional plus reset* controllers.

The main difference between proportional control and PI control is that proportional control is limited to a single final control element position for each value of the controlled variable. With a PI controller, the position of the final control element depends not only on the location of the controlled variable within the proportional band, but also on the duration and magnitude of its offset from the setpoint. When the controlled variable is essentially steady, the control point and setpoint will be the same.

The PI control mode can be adversely affected by sudden large error signals. This can be caused by a large demand deviation, a large setpoint change, or during initial startup. Such a large sustained error signal can eventually cause the controller output to drive to its limit. The result is called *reset windup*. This situation causes loss of control for a period of time.

1.2.3 Proportional + Integral + Derivative (PID) Control

Proportional + integral + derivative (PID) control is the most sophisticated form of loop control. It adds a derivative function to the proportional and integral functions used in PI control. (Derivative control is also called *rate control*.) It provides an output that is proportional to the rate of change of error. This means that it acts only when the error is changing with time. As a result, the more quickly the control point changes, the greater the corrective action that the derivative function applies. If the control point moves rapidly away from the setpoint, the derivative function outputs a corrective action to bring the control point back more quickly than would occur using the proportional or integral functions alone. As the control point moves toward the setpoint, the derivative function reduces the corrective action to slow down the approach to setpoint, which reduces the possibility of overshoot. Derivative control can never be used on its own because it cannot detect a steady (unchanging) error condition.

PID control provides the most accurate and stable control when applied in systems that have a relatively small mass and those that react quickly to changes in energy added to the process. PID control is recommended in systems where the load on the process changes often, and the controller is expected to compensate automatically due to frequent changes in setpoint, the amount of energy available, or the mass to be controlled.

1.2.4 Terms and Definitions Associated with Loop Tuning

Generally, in the context of loop tuning, the system's behavior is represented by a graph in which the x-axis (the horizontal line) represents time and the y-axis (the vertical line) represents the controlled variable. A typical graph (*Figure 1*) shows how the variable behaves after a process disturbance. This disturbance is sometimes called a *step change* because some aspect of the process suddenly jumps up or down, much like a stair step, to a different value. The step change causes the control system to react as it attempts to bring the controlled variable back to the setpoint.

Besides the process itself changing, a step change can also be created by altering the setpoint. For example, if the control operator suddenly changes a thermostat to a different temperature, the control system will see this alteration as a step change and attempt to bring the controlled variable into line with the new setpoint. When tuning a control loop, suddenly altering the setpoint is a common way of deliberately creating a step change to see how the control system reacts. This action is sometimes called a bump test.

Time is a crucial element when working with control systems. While the x-axis represents time, it may not be scaled into a standard measurement of time such as minutes or seconds. Instead, it may be divided into generic time units. *Figure 1* takes this approach. As long as you know what a time unit represents, however, you can easily convert them into real time values by multiplying by the amount of actual time units they represent.

There are certain key terms associated with these graphical models. Since you will be encountering them many times as you learn how to tune PID controllers, the ability to recognize them is important.

- Deadtime – This is the time interval between the point when the step change occurs and the point when the controlled variable starts to change in response. During the deadtime, the controlled variable will be changing but the controller will not be making any attempts to respond to it. Deadtime is present in all control systems. It can be minimized but never totally eliminated. It is the most common cause of

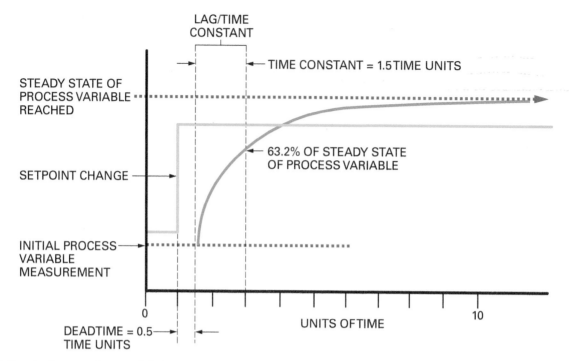

Figure 1 Graph of a control system's behavior.

many closed loop performance problems, especially if it's excessive. Deadtime is sometimes called *delay*. The deadtime in *Figure 1* happens to be 0.5 time units long.
- Lag – This is the tendency of the output to lag behind the input when the system is changing. First-order lag is the most common lag in process systems. Higher orders are harder to deal with, and are not discussed here. Many processes can be approximated as a combination of first-order lag plus deadtime (FOLPDT).
- Steady state – The steady state condition is the point at which the controlled variable is no longer changing significantly but is holding steady near the setpoint. While it may oscillate around the setpoint by a small amount, the difference between it and the setpoint is small enough that it is not considered an error. The goal of all control systems is to maintain a steady state condition.
- Time constant – A time constant is an amount of time that is unique to a particular process and control system. More importantly, it helps describe the speed at which the system responds to a disturbance. All control systems take a certain amount of time to reach the steady state condition from the point at which they start reacting (the end of the deadtime). A single time constant is the time required for a process change to reach 63.2 percent of the total change. In the graph (*Figure 1*), the time constant is 1.5 time units. A full response from the end of the deadtime to arrival at steady state requires five time constants to pass (7.5 time units in the graph). Depending on the complexity and nature of the controlled process, time constants can be difficult to measure or estimate. A complex system with several lags in series has a time constant equal to the sum of the individual time constants.

1.2.5 Process Type Identification

Processes fall into one of two general categories: *non-integrating processes* and *integrating processes*. Understanding the difference is crucial because each type behaves very differently in response to the controller. When tuning a loop, one should be aware of the process type, since tuning generally involves manipulating the process to gather information. Failure to understand a process's overall behavior could easily result in undesirable behavior, or even a serious accident.

A non-integrating process (also known as a *self-regulating process*) is one that responds to a controller output change (ΔCO) by moving toward a new process value (PV). For a given ΔCO, the process value will change by a specific amount. It will show some amount of deadtime before it starts changing, and it will take a certain amount of time to reach the new steady state value. Consequently, it will have a definite time constant. But the key idea is that it *will* reach a steady state condition and level off at that point.

This is the reason that these processes are considered *self-regulating*. Even if the control output remains in its new position, the process will not keep changing without limit.

A good example of a non-integrating process is a tank with a heater that it set to output a particular temperature. No matter how long the heater is on, the highest temperature that the liquid can reach is the heater's temperature. *Figure 2* shows the way that this kind of process will respond to a change in the controller output. As you can see, the process value ramps upwards but eventually levels off after five time constants have passed. Thus, after the ΔCO in a non-regulating process, the controller can remain at the new output, and the process will level off on its own.

> **CAUTION**
> Although non-integrating processes are self-regulating, they can still reach levels that are dangerous and potentially destructive if the controller is set to produce too large a ΔCO. When manually manipulating a non-integrating process, always use caution so that you don't inadvertently cause the process output to exceed a limit, possibly causing a plant trip or even damage to the equipment.

An integrating process, in contrast, is *not* self-regulating. Once the controller output changes, the process will begin to react and will keep changing as long as the controller output is in the new position. Of course, a process that keeps changing will eventually reach dangerous or damaging levels. For this reason, the controller output must be changed to the new level only for as long as required to produce the desired change. It must then return to a neutral level, after which the process will eventually level off at a new equilibrium condition.

Figure 3 shows the behavior of and parameters associated with an integrating process. Notice that just as with the non-integrating process, there is a deadtime. Again, this is the interval between the point when the control output changes and the point when the controller starts responding to the change. However, the time constant is determined differently. Notice the angled dashed line that runs alongside the curve; this represents the process variable's maximum slope. It intersects the horizontal line that marks the curve's amplitude at the end of the deadtime. A vertical line marks this intersection point and meets the curve. The time interval between this vertical line and the one that marks the end of the deadtime is the process time constant.

A good example of an integrating process is a tank with a drain and an inlet pipe. Liquid enters and leaves the tank at specific rates. As long as these stay matched, all will be well and an equilibrium condition will be sustained. But if the input flow value is increased above the output value (causing the liquid to enter at a faster rate than it leaves), the liquid level will start to rise, and will keep rising until the tank overflows.

Typically, therefore, changes in the controller output are made for a period of time sufficient to produce the desired change. The controller output is then returned to the previous state, and the process value will reach equilibrium at its new value. In the tank example, if the goal is to increase the liquid level in the tank, the inlet flow will be increased for sufficient time to cause the new equilibrium point to be reached. Once the

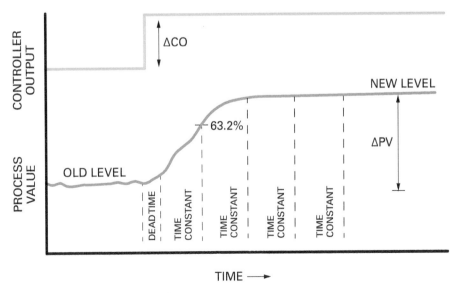

Figure 2 Process response curve of a non-integrating process.

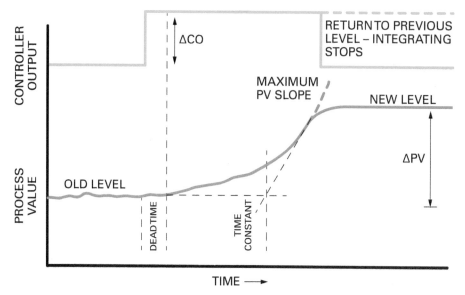

Figure 3 Process response curve of an integrating process.

desired liquid level has been reached, the inlet flow must be brought back to the steady state condition to prevent the tank from overflowing.

In many real-world situations, processes have both integrating and non-integrating characteristics, but one will tend to dominate the total picture. This is what you must identify before loop tuning. When tuning control loops, it's typical to manipulate the process in order to observe its behavior. This may be done with the controller in automatic (closed loop) or manual (open loop) modes. Regardless of the method employed, you should always understand the process that you are manipulating before making any changes. An integrating process can easily reach excessive levels if the control output is changed for too long. A non-integrating process can reach an excessive level if the control output is changed by too large a value.

Additional Resources

Good Tuning: A Pocket Guide, G. K. McMillan. Second Edition. 2005. Research Triangle Park, NC: International Society of Automation.

The International Society of Automation, www.isa.org

1.0.0 Section Review

1. A control system can become unstable at the _____.
    a. phase-shift point
    b. resonant amplitude
    c. critical frequency
    d. deadtime band

2. Oscillations in a process variable will die out if the control system's gain is set to _____.
    a. 50 percent
    b. less than 1
    c. 75 percent
    d. greater than 1

3. How many time constants are required for a process variable to return to steady state?
    a. 1
    b. 2
    c. 5
    d. 10

Section Two

2.0.0 Tuning Methods

Objective

State the basic equations needed for loop tuning and describe various loop tuning processes.
a. State the basic equations needed for loop tuning.
b. Describe open loop tuning processes.
c. Describe closed loop tuning processes.
d. Describe a visual loop tuning process.

Performance Tasks

1. Perform open loop tuning.
2. Perform closed loop tuning.
3. Perform visual loop tuning.

Trade Terms

Dampened oscillation method: A closed loop tuning method in which calculations are based on the period of the process response curve when its oscillations have been adjusted to die out.

Differentiator: A control circuit or mechanism that outputs a signal based on the derivative of the input signal.

Integrator: A control circuit or mechanism that outputs a signal based on the integral of the input signal.

Over-the-hump time: The time that it takes a waveform to reach its peak amplitude and start moving downwards.

Reaction rate method: An open loop tuning method in which calculations are based on the process reaction rate.

Repeats per minute (rpm): A measure of how fast a signal is being integrated.

Saturate: A condition in which a control element moves to its maximum position in response to an excessive control signal value.

Stiction: The tendency for a valve stem to stick due to static (non-moving) friction and not respond smoothly to its actuator. Stiction tends to cause a response that jumps between positions rather than moving smoothly.

Time constant method: An open loop tuning method in which calculations are based on the process response curve's time constant.

Ultimate period method: A closed loop tuning method in which calculations are based on the oscillation period that occurs when the controller is set to its ultimate gain or ultimate proportional band.

There are three basic approaches to tuning loops: open loop, closed loop, and visual tuning. Several methods can be used within each of these categories. Open loop tuning methods are based on there being no feedback. These methods do not even require that the controller be installed. On the other hand, in the closed loop tuning methods, the controller plays a crucial role and is constantly receiving feedback from the process. Visual loop tuning is the simplest method. This section examines each approach and some of the equations and concepts applied in order to determine the optimal controller settings.

2.1.0 Tuning Equations

The equations discussed here are the most common ones with which you'll need to be familiar. Each has a basic function in determining the best settings to obtain a stable process.

2.1.1 Energy Balance

The *energy balance equation* is used as a simple method to show how a process system receives energy and how it loses energy. In order for any process system to be stable, the equation must be balanced. The equation is written as follows:

$$E_{out} = E_{in} - E_{lost} - E_{stored}$$

Where

E_{out} = total energy leaving the process system
E_{in} = total energy coming into the process system
E_{lost} = amount of energy that the process system loses to its surroundings
E_{stored} = amount of energy remaining in the process system

This energy can be in the form of heat, pressure, or flow. As an example, consider water entering the shell side of a heat exchanger. Assume that its temperature increases. Due to the temperature rise, the energy input (E_{in}) to the heat exchanger is increased. This increases the energy transfer to the secondary fluid, raising its temperature. This is the energy output (E_{out}). At the same time, more energy is being lost to the surroundings (E_{lost}) because of the increased temperature differential. Also, the internal components of the heat exchanger are absorbing and storing energy (E_{stored}).

These quantities continue to change until equilibrium is reached. At that time, the increased energy input is balanced by an increase in the energy output, the energy lost, and the energy stored. The net result is increased temperature of the secondary fluid output, greater temperature loss to the surroundings, and increased stored energy within the heat exchanger. It's important to remember that all of the energy must go somewhere. Energy never simply vanishes or appears out of nowhere; it always goes somewhere, and the total quantity always remains the same.

The energy balance equation applies to all systems. It is important to note that a system will only be stable once the equation is balanced.

2.1.2 Time Constant

The time constant of a process is associated with the time that the process will take to respond to a disturbance. As you learned earlier, the time constant is the time required for a process to reach 63.2 percent of its total change in returning to steady state. A full response requires five times this value (five time constants). A simple formula for determining the time constant for a pneumatic or fluid system is as follows:

$$TC = V \div F$$

Where

TC = time constant
V = volume or capacity of the system
F = flow

This equation is a basis for all open loop tuning methods, which will be discussed later in this module. The following example illustrates this equation.

If the fluid flow through a system is 100 gpm (379 lpm) and the capacity of the system is 1,000 gallons (3,790 L), how long would it take for this system to settle into a steady state if a disturbance were introduced? You can find the answer by applying the time constant equation as follows (US measurements are used in this example, but their metric equivalents can also be used to obtain the correct metric answer):

$$TC = V \div F$$
$$TC = 1,000 \div 100$$
$$TC = 10 \text{ minutes}$$

Because five time constants are required for the process to return to steady state, this number must be multiplied by 5. The outcome is that it will require 50 minutes ($10 \times 5 = 50$) for the process to respond to a disturbance and return to steady state conditions.

2.1.3 Complete Response

To find out by what percent a process has changed in a given time period, use the *complete response equation*. It is used when performing open loop tuning and is expressed as follows:

$$CR = A[1 - e^{(-t/TC)}]$$

Where:

CR = complete response of the controlled variable at any given time (%)
A = amplitude of change applied to the controlled variable (%)
e = base of the natural logarithm, 2.718 (no units)
t = given time
TC = time constant

If you know the time constant, you can find the value of the controlled variable for any given point in time. For example, if the time constant (TC) is one minute, and the amplitude of total change (A) is 100%, what is the value of the controlled variable after 2.5 minutes? You can find the answer by applying the complete response equation as follows; a scientific calculator can be used to help make these calculations:

$$CR = A[1 - e^{(-t/TC)}]$$
$$CR = 100\%[1 - 2.718^{(-2.5/1)}]$$
$$CR = 100\%(0.917)$$
$$CR = 91.7\%$$

The process has completed approximately 92% of its total change in 2.5 minutes.

2.1.4 Process Gain

Knowing the process gain can be helpful in tuning, particularly in open loop situations. The gain is simply the output change divided by the input change. The equation is written as follows:

$$K = \text{change in output} \div \text{change in input}$$

Where

K = process gain

The changes in input and output are both normally expressed as a percentage of the span. This makes process gain a unitless number that describes the percentage change in output that can be expected following a known percentage change in the input. The use of the percent of span notation allows the comparison of gains for processes with different ranges of operation.

The process illustrated in *Figure 4* (A) is a surge tank. As long as the water flow into the tank is the same as the water flow out of the tank, the water level remains constant. The span of the inlet flow is 80 gpm (303 lpm), and the span of the tank level is 20' (6.1 m). Imagine that at some point in time, the inlet flow is suddenly changed from 60 gpm (227 lpm) to 80 gpm (303 lpm). The response of the surge tank to the change in inlet flow is shown in *Figure 4* (B). The 20 gpm (76 lpm) change has been balanced by an increase of 4' (1.2 m) of tank level. Since this is an integrating process, sustaining the new input level indefinitely will cause the tank to overflow. Once the new level has been reached, the input and output flows must become equal again. The gain of this process can then be determined as follows.

Determine the change in output as a percentage of the span:

Change in level =
$$\frac{\text{changed level} - \text{initial level}}{\text{span}} \times 100 =$$
$$\frac{14' - 10'}{20} \times 100 =$$
$$\frac{4}{20} \times 100 = 20\%$$

Determine the change in input as a percentage of the span:

Change in inlet flow =
$$\frac{\text{changed flow rate} - \text{initial flow rate}}{\text{span}} \times 100 =$$
$$\frac{80\text{gpm} - 60\text{gpm}}{80\text{gpm}} \times 100 =$$
$$\frac{20}{80} \times 100 = 0.25 \times 100 = 25\%$$

Using these calculations, find the process gain:

K = change in output ÷ change in input
K = 20% ÷ 25%
K = 0.8

The gain of this process is 0.8. That means that for every 1% change of the span of the inlet flow, the level will change 0.8% of its span. In physical units, an 8 gpm (30 lpm) change in inlet flow, which is 10% of the span, will cause a 1.6' (0.5 m) change in the surge tank level (8% of span).

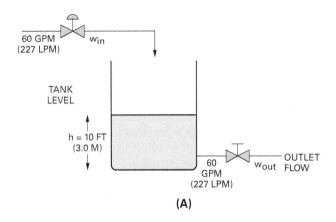

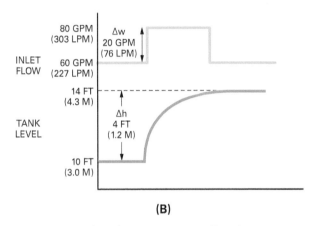

Figure 4 Surge tank and its response to a disturbance.

2.1.5 Proportional Band

The input band over which the controller provides a proportional output is called the *proportional band*. It is defined as the change in input required to produce a full-range change in output due to proportional control action, and is shown by the following equation:

PB =
(change in input ÷ change in output) × 100%

Consider a situation in which a 100% input change causes a 100% output change. The proportional band is calculated as follows:

PB = (100% ÷ 100%) × 100%
PB = 1 × 100%
PB = 100%

Gain and proportional band are inversely related, as illustrated by the following pair of equations:

K = 100% ÷ PB
and
PB = 100% ÷ K

This inverse relationship is important to remember. Some proportional controllers have an adjustment that is expressed in units of gain, while others have the adjustment expressed in units of percent proportional band.

As an example, consider a valve with a 4" (10 cm) stroke that controls the liquid level in a tank. A float sensor provides feedback and controls the valve. A 4" (10 cm) level change causes a 2" (5 cm) change in valve position. Thus, you can determine that the 4" change in level is 100% of the span, and the 2" change in valve position is 50% of the span. Using this information, you can calculate the proportional band and the gain.

Find the proportional band:

$$PB = (\text{change in input} \div \text{change in output}) \times 100\%$$
$$PB = (100\% \div 50\%) \times 100\%$$
$$PB = 2 \times 100\%$$
$$PB = 200\%$$

Find the gain:

$$K = \text{change in output} \div \text{change in input}$$
$$K = 2" \div 4"$$
$$K = 0.5$$

According to these calculations, the proportional band is 200%, and the gain is 0.5.

2.1.6 Integral Time

A device that performs the mathematical function of integration is called an integrator, and the mathematical result of integration is called the *integral*. The integrator provides a constant that specifies the function of integration. It also provides a linear output with a rate of change that is directly related to the amplitude of the step change input. The equation for this rate of change is written as follows:

$$P(t) = (K_i)(\Delta i/t)(T)$$

Where

$P(t)$ = output rate of change
K_i = integral time constant
$\Delta i/t$ = percent change in input with respect to time
T = time that change in input lasts

For example, consider an input ramp that changes at a rate of 3 percent per second and lasts for 10 seconds. The integral time constant is 2 seconds. What is the output of the integrator after 10 seconds? The output rate of change is calculated as follows:

$$P(t) = (K_i)(\Delta i/t)(T)$$
$$P(t) = (2s)\,3\% \div 1s)(10s)$$
$$P(t) = (2s)(3)(10s) = 6(10s)$$
$$P(t) = 60\%$$

Repeats per minute (rpm) is also a measure of how fast a signal is being integrated. Instead of expressing it in terms of time, it is expressed in terms of repeats. The repeats refer to how often a signal is integrated. The higher the rpm, the greater the rate of integration is.

2.1.7 Derivative Time

A device that produces a derivative signal is called a differentiator. *Figure 5* shows the input-versus-output relationship of a differentiator.

The differentiator provides an output amplitude that is directly related to the rate of change of the input and is a constant that specifies the function of differentiation. The output is expressed by the following equation:

$$P(d) = (T_d)(\Delta i/t)$$

Where

$P(d)$ = derivative output
T_d = derivative time
$\Delta i/t$ = percent change in input with respect to time

The derivative constant is expressed in units of seconds. As an example, a change in input is 40 percent over a time of 10 seconds (40%/10s) and a derivative time of 2 seconds. What is the derivative output?

$$P(d) = (T_d)(\Delta i/t)$$
$$P(d) = (2s)\,40\% \div 10s)$$
$$P(d) = (2s)(4)$$
$$P(d) = 8\%$$

The differentiator equation suggests that a step change input would theoretically produce an infinite output. Since no device can produce an infinite output in reality, the device will rapidly saturate to its maximum output value when a step change is applied to the input. Increasing the derivative constant provides a larger output amplitude for a given input rate of change.

Basically, derivative time (T_d) determines how strongly the differentiator is allowed to affect the control system. The larger the number, the more affect it has on the process. For example, if a process changes from 0 to 40 percent in one minute, the addition of a derivative time of one minute would cause the same process to go to 40 percent immediately, reaching a level of 80 percent after

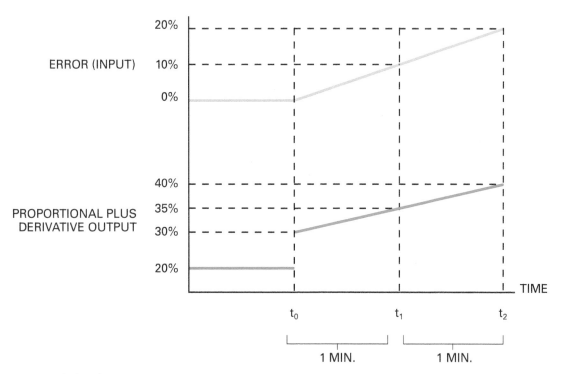

Figure 5 Proportional plus derivative output.

one minute. Determining an appropriate T_d value is an important aspect of tuning, since it affects the control system's stability and behavior.

2.1.8 PID Loop

The *PID loop equation* determines the output from a PID controller. The components of the equation are the proportional output, the integral output, and the derivative output. The equation that describes the PID controller output at a given time during a linear rate of change error input is as follows:

$$P_{PID}(t) = K_p(\Delta E_p/\Delta t)(t) + \tfrac{1}{2}K_pK_i(\Delta E_p/\Delta t)(t^2) + K_pT_d(\Delta E_p/\Delta t) + P_0$$

Where

$P_{PID}(t)$	=	proportional + integral + derivative controller output at a specified time
$K_p(\Delta E_p/\Delta t)(t)$	=	*proportional* action to a linear rate of change
$\tfrac{1}{2}K_pK_i(\Delta E_p/\Delta t)(t^2)$	=	*integral* action to a linear rate of change
$K_pT_d(\Delta E_p/\Delta t)$	=	*derivative* action to a linear rate of change
P_0	=	controller output at the start of the time period of change

Figure 6 is a graphical representation of this kind of control and its equation in action. At time t_0, the proportional and integral sections are just beginning to produce output amplitudes. At this time, their outputs are essentially 0 percent. The derivative section provides an output at the instant the proportional section output has a rate of change. In this example, a 10 percent derivative section output is produced and added to the 30 percent initial controller output for a total controller output of 40 percent at time t_0.

2.2.0 Open Loop Tuning

Open loop tuning is comparatively simple and involves a single basic test. The controller itself is not required for the test. The results of the test describe the process characteristics. From the process characteristics, you obtain data that can be inserted into appropriate equations, which yield the necessary initial settings for the controller. Since most processes are somewhat complex, tuning methods are approximations that require tweaking later on.

2.2.1 Procedure Overview

For open loop tuning methods, you will need to obtain the necessary process information from the process reaction curve. To generate this curve, you must introduce a single disturbance (bump) into the control system. Typically, a supply disturbance works well. A sudden setpoint change is a good way to provoke this situation. *Figure 7* shows the block diagram for a process being tuned by the open loop method.

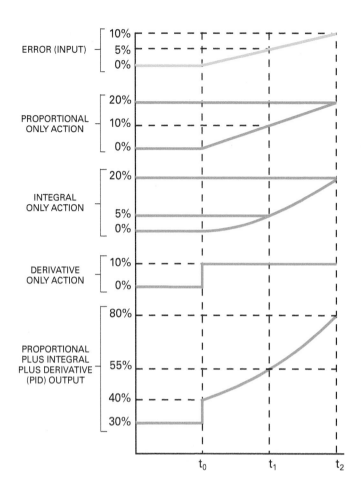

Figure 6 PID output.

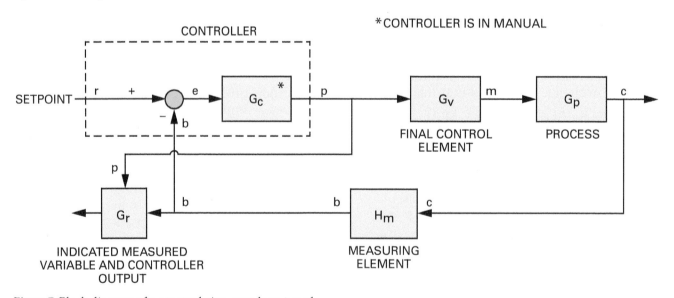

Figure 7 Block diagram of a process being open loop tuned.

While this diagram looks similar to that of a process being controlled by a closed loop system, the major difference is that the controller is in manual mode. With this condition, the control loop is effectively broken because the automatic signal does not reach the final control element. The system cannot respond to changes in the controlled variable. The other difference is that, besides indicating the measured variable, the recorder indicates the controller output rather than its setpoint. You can use an arrangement of this type to obtain the process reaction curve.

To determine the process reaction curve, bring the process to a steady state level. Ideally, the steady state level should be one where the controlled variable is at the setpoint. You may then place the controller in manual mode. Introduce a supply disturbance by changing the manual output of the controller. The response of the process to the supply disturbance is recorded by the recorder that receives the measured variable signal. The output of the controller is also recorded to provide a reference for when the disturbance was initiated. After the process reaction curve has been recorded, you should return the controlled variable to the setpoint by manual operation. It is easier to obtain a process reaction curve than it is to determine the ultimate gain.

Figure 8 shows a typical process reaction curve for a process that has been open loop tested. The curve is one for a multiple capacity process. It is simplified using the single capacity with deadtime approximation.

2.2.2 Time Constant Method

You can employ one of two methods to extract the information provided by the single capacity with deadtime approximation. The methods differ only in the way that the time constant is determined.

The first method is shown in *Figure 9* and is referred to as the time constant method. The information required from the process reaction curve is the time constant, process gain, and deadtime. The time constant is the time that it takes the process to reach 63.2 percent of its total change, as shown in the diagram. Recall that the process gain is the change in the output divided by the change in the input. The change in output is the change in the measured variable; this change must be converted to a percentage of the span first.

The reason that percentage of span is used, rather than the units of the measured variable, is because a change of 20°F (11°C) for a 100°F (55°C) span is greater than a change of 20°F (11°C) for a 200°F (110°C) span, because it is larger in comparison to the span. Therefore, a 20°F (11°C) change in the measured variable does not indicate the magnitude of the change as effectively as the percentage of span does. A 20°F (11°C) change for a 100°F (55°C) span is 20% of span, whereas a 20°F (11°C) change for a 200°F (110°C) span is only 10% of span. The 20 percent span change is a larger change with respect to the measuring span of the instrument than the 10 percent change, even though both changes are 20°F (11°C). The input change is the percent of span change in the controller output. Using the percentage of span requirement for the output change, the process gain may be written as follows:

$$K = \text{change in measured variable (\%)} \div \Delta CO\ (\%)$$

Where

K = process gain

You can determine the deadtime (t_d) using the graph; the deadtime is shown in *Figure 9*. Use this information and the following equations to obtain the needed settings to tune the controller:

Proportional

$K_p = t \div (t_d K)$
$PB = 100\% \div K_p$

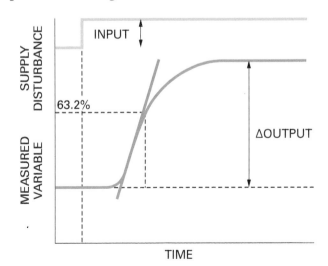

Figure 8 Process reaction curve for an open loop tested process.

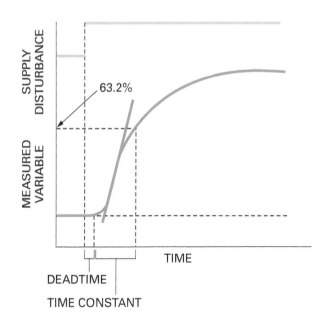

Figure 9 Time constant method for open loop tuning.

Proportional + Integral

$K_p = 0.9t \div (t_d K)$
$PB = 100\% \div K_p$
$T_i = 3.33 t_d$
$rpm = 1 \div 3.33 t_d$

Proportional + Integral + Derivative

$K_p = 1.2t \div (t_d K)$
$PB = 100\% \div K_p$
$T_i = 2.0 t_d$
$rpm = 1 \div 2.0 t_d$
$T_d = 0.5 t_d$

As an example, consider the results of the testing shown in *Figure 10*. The total change in the measured variable is 40°F (22°C). The output of the controller has a range of 3 to 15 psi (21 to 103 kPa). The change in the output of the controller is 1 psi (6.9 kPa). To determine the time constant, you need to know the value of the measured variable at 63.2 percent of the total change. Since the total change is 40°F (22°C), 63.2 percent of this change is 25.3°F (13.9°C). The 63.2 percent value of the measured variable occurs when the temperature reaches 65.3°F (18.5°C). This value is plotted on the vertical axis in the figure. The value of the time constant is read from the graph. For this example, the time constant is 1.6 minutes. The deadtime can also be read from the graph, and is 0.2 minutes.

The only piece of information remaining is the process gain. Since the span is 100°F (55°C), and the total change is 40°F (22°C), the change is 40 percent of the span. The controller output has changed 1 psi (6.9 kPa) of its 12 psi (83 kPa) span or 8.33 percent of span change. The process gain can be determined with the following equation:

$$K = \text{change in measured variable (\%)} \div \Delta CO\ (\%)$$
$$K = 40\% \div 8.33\% = 4.80$$

Now that you have determined the necessary information using the process reaction curve, you can use it to calculate the following settings (formulas to be used are presented in bold):

Proportional

$K_p = t \div (t_d K)$
$K_p = 1.6 \text{min} \div (0.2 \text{min} \times 4.80)$
$K_p = 1.67$
$PB = 100\% \div K_p$
$PB = 100\% \div 1.67$
$PB = 60\%$

Proportional + Integral

$K_p = 0.9t \div (t_d K)$
$K_p = (0.9 \times 1.6 \text{min}) \div (0.2 \text{min} \times 4.80)$
$K_p = 1.5$
$PB = 100\% \div K_p$
$PB = 100\% \div 1.5$
$PB = 67\%$
$T_i = 3.33 t_d$
$T_i = 3.33(0.2 \text{min})$
$T_i = 0.666 \text{min}$
$rpm = 1 \div 3.33 t_d$
$rpm = 1 \div (3.33 \times 0.2 \text{min})$
$rpm = 1.5 \text{ repeats/min}$

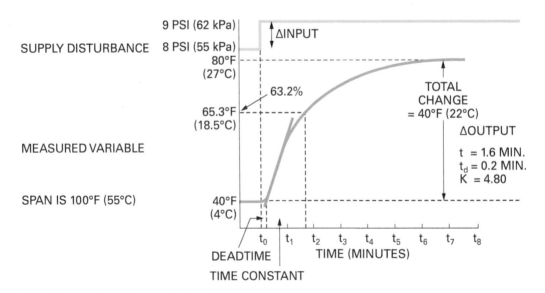

Figure 10 Test data for open loop time constant method.

Proportional + Integral + Derivative

$$K_p = 1.2t \div (t_d K)$$
$$K_p = (1.2 \times 1.6\text{min}) \div (0.2\text{min} \times 4.80)$$
$$K_p = 2.0$$
$$PB = 100\% \div K_p$$
$$PB = 100\% \div 2.0$$
$$PB = 50\%$$
$$T_i = 2.0 t_d$$
$$T_i = 2.0 \times 0.2\text{min}$$
$$T_i = 0.4\text{min}$$
$$\text{rpm} = 1 \div 2.0 t_d$$
$$\text{rpm} = 1 \div (2.0 \times 0.2\text{min})$$
$$\text{rpm} = 2.5 \text{ repeats/min}$$
$$T_d = 0.5 t_d$$
$$T_d = 0.5 \times 0.2\text{min}$$
$$T_d = 0.1\text{min}$$

2.2.3 Reaction Rate Method

The second method of open loop tuning (shown in *Figure 11*) is known as the reaction rate method. In this method, the information required from the process reaction curve is the slope of the maximum rate of rise line, which is termed the *reaction rate* (R_r). Deadtime is the other required piece of information. The reaction rate, R_r, is a change in the measured variable in percentage of span divided by a change in time. However, the reaction rate depends on more than the process. It is also affected by the magnitude of the change in the disturbance. A large disturbance produces a steeper reaction rate. To make the reaction rate independent of the magnitude of the disturbance, the magnitude of the change in the measured variable is divided by the disturbance. Normally, the disturbance is a change in the controller output in percentage of span. In equation form, it is stated as follows:

$$R_r = \frac{\text{changed in measured value}}{\text{change in time} \times \Delta CO}$$

Where

R_r = reaction rate
ΔCO = change in controller output

As before, the deadtime is read directly from the graph. With the reaction rate and the deadtime, the following equations can be used to determine the controller tuning settings:

Proportional

$$K_p = 1 \div (R_r t_d)$$
$$PB = 100\% \div K_p$$

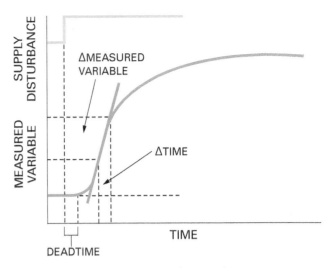

Figure 11 Reaction rate method for open loop tuning.

Proportional + Integral

$$K_p = 0.9 \div (R_r t_d)$$
$$PB = 100\% \div K_p$$
$$T_i = 3.33 t_d$$
$$\text{rpm} = 1 \div 3.33 t_d$$

Proportional + Integral + Derivative

$$K_p = 1.2 \div (R_r t_d)$$
$$PB = 100\% \div K_p$$
$$T_i = 2.0 t_d$$
$$\text{rpm} = 1 \div 2.0 t_d$$
$$T_d = 0.5 t_d$$

Figure 12 gives an example of this tuning method. As shown, the measured variable changes by 10°F (6°C) over a 100°F (55°C) span. The controller changes 1 psi (6.9 kPa) over a 3 to 15 psi (21 to 103 kPa) range, which is equal to a 12 psi (82 kPa) span. So, the change in the measured variable in percent of span is 10% and the change in controller output in percent of span is 8.33%. The time period for the 10°F (6°C) change is 0.5 min; this time is determined by using the maximum rate of rise line. The deadtime is read directly from the graph, and equals 0.2 min. The reaction rate can be determined with the following equation:

$$R_r = \frac{\text{changed in measured value}}{\text{change in time} \times \Delta CO}$$
$$R_r = \frac{20\%}{05 \times 8.33\%}$$
$$R_r = \frac{10\%}{4.165}$$
$$R_r = 2.4\%/\text{min}$$

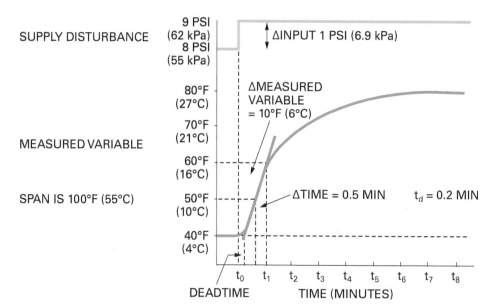

Figure 12 Test data for open loop reaction rate method.

Now that you have determined the necessary information using the process reaction curve, you can use it to calculate the following settings (formulas to be used are presented in **bold**):

Proportional

$$K_p = 1 \div (R_r t_d)$$
$$K_p = 1 \div (2.4\%/min \times 0.2min)$$
$$K_p = 2.1$$
$$PB = 100\% \div K_p$$
$$PB = 100\% \div 2.1$$
$$PB = 48\%$$

Proportional + Integral

$$K_p = 0.9 \div (R_r t_d)$$
$$K_p = 0.9 \div (2.4\%/min \times 0.2min)$$
$$K_p = 1.88$$
$$PB = 100\% \div K_p$$
$$PB = 100\% \div 1.88$$
$$PB = 53\%$$
$$T_i = 3.33 t_d$$
$$T_i = 3.33 \times 0.2min$$
$$T_i = 0.666min$$
$$RPM = 1 \div 3.33 t_d$$
$$RPM = 1 \div (3.33 \times 0.2min)$$
$$RPM = 1.5 \text{ repeats/min}$$

Proportional + Integral + Derivative

$$K_p = 1.2 \div (R_r t_d)$$
$$K_p = 1.2 \div (2.4\%/min \times 0.2min)$$
$$K_p = 2.5$$
$$PB = 100\% \div K_p$$
$$PB = 100\% \div 2.5$$
$$PB = 40\%$$

$$T_i = 2.0 t_d$$
$$T_i = 2.0 \times 0.2min$$
$$T_i = 0.4min$$
$$rpm = 1 \div 2.0 t_d$$
$$rpm = 1 \div (2.0 \times 0.2min)$$
$$rpm = 2.5 \text{ repeats/min}$$
$$T_d = 0.5 t_d$$
$$T_d = 0.5 \times 0.2min$$
$$T_d = 0.1min$$

The major difference between the two methods for open loop tuning is the information that must be gathered to set the controller adjustments. However, both methods of open loop control tuning yield good results. As with all tuning methods, a certain amount of trimming is usually required to bring the controller to the best condition.

2.3.0 Closed Loop Tuning

Closed loop tuning accomplishes the same basic goal as open loop. The crucial difference is that the controller is active in the process. Like open loop tuning, closed loop methods require you to introduce a disturbance in the process and note its effect. What is different is that the process must be disturbed more significantly and for a longer period. For this reason, closed loop methods can only be used on processes that are tolerant to these conditions. The following sections describe different ways to accomplish this kind of tuning.

2.3.1 Ultimate Period Method

The **ultimate period method** for tuning controllers was first proposed by Ziegler and Nichols in the early 1940s. The term *ultimate* is used with this method because it requires the determination of the ultimate gain (or ultimate proportional band) and the ultimate period. The ultimate gain (K_u) or ultimate proportional band (PB_u) is the maximum value of gain (or proportional band) at which the system is stable. The ultimate period is the period of response with the gain or proportional band set at its ultimate value. To determine the ultimate gain or proportional band and the ultimate period, operate the controller in proportional mode alone. *Figure 13* shows the response of a control system that is at the ultimate gain or proportional band. As shown, the controlled variable is oscillating at a constant amplitude and frequency. This is the critical frequency.

A 360-degree phase shift exists in the control system. The amplitude of the oscillations is constant. Therefore, the gain of the system is 1 and is at the ultimate gain (or proportional band) level. *Figure 13* also shows the ultimate period (P_u). The ultimate period, in this case, is the time period between two successive peaks in amplitude of the controlled variable. In other words, it is the time period from when the controlled variable is at its maximum amplitude to when it goes back down and then returns to its maximum amplitude again. These two pieces of information can then be inserted into appropriate formulas that provide tuning information. To find this information, manipulate the control system and observe the process behavior.

The ultimate period method is performed with the controller in service. Obtain the required information for tuning the controller by observing the waveform on a recorder connected to the loop. The first step is to set the controller for proportional control action only. This is required so that only one of the three possible control actions will cause the system to break into continuous oscillations. The formulas used depend on this condition being true. Achieve proportional-only control by setting the reset adjustment to either minimum repeats per minute or maximum integral time. Set the rate time to minimum, and set the controller gain at a low value. This combination of settings is the same as a wide proportional band. Make sure that these adjustments are done with the controller in manual mode.

After placing the controller back into automatic mode, introduce a disturbance. The easiest way to accomplish this is to introduce a supply disturbance, which you can do by changing the setpoint. On the recorder, observe the effect of a supply disturbance on the measured variable. Generally, the response is similar to Curve A in *Figure 14*; the oscillations tend to decrease as time continues. In some cases, the measured variable responds like Curve B. This curve shows an unstable response with the amplitude growing over time. If you encounter this kind of response, decrease the gain (or widen the proportional band). You must achieve a response like the one shown in Curve A (*Figure 14*) before continuing.

Once you've achieved this response, increase the gain (or narrow the proportional band) and disturb the process until you achieve continuous, constant-amplitude oscillations. Between each adjustment and disturbance, it may be necessary to stabilize the process. Once you reach the desired point, record the gain or proportional band setting. This value is the ultimate gain or ultimate proportional band. You can then measure the ultimate period from the information captured on the recorder.

Using the ultimate gain or proportional band value and the ultimate period value, you can calculate the appropriate controller settings using the following equations:

Proportional

$K_p = 0.5K_u$
$PB = 2PB_u$

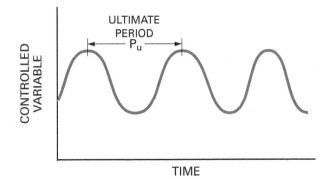

Figure 13 Control system at the ultimate gain or proportional band.

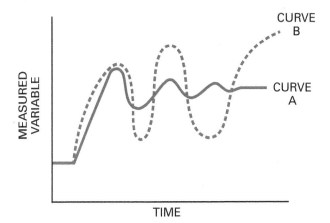

Figure 14 Responses of the measured variable to a supply disturbance.

Proportional + Integral

$K_p = 0.45K_u$
$PB = 2.2PB_u$
$T_i = P_u \div 1.2$
$RPM = 1.2 \div P_u$

Proportional + Derivative

$K_p = 0.6K_u$
$PB = 1.66PB_u$
$T_d = P_u \div 8$

Proportional + Integral + Derivative

$K_p = 0.6K_u$
$PB = 1.66PB_u$
$T_i = P_u \div 2$
$RPM = 2 \div P_u$
$T_d = P_u \div 8$

Note that these equations are not perfect. A certain degree of trimming will be necessary to achieve the best tuning. However, the values that these equations give will help achieve a well-tuned system.

As an example, consider the test results shown in *Figure 15*. The sustained continuous oscillation occurs for this process when the controller's gain is set to 4. This is equivalent to a proportional band of 25%. A measurement of the time between adjacent maximum values of the controlled variable gives an ultimate period of 16 minutes.

Using the test data in *Figure 15*, you can calculate the following settings (formulas to be used are presented in **bold**):

Proportional

$\mathbf{K_p = 0.5K_u}$
$K_p = 0.5(4)$
$K_p = 2$
$\mathbf{PB = 2PB_u}$
$PB = 2(25\%)$
$PB = 50\%$

Proportional + Integral

$\mathbf{K_p = 0.45K_u}$
$K_p = 0.45(4)$
$K_p = 1.8$
$\mathbf{PB = 2.2PB_u}$
$PB = 2.2(25\%)$
$PB = 55\%$
$\mathbf{T_i = P_u \div 1.2}$
$T_i = 16\text{min} \div 1.2$
$T_i = 13.3\text{min}$
$\mathbf{rpm = 1.2 \div P_u}$
$rpm = 1.2 \div 16\text{min}$
$rpm = 0.075 \text{ repeats/min}$

Proportional + Derivative

$\mathbf{K_p = 0.6K_u}$
$K_p = 0.6(4)$
$K_p = 2.4$
$\mathbf{PB = 1.66PB_u}$
$PB = 1.66(25\%)$
$PB = 41.5\%$
$\mathbf{T_d = P_u \div 8}$
$T_d = 16\text{min} \div 8$
$T_d = 2\text{min}$

Proportional + Integral + Derivative

$\mathbf{K_p = 0.6K_u}$
$K_p = 0.6(4)$
$K_p = 2.4$
$\mathbf{PB = 1.66PB_u}$
$PB = 1.66(25\%)$
$PB = 41.5\%$
$\mathbf{T_i = P_u \div 2}$
$T_i = 16\text{min} \div 2$
$T_i = 8\text{min}$
$\mathbf{rpm = 2 \div P_u}$
$rpm = 2 \div 16\text{min}$
$rpm = 0.125 \text{ repeats/min}$
$\mathbf{T_d = P_u \div 8}$
$T_d = 16\text{min} \div 8$
$T_d = 2\text{min}$

2.3.2 Dampened Oscillation Method

The dampened oscillation method is a modification of the ultimate period method. This method was proposed by P. Harriott in the 1960s. It was developed because some processes cannot tolerate continuous oscillations. For this method, put the controller in proportional mode alone and set the reset and rate portions to minimum values. Then adjust the gain or proportional band using

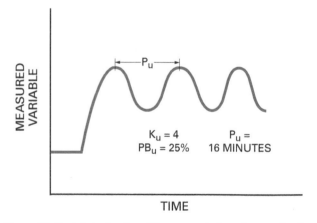

Figure 15 Test data for the ultimate period tuning method.

similar steps to those used in the ultimate period method until you obtain a response that resembles the one shown in *Figure 16*. The data obtained by this method is the period of the oscillation (P). Once you have this value, use the following equations to calculate the reset (integral) and rate (derivative) adjustments:

Reset

$$T_i = P \div 1.5$$
$$rpm = 1.5 \div P$$

Rate

$$T_d = P \div 8$$

After you make the reset and rate adjustments, readjust the gain or proportional band to provide the correct response of the measured variable (*Figure 16*).

2.4.0 Visual Loop Tuning

Visual loop tuning involves making changes while monitoring the effects of the changes on the process. It does not require formulas or complex mathematics. However, it should only be performed by experienced technicians.

> **CAUTION:** Do not attempt visual tuning unless you are completely knowledgeable and have a thorough understanding of the risks involved with varying the process with tuning experiments. If you lack this experience, obtain help from an experienced operator.

2.4.1 Incremental Changes

Changes of approximately 40 percent in gain, derivative time, and integral time almost always lead to a significant and visible change in closed loop response. A good start is to increase gain in a sequence with values of 1, 1.4, 2, 2.8, 4, 5.6, 8, and 10.

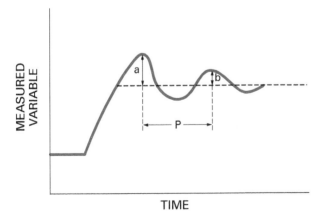

Figure 16 Test data for the dampened oscillation method.

Even though the last step (10) does not represent a 40 percent incremental increase of the previous value, it allows the same sequence to be repeated through the next decade (10, 14, 20, 28, 40, 45, 80, and 100). Since the percent change is constant for each adjustment, you can learn to anticipate the magnitude of change in response that will occur, whether the gain step is starting from 0.1 or 100.

These 40 percent stepped-increment values are usually fine enough to predict the response to the change. In some situations, you may want to finish by splitting a pair of these values. Keep in mind that such precise tuning may sometimes create problems if the process shifts slightly and becomes less stable. Successful tuning is based on having a clear understanding of when you are getting close to optimal control or when you are nearing the point of driving the process into instability.

2.4.2 Apparent Instability

When attempting to visually tune a loop having an apparent instability, it is necessary to first make sure that the cause of the loop oscillations is inherent to the loop that you are tuning and that it is not being caused by another process. After receiving permission from the unit operator, one way to do this is to place the loop controller for the loop that you are tuning in manual mode. If the process tends to stabilize in the manual mode, the controller most likely requires tuning.

> **CAUTION:** Never take any action involving live processes and instrumentation without explicit awareness of and permission from the unit operator. Furthermore, when you are communicating information, be certain that the operator is paying full attention.

You can do visual tuning of a loop having an apparent instability by incrementally stepping the output in manual mode, and then immediately setting the controller back to automatic mode while observing the trends of controller output and the feedback from the process variable. Modern smart control loops and those that are part of a DCS make this easy. With older single-loop controllers, you may need to watch indicating gauges or displays.

If the controller that you are visually tuning is equipped with setpoint tracking in which the setpoint varies with the process, disable this feature so the setpoint remains stationary during the visual tuning process.

One safer method to stop oscillations is to increase the integral time, thereby reducing the integral gain. This is especially true when the integral time is less than one-half of the oscillation period. With incremental increases in the integral time, you should obtain a smooth sinusoidal oscillation. After this point occurs, any additional incremental increases should result in an increased oscillation period without any increase in the amplitude. Following this, the integral time should continue to be increased until the oscillation damps out or stops lengthening in period.

Normally, if the integral time is longer than the oscillation time, the integral action does not play much of a role in the oscillation period and is minimally involved in the oscillation. An exception to this occurs in linear integrating processes such as level controls. In such a case, a 90-degree (or more) phase shift is produced over a wide range of period, and the additional phase shift of integral action can push the total phase shift to the point where it will sustain oscillation. In these types of processes, the integral effect that causes the process to return to setpoint is performed by the process itself. For this reason, the integrating function of the controller must be disabled by setting it well above the closed loop response period of the process, thus preventing the integral function from contributing any phase shift to the process control.

Once the integral time is set long enough that it is not involved in the oscillation, it is safe to try lowering the gain until the oscillation is damped out. Following this, you can try stepping the output as mentioned earlier, without anticipating that a sustained oscillation will result.

2.4.3 Sluggish Response

Two conditions that usually exist in a loop that is experiencing sluggish response are a lack of derivative action and a long integral time relative to the process response time. This section examines tuning areas and procedures that are intended to improve sluggish response.

A good starting point in addressing sluggish response is first to adjust the gain by changing the controller output in steps. You can do this by placing the controller in manual mode and then stepping the output by whatever is allowable. If you are not sure what is allowable, a good practice is to start with very small steps, letting the process operator know what you are doing. Knowing what is permissible lets you step the output in somewhat larger steps. The advantage of making larger steps is that any minor nonlinearities and noise tend to have less impact on the trends.

After making a stepped change to the output, put the controller back in automatic mode and watch the effects on the process. Also, note the controller's reaction to the change in output. The objective is to apply as much gain as you can without producing a series of echoes of the original response.

If the process reaction to the stepped change in the output causes the process response to swing more than one cycle before approaching an equilibrium value, the gain is most likely set too high. Try lowering the gain by about 40 percent. On the other hand, if the controller output and process response settle at new values without producing anything resembling a sinusoidal swing, then the gain is probably still set too low. Try raising it by 40 percent.

Repeat this process, alternating the direction of the output step each time until a gain setting is reached that causes approximately one cycle of output swing before the output reaches an equilibrium value. This gain value produces a result similar to that obtained with the ultimate period method.

> **NOTE**
> A process oscillation that is not centered across the setpoint value is a sign that the gain is too high.

While in manual mode, and only after the gain is set at its optimum value, shorten the integral time on the controller in an effort to cause the output value to ramp back to setpoint. The effects of stepping the output and allowing the gain to respond to the changes in output will cause the automatic reset action of the integral to attempt to find the original output value that puts the process exactly back on setpoint. A reliable assumption is that the integral time setting will be on the order of the closed loop time constant of the process with the gain optimized by the previous gain adjustment process.

However, if you want to see the integral effect clearly, start the integral time at approximately double that time and then reduce the integral time in 4 percent steps. This will cause the ramp time back to setpoint to be reduced in 40 percent steps, but the peak disturbance will reduce or increase slightly, depending on the deadtime relative to the dominant closed loop time constant.

The best range of integral time is reached when the process ramps back to the setpoint about half as fast as it moved away from the setpoint from the output step. An integral time that is too short is indicated by an oscillation that is centered across the setpoint with a period that lengthens as the integral time is lengthened.

Derivative action is often avoided in some controllers because of the controller's adverse response to it. However, in controllers that respond well to derivative action and whose process demands using it, the benefits of derivative action are greater stability and faster response. In order to determine if the process that is being controlled would benefit from derivative action, consider the following conditions:

- Having arrived at reasonable settings for gain and integral time, the process should display a deadtime, acceleration time to full rate of change time, inflection, over-the-hump time, and settling time back to setpoint.
- Observe the inflection point between the accelerating phase and the reverse curvature as the process goes over the hump. The greater the acceleration time relative to the over-the-hump time, the more benefit the process gets from the derivative action.
- The only adverse ingredient in the loop that prevents the process from getting the full benefit of derivative action is the deadtime. If it is much longer than the accelerating phase, it will cause all the output effects from the derivative action to occur after the point at which they would have done some good. This results in the controller storing echoes of the step response in the process deadtime delay.

In summary, derivative action is beneficial whenever the accelerating phase of the process step response is a significant fraction of both the deadtime and the closed loop over-the-hump time. If the control loop exhibits a jerky output or sinusoidal oscillations that make the process measurement look like a flight of stairs as it goes over the hump, it is an indication that the derivative action in a control loop should be reduced.

The visual tuning methods described up to this point test the response of a process to output steps (load disturbances) having a constant setpoint. When a controller has its setpoint delivered as the output of another controller (in other words, cascaded) there is an effective increase in loop gain. As the inner, or slave, controller is driving the process toward its setpoint, the outer, or master, controller may be driving the setpoint toward the process value. This can be compared somewhat to shooting at a target that is rapidly approaching the end of a gun, which can result in overshooting the target.

Also, it is less important that the slave controller reach perfection in the shortest possible time because its setpoint will not be there by the time the output of the controller gets the process there. Therefore, inner cascade loops tend to work best with somewhat less gain and somewhat longer integral time than if the loop were shooting for a fixed setpoint.

An example where visual tuning might be more effective than the calculated methods of tuning is in a loop where stiction is present in the control valve. Many of the calculated approaches rely on linear response in order to be effective. Stiction usually causes a nonlinear effect on the process.

Stiction may also produce an oscillation in the process that has a period proportional to integral time; but when tuning this loop, if the integral time is increased in an effort to stabilize the oscillation, the increased integral time will probably result in a longer oscillation period.

Sinusoidal oscillation, a byproduct of stiction, is not representative of either a controller that is improperly tuned or a linear process. In sinusoidal oscillation, the controller output will generally show a triangular wave as the integral action keeps ramping the output up and down. On the other hand, in a control loop with stiction present, the process variable will often indicate a form of square wave with a fairly constant amplitude. This wave is representative of the size of the sudden jump when the valve breaks loose, as shown in *Figure 17*.

> **NOTE**
> A loop with a sticking control valve should be visually tuned to limit the oscillation caused by the sudden change in valve position. In these cases, the valve should be repaired or replaced, or a valve positioner should be installed (if one is not already installed).

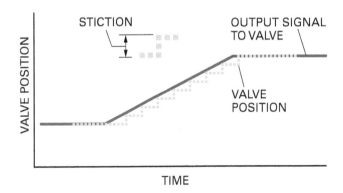

Figure 17 Waveform of a sticking control valve (stiction).

Additional Resources

Good Tuning: A Pocket Guide, G. K. McMillan. Second Edition. 2005. Research Triangle Park, NC: International Society of Automation.

The International Society of Automation, www.isa.org

2.0.0 Section Review

1. To find out by what percentage a process has changed in a time period, you should use the _____.
   a. energy balance equation
   b. time constant equation
   c. complete response equation
   d. process gain equation

2. When tuning by the open loop method, the process reaction curve is obtained by disturbing the process with _____.
   a. multiple changes to the setpoint
   b. a single change to the setpoint
   c. a switch between manual and automatic
   d. an emergency plant trip

3. When tuning by the ultimate period method, the controller must be placed in _____.
   a. proportional mode
   b. proportional + integral mode
   c. proportional + derivative mode
   d. proportional + integral + derivative mode

4. Visual tuning can be more useful than calculated tuning when which of the following is present in the process?
   a. Process disturbances
   b. Feedback
   c. PID control
   d. Stiction

SUMMARY

In process control loops, disturbances in the normal operation of a process can be caused by many things, including other processes or even faulty loop components. The response or behavior of the process control loop to the disturbance is directly related to parameters of the loop's tuning.

Many of the loops in process control have adjustable controller functions that, when used individually or in combination with each other, cause the controller's output signal to change in various ways once a disturbance occurs. There is no one common combination of controller parameters that can be applied to all processes because the variables in each process dictate specific parameters in order to function as designed and produce the desired effects. Setting these parameters for optimal process control based on the demands of the process is referred to as *loop tuning*.

Loop tuning can be as simple as increasing only one function, such as gain, to either a slightly higher or lower setting and watching the response of the controller as it adjusts its output accordingly. On the other hand, loop tuning can involve complex formulas that use loop characteristics such as deadtime, lag, and time constants. These equations must be applied in order to properly set controller functions, such as gain, integral time, and derivative, to arrive at the optimal set of parameters for a particular loop.

This module described the most common methods used in both types of loop tuning and defined the factors that apply to tuning.

Review Questions

1. Undesirable oscillations occur at the critical frequency when there is a phase shift around the loop of _____.
   a. 90 degrees
   b. 180 degrees
   c. 270 degrees
   d. 360 degrees ✓

2. Controller tuning involves setting the _____.
   a. gain to less than 1 at the critical frequency ✓
   b. I and D factors to neutral points
   c. gain to greater than 1 at the critical frequency
   d. I and D factors to equal but opposite values

3. Getting a proportional controller to work properly at startup often involves starting in _____.
   a. automatic mode
   b. loop cycling mode
   c. manual mode ✓
   d. derivative-only mode

4. Which kind of process responds to a change in the controller output by moving a specific amount and settling at a new steady state?
   a. Non-integrating ✓
   b. Open
   c. Integrating
   d. Closed

5. What is the energy balance equation's relationship to system stability?
   a. Stability will occur only when E_{in} is greater than E_{out}.
   b. Stability will occur only when the equation is balanced. ✓
   c. Stability will occur only when the equation is unbalanced.
   d. The energy balance equation is not related to system stability.

6. Which circuit or mechanism outputs a signal that is directly related to the rate of change of its input?
   a. A PI loop
   b. A differentiator ✓
   c. A P loop
   d. An integrator

7. Loop tuning usually involves getting information about the process by studying the _____.
   a. completed reaction equation
   b. new setpoint value
   c. previous cycle deadtime
   d. process reaction curve ✓

8. When tuning with the ultimate period method, what should the process reaction curve look like when the controller is successfully set to the ultimate gain / proportional band?
   a. It should show no significant oscillations.
   b. It should show oscillations that increase with time.
   c. It should show constant amplitude oscillations. ✓
   d. It should show oscillations that decrease with time.

9. A good way to check for the source of apparent instability in a loop is to _____.
   a. place the controller in manual mode ✓
   b. introduce a process disturbance
   c. switch the controller from PID to P mode
   d. switch to a non-integrating process

10. Sluggish response in a control loop is often caused by _____.
    a. an out-of-date controller
    b. a lack of derivative action ✓
    c. process disturbances
    d. supply metering

Trade Terms Introduced in This Module

Amplitude: The strength of a signal, often represented by the height of the plot in a graph.

Bump test: A method of testing the response of a control system by suddenly changing the controller setpoint to produce a process disturbance.

Critical frequency: The frequency at which a control system can become unstable if its gain is greater than 1.

Dampened oscillation method: A closed loop tuning method in which calculations are based on the period of the process response curve when its oscillations have been adjusted to die out.

Deadtime: The delay interval between the point when the controlled process variable changes and the point when the controller begins to respond to the change.

Derivative: A control method in which the controller's output depends on the rate of change of the process variable or error.

Differentiator: A control circuit or mechanism that outputs a signal based on the derivative of the input signal.

Integral: A control method in which the controller's output depends on the amount and duration of the error signal.

Integrating process: A process that changes without limit as long as the controller is outputting a value above the process equilibrium level.

Integrator: A control circuit or mechanism that outputs a signal based on the integral of the input signal.

Lag: The delay between an input change and a corresponding output response.

Non-integrating process: A process that reaches a new equilibrium level in response to a change in controller output but does not keep changing once it reaches the new level. Also known as a self-regulating process.

Oscillations: Repetitive changes in signal amplitude.

Overshoot: The tendency for the controlled variable to exceed the setpoint.

Over-the-hump time: The time that it takes a waveform to reach its peak amplitude and start moving downwards.

Reaction rate method: An open loop tuning method in which calculations are based on the process reaction rate.

Reciprocal: The value by which you can multiply a given number in order to obtain a result of 1. The reciprocal of a number is usually written as a fraction, and is equal to is 1 over that number. For example, the reciprocal of 3 is ⅓.

Repeats per minute (rpm): A measure of how fast a signal is being integrated.

Saturate: A condition in which a control element moves to its maximum position in response to an excessive control signal value.

Stability: The condition in which a control system responds to disturbances by returning quickly to the setpoint with oscillations that either remain constant or die out.

Steady state: The condition in which the controlled variable is essentially at the setpoint and is not changing significantly.

Stiction: The tendency for a valve stem to stick due to static (non-moving) friction and not respond smoothly to its actuator. Stiction tends to cause a response that jumps between positions rather than moving smoothly.

Time constant: A measurement of time that describes the response behavior of a particular control system. A single time constant is the time required for the controlled variable to change by 63.2 percent of the amount that it must change to reach steady state.

Time constant method: An open loop tuning method in which calculations are based on the process response curve's time constant.

Tuning: Adjustments made to a loop controller to produce a desired response behavior to process disturbances.

Ultimate period method: A closed loop tuning method in which calculations are based on the oscillation period that occurs when the controller is set to its ultimate gain or ultimate proportional band.

Additional Resources

This module presents thorough resources for task training. The following reference material is recommended for further study.

Good Tuning: A Pocket Guide, G. K. McMillan. Second Edition. 2005. Research Triangle Park, NC: International Society of Automation.
The International Society of Automation, **www.isa.org**

Figure Credits

©Nostal6ie/Dreamstime.com, Module Opener

Section Review Answer Key

Answer	Section Reference	Objective
Section One		
1. c	1.1.1	1a
2. b	1.1.1	1a
3. c	1.2.4	1b
Section Two		
1. c	2.1.3	2a
2. b	2.2.1	2b
3. a	2.3.1	2c
4. d	2.4.3	2d

NCCER CURRICULA — USER UPDATE

NCCER makes every effort to keep its textbooks up-to-date and free of technical errors. We appreciate your help in this process. If you find an error, a typographical mistake, or an inaccuracy in NCCER's curricula, please fill out this form (or a photocopy), or complete the online form at **www.nccer.org/olf**. Be sure to include the exact module ID number, page number, a detailed description, and your recommended correction. Your input will be brought to the attention of the Authoring Team. Thank you for your assistance.

Instructors – If you have an idea for improving this textbook, or have found that additional materials were necessary to teach this module effectively, please let us know so that we may present your suggestions to the Authoring Team.

NCCER Product Development and Revision
13614 Progress Blvd., Alachua, FL 32615

Email: curriculum@nccer.org
Online: www.nccer.org/olf

❏ Trainee Guide ❏ Lesson Plans ❏ Exam ❏ PowerPoints Other _____

Craft / Level: _____ Copyright Date: _____

Module ID Number / Title: _____

Section Number(s): _____

Description: _____

_____

_____

_____

_____

Recommended Correction: _____

_____

_____

_____

_____

Your Name: _____

Address: _____

_____

Email: _____ Phone: _____

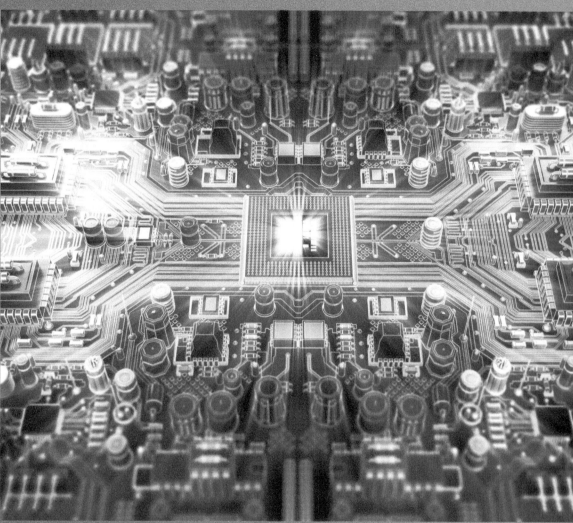

12401-16
Digital Logic Circuits

Overview

Modern instrumentation systems rely heavily on digital logic. While traditional analog systems represent information as proportional changes in voltage or current, digital systems represent information as numbers. Digital logic is inside almost every electronic device. Programmable logic controllers use digital terminology and concepts in their programming systems. Acquiring a solid understanding of digital logic technology is important for anyone in the instrumentation industry.

Module Four

Trainees with successful module completions may be eligible for credentialing through the NCCER Registry. To learn more, go to **www.nccer.org** or contact us at 1.888.622.3720. Our website has information on the latest product releases and training, as well as online versions of our Cornerstone magazine and Pearson's product catalog.

Your feedback is welcome. You may email your comments to **curriculum@nccer.org**, send general comments and inquiries to **info@nccer.org**, or fill in the User Update form at the back of this module.

This information is general in nature and intended for training purposes only. Actual performance of activities described in this manual requires compliance with all applicable operating, service, maintenance, and safety procedures under the direction of qualified personnel. References in this manual to patented or proprietary devices do not constitute a recommendation of their use.

Copyright © 2016 by NCCER, Alachua, FL 32615, and published by Pearson Education, Inc., New York, NY 10013. All rights reserved. Printed in the United States of America. This publication is protected by Copyright, and permission should be obtained from NCCER prior to any prohibited reproduction, storage in a retrieval system, or transmission in any form or by any means, electronic, mechanical, photocopying, recording, or likewise. To obtain permission(s) to use material from this work, please submit a written request to NCCER Product Development, 13614 Progress Blvd., Alachua, FL 32615.

From *Instrumentation Level Four, Trainee Guide*, Third Edition. NCCER.
Copyright © 2016 by NCCER. Published by Pearson Education. All rights reserved.

12401-16
DIGITAL LOGIC CIRCUITS

Objectives

When you have completed this module, you will be able to do the following:

1. Identify and describe the basic concepts and elements of digital logic circuits.
   a. Define digital logic technology and terminology.
   b. Identify and describe AND, OR, XOR, and NOT gates.
   c. Identify and describe NAND, NOR, and XNOR gates.
   d. Describe combination logic and its purposes.
2. Identify and describe memory elements and their function in digital circuits.
   a. Identify and describe the basic flip-flop design.
   b. Explain the operation of clocked logic and clocked flip-flops.
   c. Describe the function of various types of registers.
3. Describe counters and their function in digital circuits.
   a. Describe the numbering systems related to digital circuits.
   b. Describe the function of binary counters.
   c. Describe the function of other types of counters.
4. Describe the function of arithmetic elements and decoders.
   a. Describe the function of basic arithmetic elements.
   b. Describe the function of decoders.

Performance Task

Under the supervision of the instructor, you should be able to do the following:

1. Create the truth table that describes the behavior of an instructor-supplied schematic.

Trade Terms

5V logic	Counter	Logic level	Propagation delay
AND gate	Decoder	NAND gate	Register
BCD number	Edge-triggered	Nanosecond (ns)	Timing diagram
Binary number	Flip-flop	Negative-going	Truth table
Bit	Gate	NOR gate	Word
Buffer gate	Hexadecimal number	NOT gate	XOR gate
Clock	Level-triggered	OR gate	XNOR gate
Combination logic	Logic family	Positive-going	

Industry Recognized Credentials

If you are training through an NCCER-accredited sponsor, you may be eligible for credentials from NCCER's Registry. The ID number for this module is 12401-16. Note that this module may have been used in other NCCER curricula and may apply to other level completions. Contact NCCER's Registry at 888.622.3720 or go to **www.nccer.org** for more information.

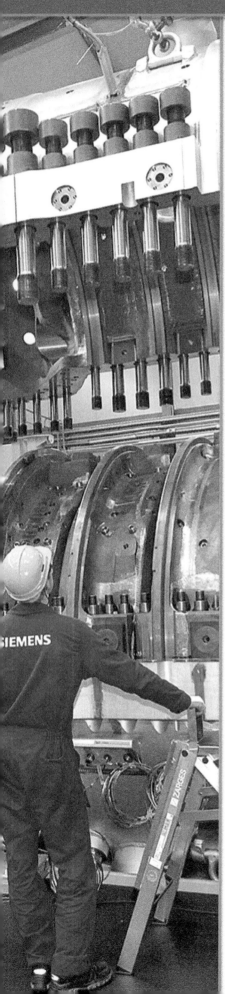

Contents

1.0.0 Digital Logic Fundamentals 1
   1.1.0 Digital Technology 1
      1.1.1 Power Supply 1
      1.1.2 Logic Levels 2
      1.1.3 Gates 2
      1.1.4 Schematic Symbols 2
   1.2.0 Basic Gates 2
      1.2.1 AND Gates 2
      1.2.2 OR Gates 4
      1.2.3 XOR Gates 4
      1.2.4 NOT Gates 5
   1.3.0 Modified Gates 5
      1.3.1 NAND Gates 5
      1.3.2 NOR Gates 5
      1.3.3 XNOR Gates 6
   1.4.0 Combination Logic 6
      1.4.1 Analyzing a Combination Logic Circuit 6
      1.4.2 Example Combination Logic Circuit 7
      1.4.3 Issues Involving Combination Logic 8
      1.4.4 Buffer Gates 9

2.0.0 Digital Memory Circuits 11
   2.1.0 Basic Memory Elements 11
      2.1.1 The RS Flip-Flop 11
      2.1.2 The NAND RS Flip-Flop 12
   2.2.0 Clocked Logic 13
      2.2.1 Clock Signals 13
      2.2.2 Timing Diagrams 13
      2.2.3 The Clocked RS Flip-Flop 13
      2.2.4 The D Flip-Flop 15
      2.2.5 The JK Flip-Flop 16
   2.3.0 Larger Memory Devices 16
      2.3.1 Basic Registers 17
      2.3.2 Specialized Registers 17

3.0.0 Counter Circuits 19
   3.1.0 Digital Numbers 19
      3.1.1 Binary Numbers 19
      3.1.2 Hexadecimal Numbers 20
   3.2.0 Counter Circuits 20
      3.2.1 Binary Counters 20
   3.3.0 Other Counters 20
      3.3.1 BCD Counters 20
      3.3.2 Up/Down Counters 20

4.0.0		Arithmetic Circuits and Decoders	22
	4.1.0	Arithmetic Circuits	22
		4.1.1 The Principles of Addition	22
		4.1.2 A Simple Adder	22
		4.1.3 Inside the Box	22
	4.2.0	Decoders	22
		4.2.1 Basic Decoder	23
		4.2.2 Complex Decoder	24

Figures and Tables

Figure 1	Example output and input logic levels comparison	3
Figure 2	Traditional and rectangular schematic symbols of the same gate	3
Figure 3	AND gate schematic symbols	3
Figure 4	AND gate equivalent circuit	3
Figure 5	OR gate schematic symbols	4
Figure 6	OR gate equivalent circuit	4
Figure 7	XOR gate schematic symbols	5
Figure 8	NOT gate schematic symbols	5
Figure 9	NOT gates used with other symbols	5
Figure 10	NAND gate schematic symbols	6
Figure 11	NOR gate schematic symbols	6
Figure 12	XNOR gate schematic symbols	6
Figure 13	An example combination logic circuit	7
Figure 14	Furnace safety circuit block diagram and schematic	8
Figure 15	Example of a multi-input AND gate	9
Figure 16	Buffer gate schematic symbols	9
Figure 17	Schematic symbol for an RS flip-flop	11
Figure 18	An RS flip-flop made from two NOR gates and its response to different inputs	12
Figure 19	An RS flip-flop made from a pair of NAND gates	13
Figure 20	A typical clock signal	14
Figure 21	Edge versus level triggering	14
Figure 22	Typical timing diagram	14
Figure 23	A clocked NOR RS flip-flop	15
Figure 24	Timing diagram for the clocked NOR RS flip-flop	15
Figure 25	Level-triggered and edge-triggered D flip-flops	15
Figure 26	Timing diagram for the level-triggered D flip-flop	16
Figure 27	Timing diagram for the edge-triggered D flip-flop	16
Figure 28	Schematic symbol for a JK flip-flop	17
Figure 29	Timing diagram for a JK flip-flop in toggle mode	17
Figure 30	A four-bit register made from D flip-flops	17
Figure 31	Schematic and timing diagram for a simple binary counter	21
Figure 32	Decimal and binary addition compared	22
Figure 33	Schematic symbol of a simple binary adder	22
Figure 34	Schematic of an eight-bit binary adder made up of simple adders	23

Figures and Tables (continued)

Figure 35   Schematic of the simple binary adder 23
Figure 36   Schematic symbol of a simple decoder 24
Figure 37   Schematic showing the inner workings of a simple decoder 25
Figure 38   A seven-segment LED display 26
Figure 39   Schematic symbol of a seven-segment display decoder 26

Table 1   Example Truth Table for a Three-Input Gate 3
Table 2   Two-Input AND Gate Truth Table 4
Table 3   Two-Input OR Gate Truth Table 4
Table 4   Two-Input XOR Gate Truth Table 4
Table 5   NOT Gate Truth Table 5
Table 6   Two-Input NAND Gate Truth Table 6
Table 7   Two-Input NOR Gate Truth Table 6
Table 8   Two-Input XNOR Gate Truth Table 6
Table 9   Empty Truth Table for a Four-Input Circuit 7
Table 10   Intermediate Truth Tables 7
Table 11   Final Truth Table 7
Table 12   NOR-Based RS Flip-Flop Truth Table 13
Table 13   NAND-Based RS Flip-Flop Truth Table 13
Table 14   Decimal, Binary, and Hexadecimal Numbers Compared 19
Table 15   Simple Binary Adder Truth Table 23

SECTION ONE

1.0.0 DIGITAL LOGIC FUNDAMENTALS

Objective

Identify and describe the basic concepts and elements of digital logic circuits.

a. Define digital logic technology and terminology.
b. Identify and describe AND, OR, XOR, and NOT gates.
c. Identify and describe NAND, NOR, and XNOR gates.
d. Describe combination logic and its purposes.

Performance Task

1. Create the truth table that describes the behavior of an instructor-supplied schematic.

Trade Terms

5V logic: A logic device that operates on +5 VDC. It has logic levels in which a logic-0 is close to 0 V and a logic-1 is close to 5 V.

AND gate: A digital gate (see *gate*) whose output is logic-1 only when both of its inputs are logic-1.

Buffer gate: A digital gate with an output that can deliver more current or a different voltage than a normal gate. Logically, it doesn't change the signal in any way, so the output equals the input.

Combination logic: A digital circuit made from a collection of gates.

Gate: A digital device whose output is the result of a logical operation performed on its inputs.

Logic family: A collection of digital circuit components that are compatible with each other.

Logic level: The voltages that represent logic-0 and logic-1 for a particular digital logic family.

NAND gate: An AND gate with an inverted output.

Nanosecond (ns): One billionth of a second (1 ns = 1/1,000,000,000 s).

Negative-going: A logic signal that's changing from logic-1 to logic-0.

NOR gate: An OR gate with an inverted output.

NOT gate: A digital gate that outputs the logical opposite of its input (also known as an inverter).

OR gate: A digital gate whose output is logic-1 if either or both of its inputs is logic-1.

Positive-going: A logic signal that's changing from logic-0 to logic-1.

Propagation delay: A delay caused by the circuits inside a digital device. It is the time interval between the moment when the inputs change and the moment when the output responds.

Truth table: A table that lists a digital circuit's output values for every possible input combination.

XOR gate: A digital gate whose output is logic-1 if exactly one of its inputs is logic-1.

XNOR gate: An XOR gate with an inverted output.

Digital technology works by representing all information as numbers. This makes digital circuits very flexible and able to handle any kind of information. By contrast, analog circuits tend to be very specialized since each type of information requires different handling methods. This is why digital technology has made such enormous changes in the world over the past few decades. Digital circuits perform three basic tasks on numbers: logical manipulation, math, and storage. Ultimately, every digital device—even the world's most powerful computer—does only those three things.

1.1.0 Digital Technology

Digital circuit components that are able to work together are grouped into collections; a collection of this kind is called a logic family. Over the years, many different families have emerged. When engineers design a piece of digital technology, they may draw from several different logic families to solve particular problems. But not all families can communicate with each other. Engineers have to know a family's electrical characteristics to make informed decisions. Similarly, instrument technicians also must have a basic understanding of the important electrical characteristics of digital technologies.

1.1.1 Power Supply

The first and most important piece of information to determine about any digital device is its power supply requirements. Digital devices all use low

 12401-16 Digital Logic Circuits · Module Four 1

voltages. However, most technologies aren't compatible unless they are using the same supply voltage. Certain specialized components are designed to tolerate a wider range of voltages. These can be used to translate between different logic families. The most popular supply voltages are +5 VDC and +3.3 VDC, although lower voltages are becoming very common in some applications. On circuit schematics, the positive power supply connection is usually designated V_{CC} or V_{DD}. The negative power supply connection is usually designated V_{SS} or GND. Other labels are possible.

In industry, the term 5V logic generically refers to any logic family that uses +5 VDC for its power supply. Be aware that not all 5V logic technologies are compatible with each other. Since they use the same power supply voltage, they won't destroy each other—but they may not communicate reliably. To determine that piece of information, you need to know something else about the family.

1.1.2 Logic Levels

Digital technologies create numbers using just two values: 0 and 1. These are represented by two discrete voltages—known as logic levels—that are spaced far enough apart to prevent confusion. A logic level can go by a variety of names. The 0 value may be called logic-0, LOW, false, or OFF. The 1 value may be called logic-1, HIGH, true, or ON. For consistency, this module will use logic-0 and logic-1.

In order for digital devices to communicate with each other, they must recognize the same logic levels. Each device will have two sets of logic levels: one for inputs, and one for outputs. This pair of levels makes up what is known as a *logic window*. As an example, a popular 5V logic family defines an output voltage in the range of 0–0.5 V as a logic-0, and a voltage in the range of 4.5–5.0 V as a logic-1. Conversely, the inputs are less rigid. Any voltage in the range of 0–1.5 V is treated as a logic-0, and a voltage in the range of 3.5–5.0 V is a logic-1. By making the input requirements looser than the outputs, the circuit becomes very reliable even when circuit conditions are less than ideal. *Figure 1* shows the relationship of the input and output logic levels for this example family.

1.1.3 Gates

The building block of all digital circuits, no matter how complex, is the gate. A gate has one or more inputs and at least one output. Circuitry inside the gate evaluates the inputs and performs a logical operation on them. The result goes to the output. A logical operation is a set of rules that define the relationship between input values and output states.

To know how a particular gate behaves, consult its truth table. This is a chart that shows all possible input patterns, along with the output that the gate will generate in response to each input pattern. The number of rows in the truth table (the possible combinations of inputs) depends upon the number of inputs. A gate with two inputs has a four-row truth table. Three inputs requires eight rows, and so on. (Every time an input is added, the number of rows in the truth table doubles.) *Table 1* shows an example truth table for a three-input gate.

1.1.4 Schematic Symbols

Digital circuit components, such as gates, are represented in schematics by symbols. Programming systems, such as those used by PLCs, may also use standard schematic symbols to represent logical functions in programs. There are currently two different types of symbols used in industry. *Figure 2* shows the same logic gate in each form; the symbol shown in (*A*) is known as the traditional style, and has been in use since the 1950s. Traditional symbols are shaped differently depending on the type of gate. The symbol shown in *Figure 2* (*B*), known as the rectangular style, was introduced in the 1980s in an attempt to make all of the symbols more uniform. Unlike traditional symbols, these are always shaped the same (as a rectangle), and the type of gate is indicated inside the rectangle. You may see either style in the workplace, although many engineers and technicians prefer the traditional symbols.

1.2.0 Basic Gates

Digital circuits are made from seven possible gates. Four of these are unique, while the other three are just variations of several of the others. The following sections examine each gate type by describing its function, showing its schematic symbol, and providing its truth table.

1.2.1 AND Gates

The AND gate is a device whose output is a logic-1 only if *both* of its inputs are also logic-1. If only one input is a logic-1 and the other is a logic-0, the output will be a logic-0. The symbols shown in *Figure 3* represent the AND gate. The two inputs, marked A and B, are on the left of the symbol, while the output, marked X, is on the right.

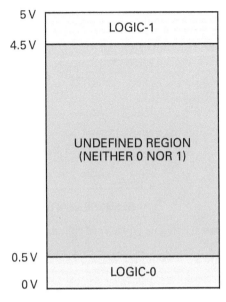
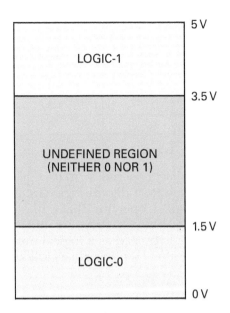

Figure 1 Example output and input logic levels comparison.

Table 1 An Example Truth Table for a Three-Input Gate

Inputs			Output
A	B	C	X
0	0	0	0
0	0	1	1
0	1	0	1
0	1	1	0
1	0	0	1
1	0	1	0
1	1	0	0
1	1	1	1

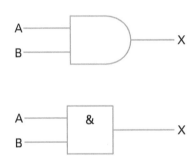

Figure 3 AND gate schematic symbols.

(A) TRADITIONAL

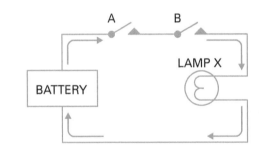

REPRESENTATIVE CIRCUIT

Figure 4 AND gate equivalent circuit.

(B) RECTANGULAR

Figure 2 Traditional and rectangular schematic symbols of the same gate.

To visualize the AND gate, use the circuit in *Figure 4* as an equivalent. In this circuit, both switches A and B must be closed for the light bulb to illuminate. If only one of the switches is closed, the light bulb will stay dark. The two switches therefore represent the AND gate's inputs, while the light bulb corresponds to the output.

Table 2 shows the truth table for a two-input AND gate. The two columns on the left show all the possible combinations of input values. The column on the right shows the corresponding output for that pattern. If you relate this table to the lamp circuit, a 0 represents an open switch or a dark lamp, while a 1 represents a closed switch or a lighted lamp.

Table 2 Two-Input AND Gate Truth Table

Inputs		Output
A	B	X
0	0	0
0	1	0
1	0	0
1	1	1

Circuit designers use a special form of algebra to calculate the behavior of digital logic circuits. Each gate has a logical operator that's like an arithmetic operator, although they are not the same. The AND gate is treated as a logical multiplication, so its equation is written out as follows:

$$X = A \bullet B \text{ or } X = AB$$

Be aware that the AND gate is *not* a multiplication circuit. Instead, it performs a logical operation that's comparable to multiplication. To stress this difference and prevent confusion, the equation should be read "X equals A *and* B" rather than "X equals A *times* B." (Note that X represents the output, and A and B represent the inputs.)

1.2.2 OR Gates

The OR gate is a device whose output is a logic-1 if any of its inputs is a logic-1. An OR gate will only output a logic-0 if *all* of its inputs are logic-0. The symbols shown in *Figure 5* represent the OR gate. To visualize the OR gate, use the circuit in *Figure 6* as an equivalent to the OR gate. Note that the switches are connected in parallel rather than in series. The bulb can now be turned on by closing either switch A, switch B, or both. *Table 3* shows the truth table for a two-input OR gate.

When expressed as an equation, the OR gate is treated as a logical equivalent of addition. Again, be aware that it is not an adding circuit. The equation is as follows:

$$X = A + B$$

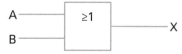

Figure 5 OR gate schematic symbols.

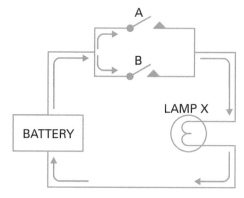

Figure 6 OR gate equivalent circuit.

Table 3 Two-Input OR Gate Truth Table

Inputs		Output
A	B	X
0	0	0
0	1	1
1	0	1
1	1	1

Like the AND function, the OR function should be read aloud differently than its arithmetic equivalent. Always say "X equals A *or* B," not "X equals A *plus* B."

1.2.3 XOR Gates

A special form of the OR gate is the exclusive OR gate, known as the XOR gate. It is a device whose output is logic-1 if just one of its inputs is logic-1. If both are logic-1 or both are logic-0, it will output a logic-0. The symbols shown in *Figure 7* represent the XOR gate. *Table 4* shows the truth table for a two-input XOR gate. Its equation is as follows:

$$X = A \oplus B$$

This equation should be read "X equals A exclusive or B."

Table 4 Two-Input XOR Gate Truth Table

Inputs		Output
A	B	X
0	0	0
0	1	1
1	0	1
1	1	0

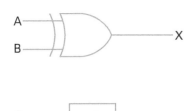

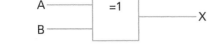

Figure 7 XOR gate schematic symbols.

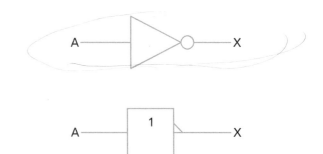

Figure 8 NOT gate schematic symbols.

Table 5 NOT Gate Truth Table

Input	Output
A	X
0	1
1	0

1.2.4 NOT Gates

The simplest gate is the NOT gate, also known as the inverter. It is different from the AND, OR, and XOR gates because it has just one input. As a result, it does not perform a decision-making function based on the combination of its inputs. Instead, it simply changes the input logic level to its opposite and sends it to the output. If the input is logic-0, the output will be logic-1. If the input is logic-1, the output will be logic-0. The symbols shown in *Figure 8* represent the NOT gate and *Table 5* shows its truth table. Its equation is as follows:

$$X = \overline{A}$$

This equation should be read "X equals A not" or "X equals A bar"; either is acceptable. The bar over a symbol generally indicates that an inversion is taking place.

NOT gates may also appear attached to other symbols, either on the inputs or outputs. They make take the form of bubbles or wedges, as *Figure 9* shows. Whenever you see either of these, assume that a NOT gate is inverting the signal at that point.

1.3.0 Modified Gates

AND, OR, XOR, and NOT gates can perform the logical functions required to create any digital circuit. But in many situations, modified versions of the first three gates (AND, OR, and XOR) can be more useful.

1.3.1 NAND Gates

The NAND gate stands for NOT AND and is simply an AND gate with an inverter attached to the output. Its output will always behave in the opposite way to an AND gate with the same inputs. For example, if an AND gate has a logic-1 on both of its inputs, its output will be logic-1, but a NAND gate will output a logic-0 for that combination.

The symbols shown in *Figure 10* represent the NAND gate. *Table 6* shows the truth table for a two-input NAND gate. Compare it to *Table 2* to be sure that you understand the relationship between AND and NAND. Its equation is as follows, and should be read "X equals A nand B":

$$X = \overline{AB}$$

1.3.2 NOR Gates

The NOR gate stands for NOT OR and is simply an OR gate with an inverter attached to the output. Its output will always behave in the opposite way to an OR gate with the same inputs. For example, if an OR gate has a logic-1 on either or both of its inputs, its output will be logic-1, whereas a NOR gate will output a logic-0 for those combinations.

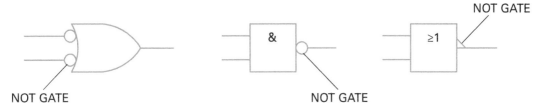

Figure 9 NOT gates used with other symbols.

Figure 10 NAND gate schematic symbols.

Table 6 Two-Input NAND Gate Truth Table

Inputs		Output
A	B	X
0	0	1
0	1	1
1	0	1
1	1	0

The symbols shown in *Figure 11* represent the NOR gate. *Table 7* shows the truth table for a two-input NOR gate. Compare it to *Table 3* to be sure that you understand the relationship between OR and NOR. Its equation is as follows, and should be read "X equals A nor B":

$$X = \overline{A+B}$$

1.3.3 XNOR Gates

The XNOR gate is simply an XOR gate with an inverter attached to the output. Its output will always behave opposite of the way an XOR gate would with the same inputs. For example, if an XOR gate has a logic-1 on either, but not both, of its inputs, its output will be logic-1. On the other hand, an XNOR gate will output a logic-0 for those combinations.

The symbols shown in *Figure 12* represent the XNOR gate. *Table 8* shows the truth table for a two-input XNOR gate. Compare it to **Table 4** to be sure that you understand the relationship between XOR and XNOR. Its equation is as follows, and is read "X equals A exclusive nor B":

$$X = \overline{A \oplus B}$$

1.4.0 Combination Logic

In most cases, individual logic gates can't do very much to solve problems. However, they can perform significant tasks when combined together. In fact, the world's most powerful computer is nothing more than many gates combined together in very complex arrangements. Circuits made up of gates are known as combination logic.

1.4.1 Analyzing a Combination Logic Circuit

Figure 13 shows a simple combination logic circuit made up of three different gates. You may encounter circuits like these in equipment schematics. For example, a machine may use a combination logic circuit in its safety interlock system. Sensors feed the circuit's inputs, and the output connects to a relay that enables or disables the machine. Only when the sensors output the right combination of logic values will the circuit allow the machine to turn on.

Figure 11 NOR gate schematic symbols.

Table 7 Two-Input NOR Gate Truth Table

Inputs		Output
A	B	X
0	0	1
0	1	0
1	0	0
1	1	0

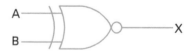

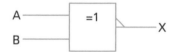

Figure 12 XNOR gate schematic symbols.

Table 8 Two-Input XNOR Gate Truth Table

Inputs		Output
A	B	X
0	0	1
0	1	0
1	0	0
1	1	1

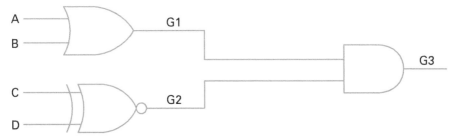

Figure 13 An example combination logic circuit.

When you encounter circuits of this kind, the first thing you must do is analyze them to understand how they behave. The best way to do this is to create a truth table for the complete circuit. Remember that truth tables can become very large if there are a lot of inputs to the circuit. For example, the circuit shown in *Figure 13* has four inputs, so its truth table contains 16 rows (*Table 9*). (Every time you add an input, the number of rows in the truth table doubles.)

When working out a large truth table, it's often easier to analyze the circuit in small pieces. For example, if you were building the truth table for *Figure 13*, first work out the smaller truth tables for gates G1 and G2. Then, use those results to work out gate G3's behavior. *Table 10* shows each small truth table, and *Table 11* shows the truth table for the whole circuit.

1.4.2 Example Combination Logic Circuit

The majority of explosions in the burner systems of furnaces can be attributed to uncontrolled ignition of combustible mixtures that have accumulated in the furnace or the related exhaust ducting systems. The National Fire Protection Association (NFPA) developed *Standard 86*, which lists required furnace safeguards as well as the

Table 10 Intermediate Truth Tables

Inputs		Output
A	B	G1
0	0	0
0	1	1
1	0	1
1	1	1

Inputs		Output
C	D	G2
0	0	1
0	1	0
1	0	0
1	1	1

Table 9 Empty Truth Table for a Four-Input Circuit

Inputs				Output
A	B	C	D	X
0	0	0	0	
0	0	0	1	
0	0	1	0	
0	0	1	1	
0	1	0	0	
0	1	0	1	
0	1	1	0	
0	1	1	1	
1	0	0	0	
1	0	0	1	
1	0	1	0	
1	0	1	1	
1	1	0	0	
1	1	0	1	
1	1	1	0	
1	1	1	1	

Table 11 Final Truth Table

Inputs				Output
A	B	C	D	G3
0	0	0	0	0
0	0	0	1	0
0	0	1	0	0
0	0	1	1	0
0	1	0	0	1
0	1	0	1	0
0	1	1	0	0
0	1	1	1	1
1	0	0	0	1
1	0	0	1	0
1	0	1	0	0
1	0	1	1	1
1	1	0	0	1
1	1	0	1	0
1	1	1	0	0
1	1	1	1	1

protection these safeguards provide. A few of the safeguards required include low-pressure fuel detectors, high-pressure fuel detectors, flame detectors, and airflow detectors.

In most furnace designs, these safeguards drive the inputs of a digital controller whose functions include controlling and monitoring the fuel and combustion conditions of the furnace. The controller normally sets the main fuel valve at a minimum setting for ignition and verifies that the flame remains lit once ignition has been detected. If for any reason any one of these conditions cannot be confirmed, the controller closes a fuel safety valve and shuts down the burner.

Figure 14 shows a simplified furnace controller circuit. The fuel pressure detectors each output a logic-1 if they detect an improper fuel pressure. If all is well, they output a logic-0. The airflow detector outputs a logic-1 if it detects insufficient airflow and a logic-0 if airflow is normal. Finally, the flame detector outputs a logic-1 when it detects a flame and a logic-0 when it doesn't. The safety valve has a digital input that causes the valve to close if it receives a logic-1. As long as it receives a logic-0, it stays open.

Basically, this circuit should only allow the valve to stay open (logic-0) if the pressure sensors and the airflow sensor are logic-0 and the flame detector is logic-1. As you can see, OR gates are used to implement the circuit. But since the flame detector is backwards compared to the other three sensors, it has to be inverted by a NOT gate first before it can be fed into an OR gate input.

1.4.3 Issues Involving Combination Logic

When working with any digital circuit, be aware that its outputs don't immediately change when new information is received. This is because the transistors making up the circuit take time to switch. There is always a delay between the moment that the inputs change and the moment when the outputs react. This is called propagation delay. In most circuits, the propagation delay is very small—usually a measure of nanoseconds. A nanosecond (ns) is a measurement of time equal to 1 billionth of a second.

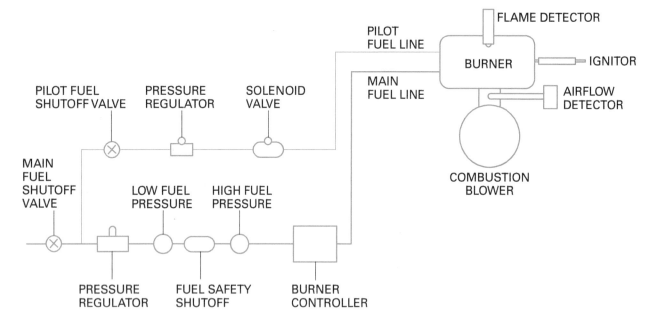

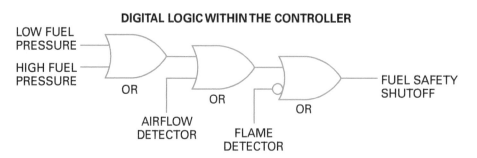

Figure 14 Furnace safety circuit block diagram and schematic.

The more gates that make up a circuit, the longer the propagation delay becomes. For small circuits, it's usually not an issue, particularly in many industrial settings where decisions don't have to be made in billionths of a second. However, it can be important in some mission-critical situations, such as a nuclear reactor shutdown system. In really large digital circuits, like those that make up a personal computer, too much propagation delay can mean a slow machine. Deciding how much propagation delay is acceptable in a given situation is the responsibility of an electrical engineer.

It is noteworthy that some digital circuits have different propagation delays for each signal direction change. For example, a positive-going signal (one changing from logic-0 to logic-1) may cause a longer propagation delay than a negative-going signal (one changing from logic-1 to logic-0). It all depends on the logic family's circuit characteristics.

One way that circuit designers can reduce propagation delay is by using gates with more than two inputs. AND, OR, XOR, NAND, NOR, and XNOR gates are available with more than two inputs (*Figure 15*). Designers can also reduce propagation delay by using mathematical tools to reduce circuits down to the smallest possible number of gates. They may also choose to reduce the circuit down to the most optimal combination of gates.

1.4.4 Buffer Gates

One issue that may come up, particularly in industrial settings, is the need for a digital output to control something that requires a larger amount of current than the output can supply. In other situations, the connected device requires different voltage levels than the output provides. Both of these problems can be solved by using a special gate called a buffer gate. Buffers are much like amplifiers. It's crucial to remember that they don't change the logic-0 or logic-1 state of the input signal; the logic value that goes in also comes out. *Figure 16* shows the schematic symbols for a buffer gate.

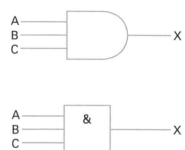

Figure 15 Example of a multi-input AND gate.

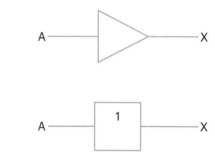

Figure 16 Buffer gate schematic symbols.

Additional Resources

Digital Fundamentals, Thomas L. Floyd. Eleventh Edition. 2015. Upper Saddle River, NJ: Prentice Hall.

1.0.0 Section Review

1. In order for digital devices to communicate with each other reliably, they must recognize the same _____.
   a. power supply voltage
   b. logic levels
   c. V_{CC} value
   d. logic process

2. The basic building block of digital technology is the _____.
   a. flip-flop
   b. binary adder
   c. gate
   d. truth table

3. A two-input digital circuit outputs a logic-1 only when both of its inputs are logic-1. What kind of gate is it?
   a. AND gate
   b. OR gate
   c. NAND gate
   d. NOR gate

4. A NOT gate is often called a(n) _____.
   a. single gate
   b. amplifier
   c. buffer
   d. inverter

5. A two-input digital circuit outputs a logic-0 only when both of its inputs are logic-1. It is a(n) _____.
   a. AND gate
   b. OR gate
   c. NAND gate
   d. NOR gate

6. Combination logic circuits experience a time period between the moment when their inputs change and the moment when their outputs react. This is called a(n) _____.
   a. fanout
   b. propagation delay
   c. gate lag
   d. edge transition

Section Two

2.0.0 Digital Memory Circuits

Objective

Identify and describe memory elements and their function in digital circuits.
a. Identify and describe the basic flip-flop design.
b. Explain the operation of clocked logic and clocked flip-flops.
c. Describe the function of various types of registers.

Trade Terms

Bit: A single logic-0 or logic-1 value. The term is a contraction of the words "binary digit."

Clock: A signal that oscillates between logic-0 and logic-1 at a particular frequency. It is used to synchronize digital circuits with each other.

Edge-triggered: A digital circuit that performs an operation when it detects a change in logic level.

Flip-flop: A digital memory circuit capable of storing a single logical value (one bit).

Level-triggered: A digital circuit that performs an operation when it detects a specific logic level.

Register: A digital memory circuit made from flip-flops and capable of storing a single word (see *word*) of information.

Timing diagram: A diagram that shows the behavior of a digital circuit's signals with respect to time.

Word: A group of bits acting together as a single unit. Words are usually (but not always) multiples of four bits.

Combination logic is only half the picture when it comes to digital technology. Although these circuits can perform amazingly complex operations, they cannot store information for later use. Once their inputs change, their outputs change as well. Without the ability to hold values and pass them on to other circuits at strategic moments, a digital system is extremely limited. Digital memory devices complete the picture by providing this essential capability.

2.1.0 Basic Memory Elements

The simplest memory device stores just a single logical value (a logic-0 or logic-1). This is known as a bit, which is a contraction of the words "binary digit." A single bit of storage may not seem like much, but it's the basis for memory devices of almost any size. A flip-flop is a digital circuit capable of storing a single bit. While there are many different types of flip-flops, they all have certain features in common.

2.1.1 The RS Flip-Flop

The simplest flip-flop, and the basis for all others, is the reset-set (RS) flip-flop. Its schematic symbol is shown in *Figure 17*. All flip-flops are similar in appearance. The outputs (labeled Q) are on the right side of the symbol, and one of the outputs has a bar over it. Recall that a bar over an output means that it's inverted. So Q and Q NOT (the one with an overbar) are outputs whose values are always the opposite of each other.

The inputs, labeled R and S, control the circuit. Activating the S input stores a logic-1 in the flip-flop. The letter S stands for "set," which is the term for a flip-flop that's storing a logic-1. A set flip-flop outputs a logic-1 on its Q output and a logic-0 on its Q NOT output. Activating the R input stores a logic-0 in the flip-flop. The letter R stands for "reset," which is the term for a flip-flop that's storing a logic-0. A reset flip-flop outputs a logic-0 on its Q output and a logic-1 on its Q NOT output.

Internally, an RS flip-flop is made either from a pair of NOR gates or a pair of NAND gates. The NOR gate version appears in *Figure 18* (A). Since this circuit is very important, it will be analyzed in considerable detail. A challenge in analyzing the RS flip-flop lies in the fact that the outputs of the circuit are connected back to the inputs. Thus, any signal that is input to the circuit goes through the circuit and returns to the input lines. As a result, the input signal has multiple effects. This method of connecting a circuit output back to its

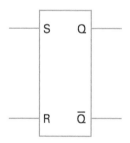

Figure 17 Schematic symbol for an RS flip-flop.

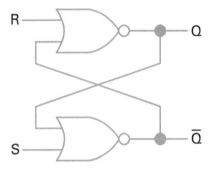

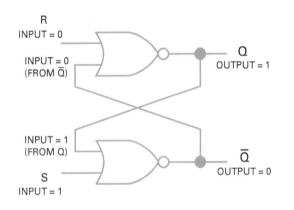

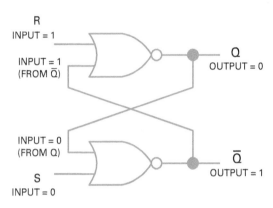

Figure 18 An RS flip-flop made from two NOR gates and its response to different inputs.

own input is called *feedback*. A feedback connection is essential for giving a logic circuit memory capability.

To analyze the RS flip-flop, begin by changing the S input to logic-1 while keeping the R input at logic-0. Refer to *Figure 18 (B)* as you trace through this scenario. Remember, a NOR gate's output is a logic-0 whenever one or both of its inputs is a logic-1. This causes the Q NOT output to change to the logic-0 state. Because the logic-0 at Q NOT is connected to one of the inputs on the upper gate, the upper gate now has two logic-0 inputs. Its output (Q) changes to logic-1.

The logic-1 output from Q is connected back to one of the inputs of the lower gate, and the lower gate now has two logic-1 inputs. The original input signal has been traced through the entire circuit in a figure-eight pattern. Since the lower NOR gate needs just a single logic-1 input to maintain its output in the logic-0 state, the logic-1 at S can change to logic-0, and the output will not change. In other words, the circuit is acting as a memory.

Now, examine what happens if the R input changes to logic-1 while the S input stays at logic-0. Do the same analysis procedure, tracing the signal around the circuit in a figure-eight pattern. If you do the analysis correctly, you'll discover that the circuit now outputs a logic-0 on Q and a logic-1 on Q NOT. In other words, the flip-flop is reset (storing a logic-0). This is illustrated in *Figure 18 (C)*.

In summary, the NOR-based RS flip-flop has the property that whenever the S input is changed to a logic-1, the circuit will become set and the Q output will be logic-1. Whenever the R input is changed to a logic-1, the circuit will become reset and the Q output will be logic-0. If both S and R are changed to logic-0, the flip-flop will simply hold its present state indefinitely. This condition is called *memory mode*.

What happens if you try to activate both S and R at the same time (make them both logic-1)? If this happens, the flip-flop behaves illogically by outputting a logic-0 on both Q and Q NOT. This condition is called the illegal mode and shouldn't be used. *Table 12* shows the truth table for the NOR RS flip-flop.

2.1.2 The NAND RS Flip-Flop

It's also possible to construct an RS flip-flop from NAND gates. *Figure 19* shows this circuit. It behaves in almost the same way as the NOR circuit, with a few crucial differences. First of all, notice that Q and Q NOT are on the opposite gates with respect to S and R compared to the NOR RS flip-

Table 12 NOR-Based RS Flip-Flop Truth Table

Inputs		Outputs		Mode
S	R	Q	Q NOT	
0	0	No change	No change	Memory
0	1	0	1	Reset
1	0	1	0	Set
1	1	0	0	Illegal

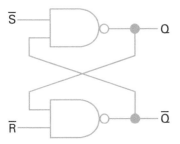

Figure 19 An RS flip-flop made from a pair of NAND gates.

flop. Also notice that the S and R inputs have bars over them. This means that they are inverted, so their behavior is opposite of the inputs on the NOR RS flip-flop. In other words, to set the flip-flop, you change the S input to logic-0 instead of logic-1. To reset it, change the R input to logic-0. *Table 13* shows the NAND RS flip-flop's truth table. Examine and compare it to *Table 12*. You may wish to take a few moments to analyze the NAND RS flip-flop in the same way as you did with the NOR version.

2.2.0 Clocked Logic

Complicated digital circuits are made up of many different subcircuits. Some of these are combination logic, while others are memory logic. Coordinating the actions of each of those can become very complicated and difficult. For this reason, a method of synchronizing the different circuits becomes an absolute necessity. A special signal called a clock solves the problem.

2.2.1 Clock Signals

A clock signal is simply a digital signal that oscillates between logic-0 and logic-1 at a particular frequency. A special circuit, called a clock genera-

Table 13 NAND-Based RS Flip-Flop Truth Table

Inputs		Outputs		Mode
S	R	Q	Q NOT	
0	0	1	1	Illegal
0	1	1	0	Set
1	0	0	1	Reset
1	1	No change	No change	Memory

tor, produces this signal. Clock frequencies can range from less than 1 Hz to 4 GHz or more. *Figure 20* shows a typical clock signal graphed with respect to time.

Digital circuits that must be synchronized are connected to the clock generator so they can monitor the clock signal. Think of the circuits like marchers in a parade, and the clock signal like a drum. As long as everyone marches to the correct beat, everything works well. Circuits that use the clock for synchronization, known as clocked logic, will activate only if the clock signal is doing a particular thing. For example, some circuits are edge-triggered. This means that they will only act if the clock signal is changing from logic-0 to logic-1 (rising) or logic-1 to logic-0 (falling).

On the other hand, some circuits are level-triggered. They act when the clock signal is at a particular logic level (logic-0 or logic-1). Edge-triggered circuits have a very small window of time to act since the clock signal changes very quickly. Level-triggered circuits have a much wider time window in which to act since the clock signal will be at each level 50 percent of the total time. Both approaches are useful, depending on the application. *Figure 21* illustrates the difference between edge and level triggering.

2.2.2 Timing Diagrams

Truth tables work very well for describing the behavior of many digital circuits. However, they don't work well for clocked logic since they cannot indicate the circuit's behavior over a period of time. For this reason, engineers and technicians use another tool, the timing diagram, to show the behavior of a clocked circuit.

A timing diagram is a kind of graph with time as the x-axis (the horizontal plane). Generally, timing diagrams include several signals stacked on top of each other so their behavior can be examined with respect to time, as well as in relation to one another. *Figure 22* shows a typical timing diagram. Notice that a clock signal is shown along with three other signals. The diagram is like a series of snapshots describing the dynamic behavior of the circuit as time goes by.

2.2.3 The Clocked RS Flip-Flop

A flip-flop is able to change any time its inputs are triggered. The basic RS flip-flop is not a clocked circuit, but in many cases, it's more useful if flip-flops are governed by a clock. Modifying the basic NOR RS flip-flop into a clocked version is quite simple.

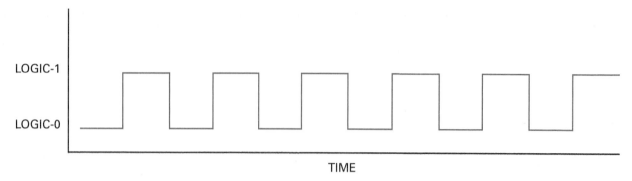

Figure 20 A typical clock signal.

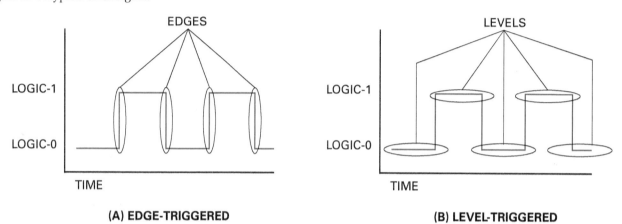

Figure 21 Edge versus level triggering.

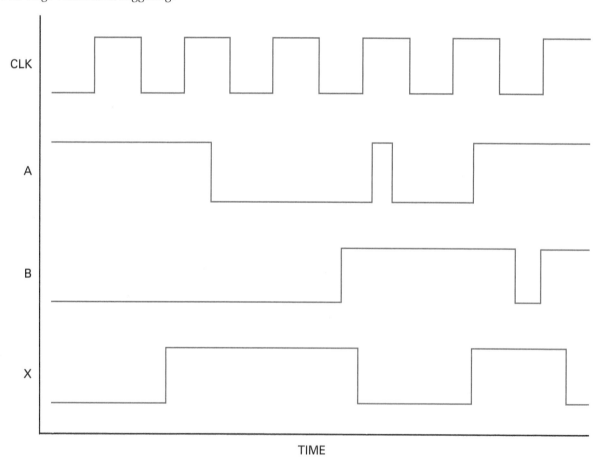

Figure 22 Typical timing diagram.

Figure 23 shows the revised RS flip-flop. Notice that it contains a pair of AND gates in front of the inputs, with the clock fed into each of them. Basically, the AND gates will block the S and R signals when the clock is logic-0. When the clock is logic-1, S and R can be used to change the flip-flop's set or reset condition. This flip-flop is therefore level-triggered; the active level is logic-1, and the inactive level is logic-0. In other words, unless the clock signal is logic-1, the flip-flop will ignore any changes that you attempt to make. *Figure 24* shows a timing diagram for this circuit.

Since the RS flip-flop is the basis for all other types, it has been given a fairly detailed analysis here. Now that you have a good understanding of this type, you can examine other varieties. They are all variations of the basic RS theme. Since some of them are quite complicated internally, the following sections examine each one as a digital black box with inputs and outputs. Timing diagrams help clarify each circuit's behavior.

2.2.4 The D Flip-Flop

Two versions of the data (D) flip-flop are shown in *Figure 25*. Examine the inputs and outputs. The D input is used both to set and reset the flip-flop (instead of a pair of inputs as on the RS flip-flop). The two Q outputs behave as expected. The second input requires some explanation. In each case, it is for the clock. However, the way the input is labeled indicates how the flip-flop responds to the clock. If the clock input is labeled EN (enable), the flip-flop is level-triggered. If the input is labeled CLK (clock), the flip-flop is edge-triggered.

Figure 26 shows the timing diagram of a level-triggered D flip-flop. In this case, the active clock level is logic-1. Notice that whenever the clock is logic-1, whatever value that is placed on the D input passes through to the outputs immediately. In other words, the flip-flop is transparent to the data. When the clock is logic-0 (inactive), the flip-flop goes into memory mode and just holds the outputs at whatever value was on the D input right before the clock changed to logic-0.

Figure 27 shows the timing diagram of an edge-triggered D flip-flop using the same data supplied in *Figure 26*. In this case, the flip-flop only permits changes to the outputs when the clock is rising (changing from logic-0 to logic-1). The rest of the time, the flip-flop ignores the D input and just keeps holding the outputs at the same value. Notice from the timing diagram that there are a lot fewer opportunities to change the

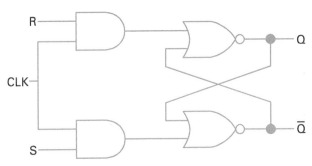

Figure 23 A clocked NOR RS flip-flop.

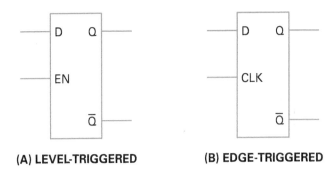

Figure 25 Level-triggered and edge-triggered D flip-flops.

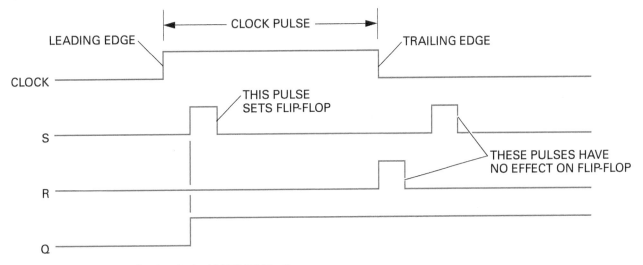

Figure 24 Timing diagram for the clocked NOR RS flip-flop.

flip-flop since the edge happens very quickly. Consequently, the two timing diagrams are quite different, even though the same data was used for each.

It is possible for a D flip-flop to respond to the logic-0 clock level instead of logic-1. It is also possible for it to respond to the falling edge (logic-1 to logic-0) instead of the rising one. The flip-flop's schematic symbol tells you how it responds to the clock. If the EN symbol has a bar over it, the flip-flop responds to the logic-0 level instead of logic-1. If an edge-triggered flip-flop's CLK symbol has a bar over it, it responds to the falling edge instead of the rising one.

2.2.5 The JK Flip-Flop

The JK flip-flop is similar to the clocked RS. It is always an edge-triggered flip-flop (never level-triggered). *Figure 28* shows its schematic symbol. Notice that instead of R and S, its inputs are labeled J and K. These letters don't stand for anything in particular; they are named this way for historical reasons. J corresponds to the set input (S) and K to the reset (R) input.

The JK flip-flop differs from the clocked RS in that, instead of having an illegal mode when you activate both J and K at the same time, the JK flip-flop does something special—it toggles. Every time it sees an appropriate clock change, the outputs switch to the opposite of their current values. *Figure 29* shows the timing diagram of a JK flip-flop in toggle mode. Toggling is a useful capability, and will be discussed in greater detail later in this module.

2.3.0 Larger Memory Devices

Flip-flops are useful as memory devices, but they are somewhat limited since they only store one bit of information. In many digital systems, it's more useful to store groups of bits organized into words. A word can be any size, although most sizes are multiples of four bits. Probably the most

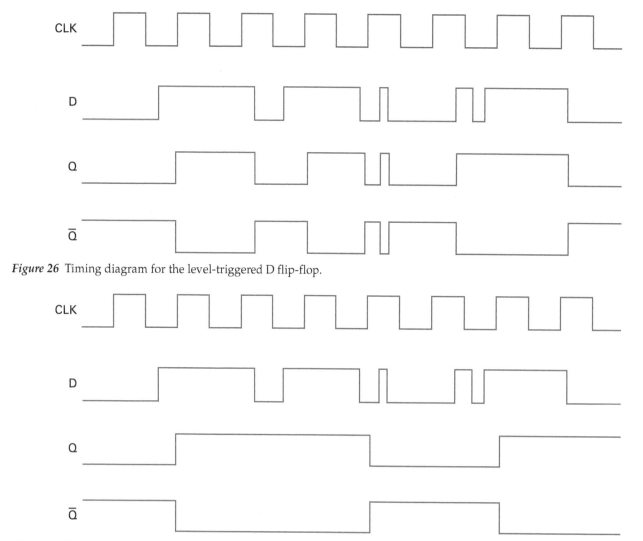

Figure 26 Timing diagram for the level-triggered D flip-flop.

Figure 27 Timing diagram for the edge-triggered D flip-flop.

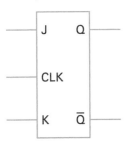

Figure 28 Schematic symbol for a JK flip-flop.

familiar word size is the byte, which is a group of eight bits. Words can represent numbers or characters, depending on the application. A register is a memory device that stores a single word. It is made up of individual flip-flops connected to work together.

2.3.1 Basic Registers

The simplest kind of register is shown in *Figure 30*. In this example, the register holds a single four-bit word. It is made up of four D flip-flops. Notice that their CLK inputs are all connected together. Doing this makes all of the flip-flops work as a single unit. When the clock rises (positive edge), whatever data appears on the four inputs, D_0 through D_3, gets stored in the flip-flops and sent to the outputs, Q_0 through Q_3. The rest of the time, the flip-flops ignore the inputs and just keep outputting their stored information. While this example register holds just four bits of information, registers can be any size; it's simply a matter of adding more flip-flops.

2.3.2 Specialized Registers

In addition to the basic register, there are many specialized varieties. Some can accept data one bit at a time and assemble it internally into a complete word. These are called *serial-in, parallel-out registers* (SIPO registers). Others can take the data word stored inside and send it one bit at a time to another device. These are called *parallel-in, serial-out registers* (PISO registers). Still others can take the data word stored inside and shift the bits to the right or left. These are called *shift registers*. All of these are used for specialized tasks. While it is helpful to be familiar with their names, it is unlikely that you'll interact with them in most instrumentation settings.

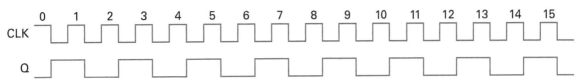

Figure 29 Timing diagram for a JK flip-flop in toggle mode.

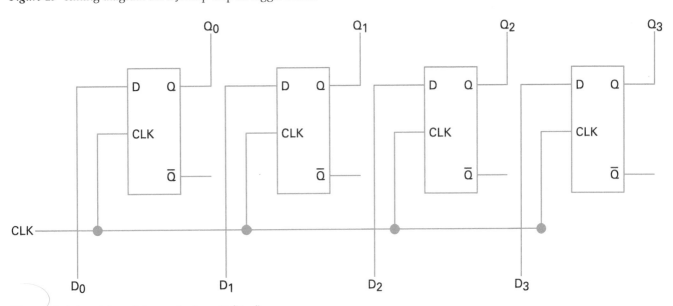

Figure 30 A four-bit register made from D flip-flops.

Additional Resources

Digital Fundamentals, Thomas L. Floyd. Eleventh Edition. 2015. Upper Saddle River, NJ: Prentice Hall.

2.0.0 Section Review

1. A flip-flop stores a single logical value known as a _____.
   a. word
   b. byte
   c. bit
   d. Q

2. To create the memory effect, digital memory circuits rely on _____.
   a. propagation delay
   b. feedback
   c. Q NOT
   d. the ripple effect

3. Clocked flip-flops that respond based on whether the clock signal is logic-0 or logic-1 are _____.
   a. positive-going
   b. negative-going
   c. edge-triggered
   d. level-triggered

4. What kind of flip-flop has a single input that can set or reset it?
   a. D
   b. JK
   c. RS
   d. Clocked RS

5. A memory device that can store a word of information is called a _____.
   a. flip-flop
   b. register
   c. byte
   d. bit

Section Three

3.0.0 Counter Circuits

Objective

Describe counters and their function in digital circuits.
a. Describe the numbering systems related to digital circuits.
b. Describe the function of binary counters.
c. Describe the function of other types of counters.

Trade Terms

BCD number: A decimal digit represented by a four-bit binary number in the range of 0000 to 1001; stands for "binary coded decimal".

Binary number: A number made up of digits that are either 0 or 1; also known as base 2.

Counter: A digital circuit that stores a value that gets increased or decreased in response to an external signal.

Hexadecimal number: A number made up of the digits 0–9 and A–F; also known as base 16.

One function that digital circuits must often perform is *counting*. A digital counter is a circuit that keeps a record of the number of times that an event occurs. Every time the event happens, the circuit either increases or decreases a stored value by a factor of one. Counters are actually a type of memory circuit, so they are based on flip-flops.

3.1.0 Digital Numbers

Almost all human cultures count and do arithmetic using the decimal (base 10) number system. Decimal numbers are made up of digits that range in value from 0 to 9. By chaining together enough digits, numbers of any size are possible.

The decimal system is commonly used for handling numbers, but other number systems do exist. Digital electronics works particularly well with the binary number system (base 2), since a binary number has just two digit possibilities: 0 and 1. These correspond neatly with the logic-0 and logic-1 values that digital circuits use.

3.1.1 Binary Numbers

The binary number system has rules that are similar to but somewhat simpler than those of the decimal system. Binary starts by representing a quantity of zero with a *0* and a quantity of one with a *1*. However, no binary symbol exists for a quantity of two. This problem is handled in the same manner as it is handled with the decimal system. No individual symbol in the decimal system represents a quantity of ten, so when you reach the maximum digit value (9), you add another digit to the left and set it to 1. Then you reset the original digit back to 0. The result is a two-digit number—10.

In binary, two is represented by resetting the original digit to 0 and putting a 1 to the left. Thus, two is represented by *10* in binary. This process can continue indefinitely. The second column of *Table 14* shows binary equivalents of the decimal numbers 0 through 15.

In most everyday situations, it won't be necessary to convert back and forth between binary and decimal numbers. When the need arises, there are a number of solutions. Many calculators and online number converters can perform this task. It is also possible to convert the numbers by hand using simple math. The steps for converting between binary and decimal numbers are described in greater detail elsewhere in this curriculum. However, just knowing the first sixteen binary numbers is likely sufficient at this stage of your training.

Table 14 Decimal, Binary, and Hexadecimal Numbers Compared

Decimal	Binary	Binary (4-bit)	Hexadecimal
0	0	0000	0
1	1	0001	1
2	10	0010	2
3	11	0011	3
4	100	0100	4
5	101	0101	5
6	110	0110	6
7	111	0111	7
8	1000	1000	8
9	1001	1001	9
10	1010	1010	A
11	1011	1011	B
12	1100	1100	C
13	1101	1101	D
14	1110	1110	E
15	1111	1111	F

Since many digital circuits work with fixed-size words, it is common practice to add leading zeros to binary numbers so they have a fixed size. This is standard practice with decimal numbers as well. For example, the number 0005 is the same as 5. The third column in *Table 14* shows the first sixteen binary numbers expressed as four-bit words. These are numerically identical to the numbers in the second column.

3.1.2 Hexadecimal Numbers

Binary numbers can be difficult to manage because the 0s and 1s are often hard to keep track of. For this reason, engineers and computer scientists started using the hexadecimal number system many years ago. This system is a method that easily converts to and from binary. Hexadecimal numbers, also known as *base 16*, use sixteen digits instead of ten (decimal) or two (binary). These digits are the familiar 0–9, along with the letters A, B, C, D, E, and F. The digits 0–9 work just as they do in decimal. The letter A stands for the number 10, B for 11, C for 12, and so on. This means that each digit position can represent a value from 0 through 15.

Many technical professionals find hexadecimal desirable because four binary bits convert perfectly into a single hexadecimal digit. For example, 0000 in binary equals 0 in hexadecimal. The binary number 1111 equals F. The fourth column in *Table 14* shows the hexadecimal equivalents of the decimal numbers 0 through 15.

3.2.0 Counter Circuits

There are many different kinds of counter circuits. Some count up, while others count down. Some can count in either direction. Many are based on a common circuit design made from JK flip-flops.

3.2.1 Binary Counters

Figure 31 shows the schematic for a simple binary counter that counts up from 0000 (0 decimal) to 1111 (15 decimal). It's made from four JK flip-flops with their J and K inputs both activated. As previously discussed, a JK flip-flop configured in this way toggles back and forth between 0 and 1 every time it sees an appropriate clock edge. Since the CLK inputs have bars over them, you know that these flip-flops trigger on the falling clock edge.

Notice that the Q output of one flip-flop feeds into the clock input of the next stage. The Q outputs are also the counter's numeric output. The counter is incremented by pulsing the CLK input on the leftmost flip-flop (labeled COUNT CLK in the diagram). *Figure 31* shows the timing diagram for this counter. Note the binary count at the bottom of the timing diagram. The counter begins at 0000, counts up to 1111, and then rolls back to 0000 before the count starts over.

Since the flip-flops are drawn from left to right on the page, they are backwards compared to the number. Essentially, flip-flop output Q_0 is actually the rightmost bit in the number, and Q_3 is the leftmost. Be sure that you understand this when you're comparing the schematic to the timing diagram and binary count values at the bottom.

3.3.0 Other Counters

The simple binary counter represents only a portion of what is available. Many other counter designs are possible, and each serves a specialized purpose. The following sections give a brief overview of a few of these counters.

3.3.1 BCD Counters

A modification of the basic binary counter is one that counts in BCD numbers. A BCD number is a binary number that ranges from 0000 to 1001 (0–9 decimal). Instead of counting 0000 through 1111, as did the binary counter, the BCD counter resets back to 0000 after it passes 1001. This behavior makes it possible to use the counter in a setting where the count needs to be turned into a decimal number. Even though the counter is still working in binary, it translates back to decimal fairly easily.

3.3.2 Up/Down Counters

There are many modifications possible for the basic binary counter. Being able to count down rather than up is one example, and some digital counters can do either. These include an input labeled DIRECTION that controls the count direction. For example, if this input is connected to a logic-0 value, the circuit counts up. If this input is changed to logic-1, the circuit counts down.

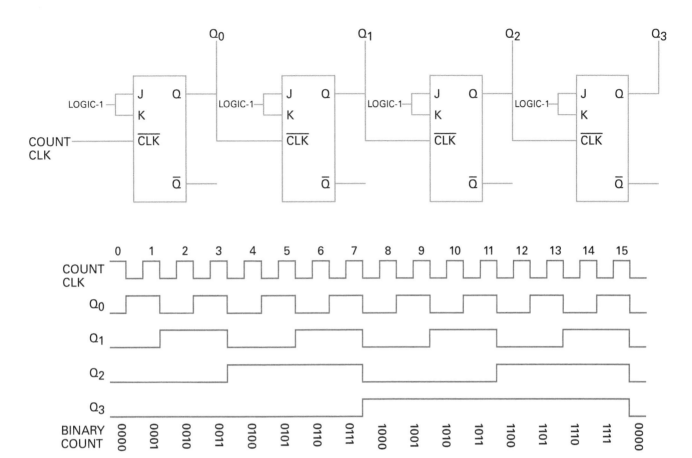

Figure 31 Schematic and timing diagram for a simple binary counter.

Additional Resources

Digital Fundamentals, Thomas L. Floyd. Eleventh Edition. 2015. Upper Saddle River, NJ: Prentice Hall.

3.0.0 Section Review

1. The binary number 0011 is equivalent to the decimal number _____.
   a. 2
   b. 3
   c. 11
   d. 12

2. The hexadecimal number C is equivalent to the decimal number _____.
   a. 2
   b. 3
   c. 11
   d. 12

3. A simple binary counter is made from _____.
   a. JK flip-flops
   b. Clocked RS flip-flops
   c. D flip-flops
   d. registers

4. The maximum BCD value is _____.
   a. 0000
   b. 1001
   c. 0110
   d. 1111

Section Four

4.0.0 Arithmetic Circuits and Decoders

Objective

Describe the function of arithmetic elements and decoders.
a. Describe the function of basic arithmetic elements.
b. Describe the function of decoders.

Trade Terms

Decoder: A digital circuit that outputs a unique pattern in response to each input combination.

While there are many different specialized digital circuits, two types come up fairly frequently in a variety of applications: the arithmetic circuit and the decoder. Arithmetic circuits perform calculations, such as addition, subtraction, multiplication, and division. Decoders translate information from one form to another. Both have applications in instrumentation.

4.1.0 Arithmetic Circuits

The arithmetic circuits in a typical computer are extremely complicated. However, all arithmetic fundamentally comes down to just one operation: addition. If you can add, you can also subtract, multiply, and divide. After all, subtracting is simply adding a negative value; multiplication is just repeated addition, and division is repeated subtraction.

4.1.1 The Principles of Addition

Decimal addition is performed one column at a time. If the sum of a column is greater than 9 (if it is a two-digit number), it generates a carry into the next column. When you add the numbers in that column, you include the carry from the previous column. Binary addition follows the same principles as basic decimal addition. The primary difference lies in the increased amount of carrying since the maximum column value is exceeded much more often. *Figure 32* illustrates decimal and binary addition, along with carrying between columns.

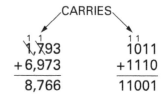

Figure 32 Decimal and binary addition compared.

4.1.2 A Simple Adder

A binary adding circuit is a combination logic circuit that has three inputs and two outputs. *Figure 33* shows the schematic symbol for a binary adder that can add two bits. The A and B inputs accept the two bits to be added together. C_{in} accepts the carry from a previous column's addition. The S output is the sum. C_{out} is the carry out to the next column.

Because adding just two bits isn't very useful, this circuit is replicated many times until there are enough bits to perform useful additions. *Figure 34* shows a binary adder that can work with two eight-bit numbers. Notice that it's just eight simple adders connected together. The C_{out} of one stage feeds the C_{in} of the next.

4.1.3 Inside the Box

Inside the box representing a simple adder, there are six gates: two XORs, three ANDs, and one OR. This is illustrated in *Figure 35*. This circuit is actually two separate circuits. The first one, made up of the two XOR gates, adds the three input bits (A, B, and C_{in}) together and generates the sum (S). The second circuit, made up of the three AND gates and the one OR gate, works out whether there needs to be a carry to the next stage. As a helpful exercise, try working through these two circuits to confirm that they work as expected. *Table 15* shows the truth table for the simple adder.

4.2.0 Decoders

Digital circuits work with binary numbers internally. However, it's often necessary for digital information to be presented in a more user-friendly

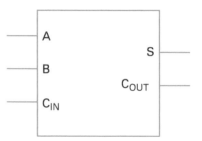

Figure 33 Schematic symbol of a simple binary adder.

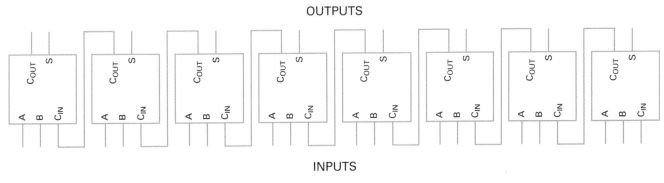

Figure 34 Schematic of an eight-bit binary adder made up of simple adders.

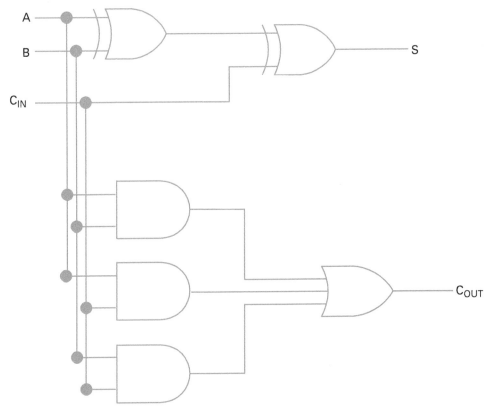

Figure 35 Schematic of the simple binary adder.

Table 15 Simple Binary Adder Truth Table

Inputs			Outputs	
A	B	C_{in}	S	C_{out}
0	0	0	0	0
0	0	1	1	0
0	1	0	1	0
0	1	1	0	1
1	0	0	1	0
1	0	1	0	1
1	1	0	0	1
1	1	1	1	1

way. Video displays and LCD panels offer a very sophisticated way to make information available to operators, but in many cases these devices would be overkill or simply too expensive for the application. Simple displays, such as status lights or number-only readouts, are often sufficient. Special circuits are required to convert binary information into other forms, such as decimal numbers or patterns of status lights. Circuits known as *decoders* perform this function, converting digital information from one form to another.

4.2.1 Basic Decoder

By definition, a decoder is a combination logic circuit with a set of inputs and a set of outputs.

The inputs usually accept a binary number. The circuit examines the number coming in from the inputs and activates a unique pattern on the outputs. Decoders can be simple or complex. *Figure 36* shows a simple decoder designed to control eight status lights on a machine panel.

Notice that the circuit has three inputs and eight outputs. A three-bit binary number ranging from 000 through 111 (0–7 decimal) comes in on the inputs. Circuitry in the decoder evaluates the number and activates the output whose number matches the binary value. For example, if a 011 (3 decimal) appears on the inputs, output O_3 changes to logic-1, while the other seven outputs remain logic-0. Each of the outputs can be connected to a status light on the machine panel. The decoder makes it possible for numeric information inside the equipment to control a simple and meaningful output.

Figure 37 shows the logic inside the decoder. Notice that it's basically a combination logic circuit made up of AND and NOT gates. As a helpful exercise, try working through its truth table and confirm that it behaves as expected.

4.2.2 Complex Decoder

Sometimes a more sophisticated decoder is needed. Many machines have numeric displays that use seven-segment LEDs (*Figure 38*). These are made up of seven bar-shaped LEDs arranged in the shape of the number 8. By lighting up spe-

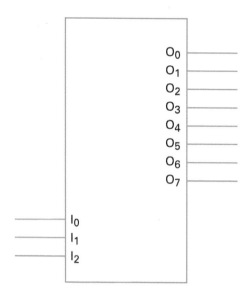

Figure 36 Schematic symbol of a simple decoder.

cific segments, the numbers 0 through 9 can be formed on the display. Controlling one of these displays requires a decoder to take a binary number between 0000 and 1001 (0–9 decimal) and translate it into the appropriate segment pattern for that number.

Figure 39 shows the schematic symbol of the decoder. Notice that it has four inputs and seven outputs. The inputs accept the binary number, while each output connects to one LED segment on the display. Internally, this decoder contains more than forty gates, so its schematic is not included here.

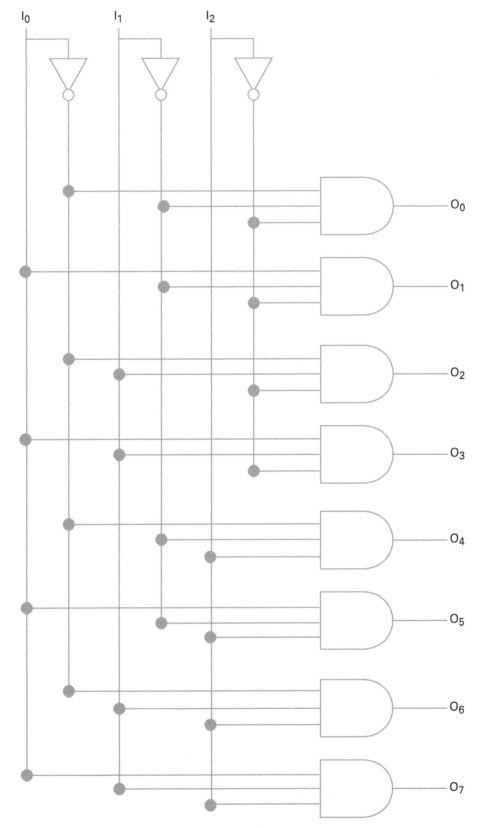

Figure 37 Schematic showing the inner workings of a simple decoder.

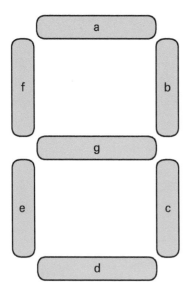

Figure 38 A seven-segment LED display.

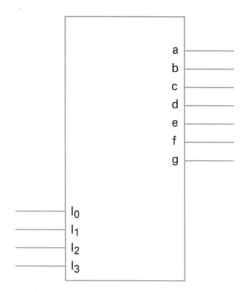

Figure 39 Schematic symbol of a seven-segment display decoder.

Additional Resources

Digital Fundamentals, Thomas L. Floyd. Eleventh Edition. 2015. Upper Saddle River, NJ: Prentice Hall.

4.0.0 Section Review

1. The A, B, and C_{in} inputs of a binary adder receive the values 0, 1, and 1. What are the circuit's output values?

   a. $S = 0$ and $C_{out} = 0$
   b. $S = 0$ and $C_{out} = 1$
   c. $S = 1$ and $C_{out} = 0$
   d. $S = 1$ and $C_{out} = 1$

2. The circuit in a binary adder that performs the summation is made from a pair of _____.

   a. AND gates
   b. OR gates
   c. XOR gates
   d. XNOR gates

3. Once a decoder examines the number coming in from its inputs, what does it activate on its outputs?

   a. A sum and a carry
   b. A unique pattern
   c. An error signal
   d. A clock signal

SUMMARY

Digital logic is the basis for many of the technological devices that we use in daily life. Similarly, many instrumentation systems rely on digital technology. All digital circuits do one of three basic things: manipulate information, perform arithmetic, or store information. Combination logic does the first two of these tasks. Using basic gates such as AND, OR, and NOT, it creates complex circuits that perform the required manipulations. Storage devices, such as flip-flops and registers, retain digital information until it's needed elsewhere. Circuits such as decoders translate information from one form into another. All digital devices, from the simplest machine controller to the largest computer, are based on these primary building blocks.

Review Questions

1. Digital technology treats all information as _____.
   a. math
   b. numbers
   c. storage
   d. specialized

2. A collection of compatible digital components make up a _____.
   a. 5V logic group
   b. logic level
   c. logic family
   d. common power supply

3. A digital component defines an output voltage between 0 and 0.8V as a logic-0, and an output voltage between 2V and 5V as a logic-1. These voltages are its _____.
   a. power supply values
   b. propagation delays
   c. logic family
   d. logic levels

4. Which type of gate outputs a logic-0 when both of its inputs are logic-0 or both are logic-1?
   a. AND gate
   b. OR gate
   c. XOR gate
   d. NOT gate

5. What kind of gate must be used if a digital output must deliver more current than is typical?
   a. An OR gate
   b. A buffer gate
   c. An inverter gate
   d. An XOR gate

6. A flip-flop that's outputting a 0 on its Q output and a 1 on its Q NOT output is said to be _____.
   a. set
   b. reset
   c. clocking
   d. toggling

7. A NOR RS flip-flop's S input is changed to logic-1 and its R input is changed to logic-0. What will its Q NOT output do?
   a. It will be logic-1.
   b. It will be logic-0.
   c. It will toggle.
   d. It will become unpredictable.

8. What is the name of a signal used to synchronize digital circuits with each other?
   a. A pulsar
   b. A sync circuit
   c. A clock
   d. A positive edge

9. A register is made of _____.
   a. RS flip-flops
   b. D flip-flops
   c. JK flip-flops
   d. clocked RS flip-flops

10. The binary equivalent of the decimal number 8 is _____.
    a. 0111
    b. 0110
    c. 1001
    d. 1000

11. The decimal equivalent of the hexadecimal number F is _____.
    a. 5
    b. 10
    c. 15
    d. 16

12. In a simple binary counter, the CLK input of each flip-flop (except for the first one) is connected to the previous flip-flop's _____.
    a. J input
    b. K input
    c. CLK input
    d. Q output

13. A useful binary addition circuit is made from many individual _____.
    a. D flip-flops
    b. simple adders
    c. JK flip-flops
    d. registers

14. A binary adder contains two separate circuits. One generates the sum, while the second generates the _____.
    a. carry
    b. BCD
    c. decoded output
    d. toggle bit

15. Converting a binary number into the pattern needed to drive a seven-segment LED display requires a circuit called a(n) _____.
    a. translator
    b. encoder
    c. decoder
    d. transcoder

Trade Terms Introduced in This Module

5V logic: A logic device that operates on +5 VDC. It has logic levels in which a logic-0 is close to 0 V and a logic-1 is close to 5 V.

AND gate: A digital gate (see *gate*) whose output is logic-1 only when both of its inputs are logic-1.

BCD number: A decimal digit represented by a four-bit binary number in the range of 0000 to 1001; stands for binary coded decimal.

Binary number: A number made up of digits that are either 0 or 1; also known as base 2.

Bit: A single logic-0 or logic-1 value. The term is a contraction of the words "binary digit".

Buffer gate: A digital gate with an output that can deliver more current or a different voltage than a normal gate. Logically, it doesn't change the signal in any way, so the output equals the input.

Clock: A signal that oscillates between logic-0 and logic-1 at a particular frequency. It is used to synchronize digital circuits with each other.

Combination logic: A digital circuit made from a collection of gates.

Counter: A digital circuit that stores a value that gets increased or decreased in response to an external signal.

Decoder: A digital circuit that outputs a unique pattern in response to each input combination.

Edge-triggered: A digital circuit that performs an operation when it detects a change in logic level.

Flip-flop: A digital memory circuit capable of storing a single logical value (one bit).

Gate: A digital device whose output is the result of a logical operation performed on its inputs.

Hexadecimal number: A number made up of the digits 0–9 and A–F; also known as base 16.

Level-triggered: A digital circuit that performs an operation when it detects a specific logic level.

Logic family: A collection of digital circuit components that are compatible with each other.

Logic level: The voltages that represent logic-0 and logic-1 for a particular digital logic family.

NAND gate: An AND gate with an inverted output.

Nanosecond (ns): One billionth of a second (1 ns = 1/1,000,000,000 s).

Negative-going: A logic signal that's changing from logic-1 to logic-0.

NOR gate: An OR gate with an inverted output.

NOT gate: A digital gate that outputs the logical opposite of its input (also known as an inverter).

OR gate: A digital gate whose output is logic-1 if either or both of its inputs is logic-1.

Positive-going: A logic signal that's changing from logic-0 to logic-1.

Propagation delay: A delay caused by the circuits inside a digital device. It is the time interval between the moment when the inputs change and the moment when the output responds.

Register: A digital memory circuit made from flip-flops and capable of storing a single word (see *word*) of information.

Timing diagram: A diagram that shows the behavior of a digital circuit's signals with respect to time.

Truth table: A table that lists a digital circuit's output values for every possible input combination.

Word: A group of bits acting together as a single unit. Words are usually (but not always) multiples of four bits.

XOR gate: A digital gate whose output is logic-1 if exactly one of its inputs is logic-1.

XNOR gate: An XOR gate with an inverted output.

Additional Resources

This module provides thorough resources for task training. The following reference material is recommended for further study.

Digital Fundamentals, Thomas L. Floyd. Eleventh Edition. 2015. Upper Saddle River, NJ: Prentice Hall.

Figure Credits

©iStockphoto.com/Henrik5000, Module Opener

Section Review Answer Key

Answer	Section Reference	Objective
Section One		
1. b	1.1.2	1a
2. c	1.1.3	1a
3. a	1.2.1	1b
4. d	1.2.4	1b
5. c	1.3.1	1c
6. b	1.4.3	1d
Section Two		
1. c	2.1.0	2a
2. b	2.1.1	2a
3. d	2.2.1	2b
4. a	2.2.4	2b
5. b	2.3.0	2c
Section Three		
1. b	3.1.1; Table 14	3a
2. d	3.1.2	3a
3. a	3.2.1	3b
4. b	3.3.1	3c
Section Four		
1. b	4.1.2	4a
2. c	4.1.3	4a
3. b	4.2.1	4b

NCCER CURRICULA — USER UPDATE

NCCER makes every effort to keep its textbooks up-to-date and free of technical errors. We appreciate your help in this process. If you find an error, a typographical mistake, or an inaccuracy in NCCER's curricula, please fill out this form (or a photocopy), or complete the online form at **www.nccer.org/olf**. Be sure to include the exact module ID number, page number, a detailed description, and your recommended correction. Your input will be brought to the attention of the Authoring Team. Thank you for your assistance.

Instructors – If you have an idea for improving this textbook, or have found that additional materials were necessary to teach this module effectively, please let us know so that we may present your suggestions to the Authoring Team.

NCCER Product Development and Revision
13614 Progress Blvd., Alachua, FL 32615

Email: curriculum@nccer.org
Online: www.nccer.org/olf

❏ Trainee Guide ❏ Lesson Plans ❏ Exam ❏ PowerPoints Other _____

Craft / Level: _____ Copyright Date: _____

Module ID Number / Title: _____

Section Number(s): _____

Description: _____

_____

_____

_____

_____

Recommended Correction: _____

_____

_____

_____

_____

Your Name: _____

Address: _____

Email: _____ Phone: _____

12406-16
Programmable Logic Controllers

Overview

Modern PLCs are powerful, flexible devices capable of interacting with thousands of inputs and outputs. Using sophisticated networking technology, they communicate over distances, controlling widely separated processes. This module describes the function and purpose of a PLC, including architecture, hardware, software and firmware, addressing and networking, and the differences between PLC programming languages. It explains how to perform basic PLC programming and how to install a PLC. A general overview of input/output (I/O) modules as well as PLC number systems is also included.

Module Five

Trainees with successful module completions may be eligible for credentialing through the NCCER Registry. To learn more, go to **www.nccer.org** or contact us at 1.888.622.3720. Our website has information on the latest product releases and training, as well as online versions of our *Cornerstone* magazine and Pearson's product catalog.

Your feedback is welcome. You may email your comments to **curriculum@nccer.org**, send general comments and inquiries to **info@nccer.org**, or fill in the User Update form at the back of this module.

This information is general in nature and intended for training purposes only. Actual performance of activities described in this manual requires compliance with all applicable operating, service, maintenance, and safety procedures under the direction of qualified personnel. References in this manual to patented or proprietary devices do not constitute a recommendation of their use.

Copyright © 2016 by NCCER, Alachua, FL 32615, and published by Pearson Education, Inc., New York, NY 10013. All rights reserved. Printed in the United States of America. This publication is protected by Copyright, and permission should be obtained from NCCER prior to any prohibited reproduction, storage in a retrieval system, or transmission in any form or by any means, electronic, mechanical, photocopying, recording, or likewise. To obtain permission(s) to use material from this work, please submit a written request to NCCER Product Development, 13614 Progress Blvd., Alachua, FL 32615.

From *Instrumentation Level Four, Trainee Guide*, Third Edition. NCCER.
Copyright © 2016 by NCCER. Published by Pearson Education. All rights reserved.

12406-16
PROGRAMMABLE LOGIC CONTROLLERS

Objectives

When you have completed this module, you will be able to do the following:

1. Define and describe PLCs and compare them to hardwired systems.
    a. Define and describe basic PLCs and systems.
    b. Compare hardwired systems to PLC systems.
2. Describe the various number systems that correspond with the digital operation of PLCs.
    a. Describe the binary number system.
    b. Describe the hexadecimal number system.
    c. Define and describe binary coding.
3. Describe and explain the function of various PLC hardware components.
    a. Describe typical power supplies.
    b. Describe and explain the operation of processors.
    c. Describe and explain the operation of I/O and communications modules.
4. Describe PLC programming concepts.
    a. Identify various programming languages used to program PLCs.
    b. Explain how ladder diagramming is used and identify the six related categories of instructions.
    c. State typical guidelines for PLC programming and installation.

Performance Tasks

Under the supervision of the instructor, you should be able to do the following:

1. Given an instructor-provided PLC diagram, identify the basic components in a PLC system.
2. Given an instructor-provided ladder diagram program, point out commonly used symbols and their meaning.
3. Implement a simple logic circuit using an instructor-provided PLC platform or simulator.

Trade Terms

Address
Analog-to-digital converter (ADC)
Control code
Data table
Digital-to-analog converter (DAC)
Encoded
Ethernet
Human-machine interface (HMI)
Input/output (I/O)
Interrupt
Isolated
Language
Memory map
Modules
Motion encoder
Nonvolatile memory
Real-time clock
Ruggedized
Scan
Serial network
Simulator
Universal serial bus (USB)
Volatile memory
Wi-Fi

Industry Recognized Credentials

If you are training through an NCCER-accredited sponsor, you may be eligible for credentials from NCCER's Registry. The ID number for this module is 12406-16. Note that this module may have been used in other NCCER curricula and may apply to other level completions. Contact NCCER's Registry at 888.622.3720 or go to **www.nccer.org** for more information.

Contents

- 1.0.0 PLC Basics 1
  - 1.1.0 Typical PLC Systems 1
    - 1.1.1 Example PLC Systems 1
  - 1.2.0 Hardwired Systems versus PLC Systems 2
    - 1.2.1 Hardwired Circuit 3
    - 1.2.2 PLC Controlled Circuit 3
    - 1.2.3 Comparison of Hardwired and PLC Systems 3
- 2.0.0 PLC Number Systems 7
  - 2.1.0 Binary Numbers 7
    - 2.1.1 Binary to Decimal Conversion 7
    - 2.1.2 Decimal to Binary Conversion 8
  - 2.2.0 Hexadecimal Numbers 9
    - 2.2.1 Hexadecimal/Binary Conversions 9
  - 2.3.0 Binary Codes 9
    - 2.3.1 ASCII 9
    - 2.3.2 Binary Coded Decimal 11
    - 2.3.3 Gray Code 11
- 3.0.0 PLC Hardware 13
  - 3.1.0 Power Supplies and Grounds 13
  - 3.2.0 Processor Modules 14
    - 3.2.1 Processor Memory 14
    - 3.2.2 Processor Scans 15
  - 3.3.0 I/O and Communications Modules 16
    - 3.3.1 I/O Modules 16
    - 3.3.2 Discrete Modules 16
    - 3.3.3 Numerical Data Interface Modules 17
    - 3.3.4 Analog Modules 17
    - 3.3.5 Special Modules 17
    - 3.3.6 Communications Modules 19
    - 3.3.7 Module Addressing 19
- 4.0.0 PLC Programming 21
  - 4.1.0 PLC Programming Languages 21
    - 4.1.1 Ladder Diagram (LD) 21
    - 4.1.2 Function Block Diagram (FBD) 21
    - 4.1.3 Sequential Function Chart (SFC) 21
    - 4.1.4 Structured Text (ST) 22
    - 4.1.5 Instruction List (IL) 22
    - 4.1.6 Development Tools 22
  - 4.2.0 Ladder Diagram Programming 22
    - 4.2.1 Relay 22
    - 4.2.2 Timer and Counter 23
    - 4.2.3 Arithmetic 25
    - 4.2.4 Data Comparison 26
    - 4.2.5 Data Transfer 26
    - 4.2.6 Program Control 26

4.3.0	Guidelines for Programming and Installation	28
4.3.1	Programming	28
4.3.2	Installation	28
4.3.3	I/O Wiring	29
4.3.4	Dynamic System Checkout	30

Figures and Tables

Figure 1	A micro PLC designed for small-scale control	1
Figure 2	A medium-scale PLC suitable for many typical applications	2
Figure 3	A large-scale PLC suitable for I/O-intense applications	2
Figure 4	Examples of HMIs	2
Figure 5	Hardwired switch/lamp circuit	3
Figure 6	PLC switch/lamp circuit	3
Figure 7	Process vat control system	4
Figure 8	Hardwired vat control	4
Figure 9	PLC vat control	4
Figure 10	Hardwired system changes	5
Figure 11	PLC system changes	5
Figure 12	PLC power supply	14
Figure 13	PLC processor with front panel controls and ports	14
Figure 14	Discrete input circuit	16
Figure 15	Input module connection diagram	17
Figure 16	Discrete output circuit	17
Figure 17	Output module connection diagram	18
Figure 18	Analog input module connection diagram	18
Figure 19	A serial communications module with multiple ports	19
Figure 20	Input contact symbols	23
Figure 21	Output coil symbols	23
Figure 22	Logic flow for NC and NO contacts	24
Figure 23	Latching/unlatching coil operation	24
Figure 24	Timers	25
Figure 25	Counter application	25
Figure 26	Arithmetic operations	25
Figure 27	Examples of data comparison instructions	26
Figure 28	BTD transfer operation	27
Figure 29	MOV and MVM transfer operations	27
Figure 30	Typical MCR instructions	28
Table 1	Hexadecimal and Binary Number Equivalents	9
Table 2	The ASCII Character Set	10
Table 3	Decimal to BCD Conversion Values	11
Table 4	Gray Code and Standard Binary Values	11
Table 5	Programming Guidelines	28

Section One

1.0.0 PLC Basics

Objective

Define and describe PLCs and compare them to hardwired systems.
a. Define and describe basic PLCs and systems.
b. Compare hardwired systems to PLC systems.

Trade Terms

Human-machine interface (HMI): Any device that allows a human to control and monitor the status of a computer or PLC.

Input/output (I/O): The component of a computer or PLC that allows it to interact with the outside world.

Isolated: Connected in such a way that a control signal does not directly interact at the electrical level with a device.

Modules: Components in a PLC system that provide particular features, such as I/O or communications.

Ruggedized: Designed to work reliably in a potentially harsh environment, such as one with unusual temperatures, electrical noise, or physical risk.

Universal serial bus (USB): A communications system built into many computers, PLCs, and portable electronic devices to allow simple, high-speed cabled connections.

Many of the PLCs currently available are simply called *programmable controllers*. This is because they perform many more functions than the relay logic functions of the original PLCs. Using special input/output (I/O) modules, they can handle motion control, drive control, and specialized process control, all of which is programmed using intuitive tools running on a notebook computer or other portable electronic device.

1.1.0 Typical PLC Systems

A programmable logic controller is a microprocessor-controlled device similar to a computer that can be programmed by the user to perform specific tasks. Unlike an ordinary personal computer, however, a PLC is ruggedized and suitable for an industrial setting. The two basic components of a PLC are the hardware and the software. The hardware includes all the physical components that make up the PLC system. The software controls the hardware, providing the intelligence and decision-making that ultimately controls industrial processes.

The term *architecture* is usually associated with building designs, but in PLCs, it refers to how the whole system is put together. Engineers or senior technicians create PLC system designs to solve particular problems. The following are a few of the things that they consider when designing and building a PLC-based control system:

- The PLC location (in the field or in a rack located in an office or dedicated room)
- The kind of power supply and backup power system required
- The type of processor and I/O modules required
- The software that will be used to program the system
- The kind of network that will handle communications among system components
- The type(s) of human-machine interface (HMI) used to interact with operators

1.1.1 Example PLC Systems

All PLC systems contain essentially the same building blocks, but a variety of scales are possible. *Figure 1*, *Figure 2*, and *Figure 3* show scales ranging from a very small system to a fairly large one. *Figure 1* shows a micro PLC that is a complete standalone unit. It includes a small number of I/O connections, both digital and analog. It can be programmed via a universal serial bus (USB) cable from a standard computer. The built-in LCD panel allows program adjustment and configuration. PLCs of this type are common in situations where some intelligence is required but a large amount of I/O is not.

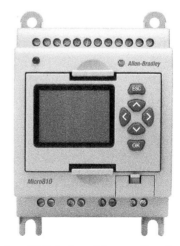

Figure 1 A micro PLC designed for small-scale control.

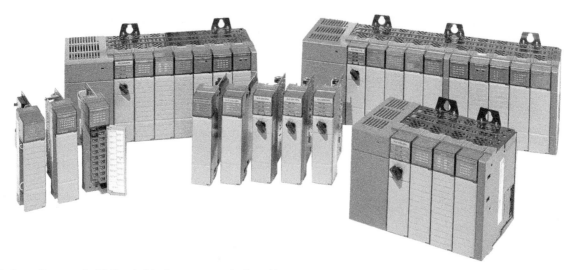

Figure 2 A medium-scale PLC suitable for many typical applications.

A typical medium-capacity PLC system is shown in *Figure 2*. The particular system is rack-mounted and can be expanded with up to several thousand I/O connections of various types. It can run fairly large programs and connects to other devices using several of the most common industrial networking systems. PLCs of this type are extremely useful and offer enough expansion capability to grow with many typical applications. *Figure 3* shows a PLC for a large-capacity, heavy-duty system. This unit has enormous expansion capability, handling tens of thousands of I/O connections. It has a large memory and so can run sophisticated control programs. The basic chassis can connect to numerous expansion chassis, each filled with modules. PLCs of this type can control the largest of processes.

Figure 3 A large-scale PLC suitable for I/O-intense applications.

Figure 4 shows two examples of HMIs, one of which is very large and sophisticated and one that is very simple. Buttons, switches, and lights wired on a panel or control box can function as HMIs as well. Each HMI must be tailored for the specific application. They must also be designed with ordinary operators in mind rather than experienced technicians and programmers. A good HMI requires very little instruction to be used effectively.

1.2.0 Hardwired Systems versus PLC Systems

The primary hardware components of a PLC are the power supply, the processor, I/O modules, and communications modules. A personal computer is often used to program the processor, but this is not considered a primary component because once the processor is programmed, the computer is disconnected.

Figure 4 Examples of HMIs.

The relationship between the processor and the I/O modules can best be seen by comparing a hardwired circuit to an identical circuit that is PLC controlled. While the example provided here is rather simple, it illustrates the key ideas, many of which scale up to very large systems.

1.2.1 Hardwired Circuit

Figure 5 is a simplified ladder diagram of a hardwired circuit used to control two lamps. Switch 1 and Switch 2 are normally open (NO) pushbutton switches. Switch 1 sends power to Lamp 1, and Switch 2 sends power to Lamp 2. When Switch 1 is closed, Lamp 1 will light. When Switch 2 is closed, Lamp 2 will light.

1.2.2 PLC Controlled Circuit

Figure 6 shows the same components connected to a PLC. It illustrates several aspects which differ from the hardwired circuit. As you can see, the switches are not connected directly to the lamps. Instead, they are connected to an input module. The lamps are connected to an output module. The input modules and the output modules do not communicate with each other directly; each is controlled by the processor.

The processor is programmed to control Lamp 1 with Switch 1 and Lamp 2 with Switch 2. This is accomplished by writing a program with a PLC development system and uploading the program into the PLC processor. The program looks very similar to a standard electrical ladder diagram.

1.2.3 Comparison of Hardwired and PLC Systems

In the hardwired system shown in *Figure 5*, the electrons flow from the power source, through the switch, to the correct indicator lamp. When the switch is opened, power is interrupted and the light goes out.

In the PLC controlled system shown in *Figure 6*, electrical power comes from the power source, flows through the switch, and enters the input module. The input module senses the presence of this voltage and, in turn, sends a signal to the processor indicating the state of that particular input.

The voltage from the switch does not come in direct contact with the processor. In other words, it is electrically isolated. This isolation is necessary due to the fact that signals coming from an industrial environment can be hostile to the processor. Many PLC input modules use devices called *optocouplers* to separate the inputs from the processor. These devices can protect the processor from voltages up to several hundred or even several thousand volts.

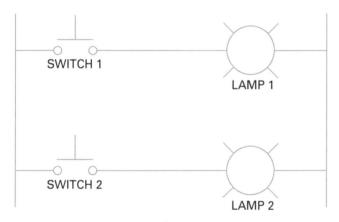

Figure 5 Hardwired switch/lamp circuit.

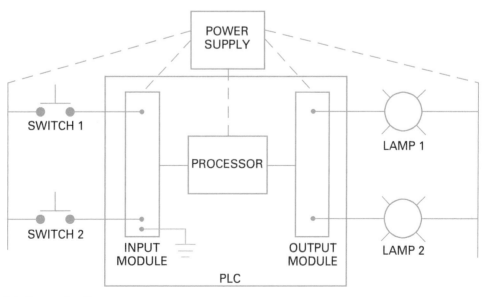

Figure 6 PLC switch/lamp circuit.

12406-16 Programmable Logic Controllers   Module Five 3

Under program control, the processor now needs to turn on the lamp connected to the appropriate output. The processor sends a signal to the output module instructing it to turn on a specific output. The output module usually isolates the controlled device from the processor by using a relay or solid-state control device to perform the actual switching action. When Switch 2 is activated, the processor completes a similar action, but this time the signal is sent to the output module terminal for Lamp 2.

An observer of both the hardwired and PLC controlled systems would not notice any difference in system operation. In both systems, Switch 1 controls Lamp 1, and Switch 2 controls Lamp 2.

The advantage of the PLC becomes evident when circuit changes are required. For example, in order to change the circuits of a hardwired system to have Switch 1 control Lamp 2 and Switch 2 control Lamp 1, it would take several minutes to rewire them and would involve moving wires. With a PLC, a simple program edit makes the change with no rewiring needed.

This simple example illustrates the basic differences between a hardwired and a PLC system. The clear advantage of the PLC will be much more apparent when applied to a larger system that has many inputs and outputs.

A practical application demonstrating the flexibility of a PLC program can be seen in *Figure 7*. In this system, a vat is filled with liquid. When temperature and pressure conditions are met, a motor comes on and operates a stirrer to mix the liquid.

Figure 8 shows the hardwired method used to control this system. A pressure switch and a temperature switch are hardwired into the system with a manual override installed.

Figure 9 shows the way the circuit would look if it were programmed into a PLC. The only hardwired connections are the temperature, pressure, and override switch inputs as well as the mixer motor output. Each input and output has a unique identifier that the program uses when it wishes to read or write the inputs or outputs. The actual identification system varies with different brands of PLCs.

It is very easy to change the circuit setup in the PLC without ever moving a wire connection. *Figure 10* shows how a traditional circuit would have to be rewired in order to make temperature a critical path for the motor to work. As you can see, the wiring must be physically changed, which could involve extensive work, depending on its location.

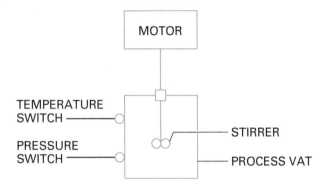

Figure 7 Process vat control system.

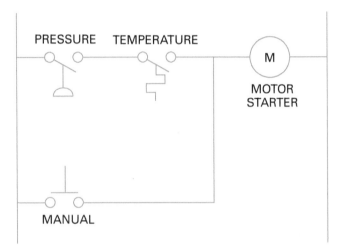

Figure 8 Hardwired vat control.

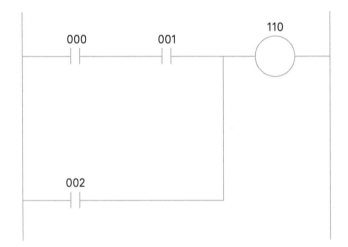

Figure 9 PLC vat control.

Figure 11 shows how the PLC-controlled circuit can be reconfigured to perform the same function. As you can see, there is no need to physically touch the wiring. All that is required is a simple programming change.

Smart HMIs

In recent years, PLC manufacturers have introduced a new and relatively inexpensive way to implement sophisticated HMIs—tablet computers and smartphones. Based around apps running on off-the-shelf portable electronic devices, these HMIs communicate over wireless networks and allow convenient interaction with control systems. A PLC technician can handle an emergency from home simply by using a smartphone to access the PLC. While traditional HMIs aren't going to disappear from the factory floor, portable ones have added a new dimension to PLC flexibility and control.

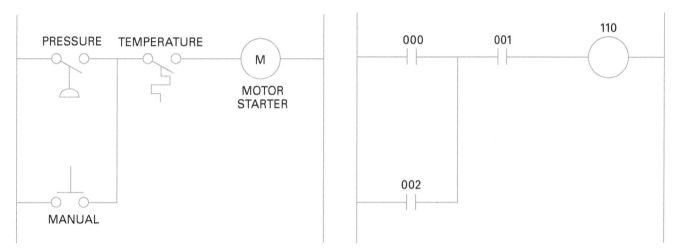

Figure 10 Hardwired system changes.

Figure 11 PLC system changes.

Additional Resources

Programmable Controllers, Thomas A. Hughes. Fourth Edition. 2004. Research Triangle Park, NC: International Society of Automation.

Programmable Logic Controllers, W. Bolton. Sixth Edition. 2015. Oxford, UK: Newnes/Elsvier.

The following websites offer resources for products and training:

Rockwell Automation, Inc., **ab.rockwellautomation.com**

AutomationDirect, **www.automationdirect.com**

1.0.0 Section Review

1. A PLC interacts with the outside world through its _____.
   a. processor
   b. power supply
   c. I/O modules
   d. programming tools

2. How does a PLC differ from an ordinary general-purpose computer?
   a. It is faster.
   b. It has more memory.
   c. It has a graphic interface.
   d. It is ruggedized.

3. The biggest difference between a large PLC and a micro PLC is the _____.
   a. power supply
   b. rack mounting
   c. number of I/O connections
   d. programming system

4. PLCs directly interact with their operators through _____.
   a. an HMI
   b. the control logic
   c. their processors
   d. a network module

5. A hardwired system controls outputs directly with its inputs. How does a PLC use its inputs to control its outputs?
   a. Through special internal wiring
   b. Through its processor
   c. With an optoisolator
   d. Using relays

Section Two

2.0.0 PLC Number Systems

Objective

Describe the various number systems that correspond with the digital operation of PLCs.
 a. Describe the binary number system.
 b. Describe the hexadecimal number system.
 c. Define and describe binary coding.

Trade Terms

Control code: A character that when received causes a device to modify its behavior in some way.

Encoded: The process of representing one kind of information as another. For example, a character encoding system represents each character with an equivalent number.

Motion encoder: An electromechanical device that converts a rotary or linear motion into a number representing the current position.

Since PLCs are microprocessor-based devices, they work with numbers in the same way as any digital circuit. While most of what technicians do with a PLC involves ordinary decimal numbers, it's useful to have some familiarity with the numbering systems that PLCs use internally. In some cases, devices that connect to a PLC communicate with it or display their results in a non-decimal format. Binary, hexadecimal, and BCD numbers are covered in another module. This section explains a bit more about how these numbering systems work.

2.1.0 Binary Numbers

Binary (base 2) uses just two symbols—0 and 1—rather than the ten used in the decimal system (0 through 9). Like any number system, as the number gets larger, more columns (or digits) are added. For example, once you reach the number 9 in decimal, you have to add another column to move to the next value, 10. Once you reach 99, you have to add another column to move to 100. Since binary numbers have fewer symbols, new columns are added fairly frequently at first.

Each column in a decimal number is ten times the "weight" of the one before it. These columns are also known as *places*. Essentially, the first column (on the right) could be called the "ones" column, the second the "tens", the third the "hundreds", and so on. Each column added to the left is an increasing power of ten. If a number is in the third column (hundreds), it is said to have a place value in the hundreds place.

10^3	10^2	10^1	10^0
1,000s	100s	10s	1s

With binary, each column added to the left is an increasing power of two.

2^7	2^6	2^5	2^4	2^3	2^2	2^1	2^0
128s	64s	32s	16s	8s	4s	2s	1s

For larger binary numbers, continue to add columns as needed. Each additional column is the next power of two greater. Even more simply, each new column is double the value of the previous one.

Understanding this basic principle, which is true for all number systems, is the key to converting numbers back and forth between bases. It is critical when dealing with multiple bases to be clear about which base(s) you're using. When more than one base is in use, it's standard practice to place a subscript that identifies the base at the end of the number. For example, the number 946_{10} is the number 946 in decimal. Similarly, $105A_{16}$ is the number 105A in hexadecimal, and 10100101_2 is the number 10100101 in binary.

2.1.1 Binary to Decimal Conversion

To convert between bases in most real-world situations, it's best to use a scientific calculator, smartphone app, or online conversion program. However, these tools may not always be available, so it's important to know how manual conversions work.

Converting from binary to decimal is fairly simple. If a 1 is present in a given position, the column value for that position is added to a running sum. If a 0 is present, ignore that column value and move on to the next position. In other words, to find the decimal equivalent of a binary number, multiply each digit by its place value's corresponding power of two, and find the sum of the results. For example, to convert the binary number 101101_2 to its decimal equivalent, use the following process:

$101101_2 =$
$(1 \times 2^5) + (0 \times 2^4) + (1 \times 2^3) + (1 \times 2^2) + (0 \times 2^1) + (1 \times 2^0) =$
$(1 \times 32) + 0 + (1 \times 8) + (1 \times 4) + 0 + (1 \times 1) =$
$32 + 0 + 8 + 4 + 0 + 1 = 45_{10}$

2.1.2 Decimal to Binary Conversion

There are two methods for converting decimal numbers to binary numbers: the subtractive method and the division method. These two methods are completely different, but each will give the same result.

To convert decimal to binary using the subtractive method, start by determining the highest possible power of two that's less than or equal to the decimal number, and subtract it from the original number. Write down a 1. If you can subtract the next power of two without producing a negative number, do so and write down a 1. If you can't subtract the next value without a negative result, write down a 0 and move on to the next lowest power of two. Repeat this subtraction process on the result of each subtraction until there is nothing left.

For example, to convert the number 47_{10} to binary using the subtractive method, use the following steps:

Step 1 Determine the largest power of two that is still less than or equal to 47. It is 32 (which is 2^5), because the next power of two (64, which is 2^6) is larger than 47.

Step 2 Subtract 32 from 47, and write down a 1. (This will be in the 32s place of the binary number.) The result is 15.

Step 3 Go to the next lowest power of two (in this case, 16), and subtract it from the result in the previous step. Since 16 cannot be subtracted from 15 without resulting in a negative number, write down a zero in the 16s place of the binary number, and carry the 15 to the next step.

Step 4 Subtract the next lowest power of two (8) from 15. The result is 7; because this is a positive number, write a one in the 8s place.

Step 5 Repeat this process until you obtain a result of zero. The ones and zeroes you recorded are the binary number. In this example, 47_{10} converts to 101111_2.

This process is further illustrated as follows; the left hand column shows the place value (power of two), the middle column shows the process of subtraction, and the right hand column shows the recorded binary digits for that power of two:

Power of 2	Subtract	Binary Digits
$2^5 = 32$	$47 - 32 = 15$	1
$2^4 = 16$	(Can't subtract 16)	0
$2^3 = 8$	$15 - 8 = 7$	1
$2^2 = 4$	$7 - 4 = 3$	1
$2^1 = 2$	$3 - 2 = 1$	1
$2^0 = 1$	$1 - 1 = 0$	1

The digits recorded on the right, when read from the *first* to *last* (top to bottom), are the resulting binary number (101111_2).

To convert decimal to binary using the division method, divide the decimal number by two and record the remainder (which will be a 0 or a 1). Then, divide the result by two and record the remainder. Repeat this process until you reach zero. The recorded digits, when arranged from the *last* one recorded to the *first*, are the resulting binary number.

For example, to convert the decimal number 351_{10} to binary using the division method, proceed as follows; remember that the final binary number will be the remainders arranged from last to first (bottom to top in the right hand column):

Divide by 2	Record Remainders
$351 \div 2 = 175, R1$	1
$175 \div 2 = 87, R1$	1
$87 \div 2 = 43, R1$	1
$43 \div 2 = 21, R1$	1
$21 \div 2 = 10, R1$	1
$10 \div 2 = 5, R0$	0
$5 \div 2 = 2, R1$	1
$2 \div 2 = 1, R0$	0
$1 \div 2 = 0, R1$	1

Arrange the remainders you recorded from last to first to find the binary number:

$$351_{10} = 101011111_2$$

2.2.0 Hexadecimal Numbers

Hexadecimal (base 16) uses sixteen digits (0–9 and A–F) to represent numbers. Each column in a hexadecimal number is a power of 16. Digital systems often use hexadecimal because groups of four binary bits convert easily into a single hexadecimal digit. Hexadecimal numbers are significantly easier to work with, since they look more like a "normal" number and are shorter than binary. *Table 1* shows the relationship of each hexadecimal digit to a four-bit binary equivalent.

2.2.1 Hexadecimal/Binary Conversions

To convert a binary number into hexadecimal, divide it into groups of four bits, starting from the right side. If the number of bits isn't evenly divisible by four, add sufficient leading zeroes to the left side of the binary number to make it so.

Next, using *Table 1*, convert each group of four bits into the equivalent hexadecimal digit. After a while, you'll have these memorized and won't have to use the table.

For example, to convert the binary number 1010011011100010_2 to hexadecimal, proceed as follows:

$$1010\ 0110\ 1110\ 0010$$
$$1010 = A$$
$$0110 = 6$$
$$1110 = E$$
$$0010 = 2$$
$$1010011011100010_2 = A6E2_{16}$$

To convert a hexadecimal number to binary, use *Table 1* to convert each hexadecimal digit to its four-bit binary equivalent. (Do not drop leading zeroes.)

For example, to convert the hexadecimal number $E59C_{16}$ to binary, proceed as follows:

$$E = 1110$$
$$5 = 0101$$
$$9 = 1001$$
$$C = 1100$$
$$E59C_{16} = 1110010110011100_2$$

2.3.0 Binary Codes

All microprocessor devices need to be able to communicate with other electronic devices as well as with humans. In most cases, PLCs communicate with other devices in binary, although there are a few exceptions. But human communication is easier if internal binary data is encoded into a form that's more human-accessible. There are a number of industry-wide standards for binary coding; the most common are ASCII, binary coded decimal (BCD), and Gray code.

Table 1. Hexadecimal and Binary Number Equivalents

Hexadecimal	Binary
0	0000
1	0001
2	0010
3	0011
4	0100
5	0101
6	0110
7	0111
8	1000
9	1001
A	1010
B	1011
C	1100
D	1101
E	1110
F	1111

2.3.1 ASCII

One of the most useful ways to enhance human-computer communication is to be able to work with letters, numbers, and symbols—in other words, text. While computers internally handle all information as binary numbers, it is helpful if HMI devices can work with text. For this reason, schemes that represent text as binary numbers were developed in the early days of computers. One of the oldest and still most commonly used is ASCII.

ASCII stands for American Standard Code for Information Interchange. It is a seven or eight-bit binary code in which each number represents a standard character of some kind. Early ASCII systems used seven bits, giving a total of 128 possible characters. Not all of these corresponded to visible characters. Instead, some acted as a control code that modified the behavior of certain types of devices. Almost all of the control codes defined in ASCII pertain to long-obsolete equipment and no longer perform their original functions. Many computer manufacturers later added an eighth bit to the ASCII standard, which made an additional 128 characters available. Many of these were used for graphics and foreign language alphabets.

Today, virtually all devices that communicate in ASCII use the eight-bit "extended" version. The first 32 codes are control codes. Some HMI devices modify their behavior when they receive one of these. For example, a simple LCD text display might switch to bold text when it receives the code 00000010_2 (2_{10}), which is called CTRL-B. The next 95 codes include the letters of the alphabet, numbers, and various symbols.

For example, 01000001_2 (65_{10}) is the code for the letter A. The code 01111111_2 (127_{10}) is called DEL, which stands for delete. The backspace key on many keyboards sends this code when pressed. The final 128 codes are not standardized and can represent a wide variety of symbols. *Table 2* shows the first 128 characters of the ASCII character set. A device that communicates in ASCII sends or receives data in binary form. The device translates into or out of ASCII as required. For example, if you hit the space bar on an ASCII keyboard, it will transmit a 00100000_2 (32_{10}), which is the code for space. Similarly, if you send a 00100100_2 (36_{10}) to an LCD text panel, it will display the $ character.

Table 2 The ASCII Character Set

Number	Character	Number	Character	Number	Character	Number	Character
0	CTRL-@	32	SPACE	64	@	96	`
1	CTRL-A	33	!	65	A	97	a
2	CTRL-B	34	"	66	B	98	b
3	CTRL-C	35	#	67	C	99	c
4	CTRL-D	36	$	68	D	100	d
5	CTRL-E	37	%	69	E	101	e
6	CTRL-F	38	&	70	F	102	f
7	CTRL-G	39	'	71	G	103	g
8	CTRL-H	40	(	72	H	104	h
9	CTRL-I	41	)	73	I	105	i
10	CTRL-J	42	*	74	J	106	j
11	CTRL-K	43	+	75	K	107	k
12	CTRL-L	44	,	76	L	108	l
13	CTRL-M	45	-	77	M	109	m
14	CTRL-N	46	.	78	N	110	n
15	CTRL-O	47	/	79	O	111	o
16	CTRL-P	48	0	80	P	112	p
17	CTRL-Q	49	1	81	Q	113	q
18	CTRL-R	50	2	82	R	114	r
19	CTRL-S	51	3	83	S	115	s
20	CTRL-T	52	4	84	T	116	t
21	CTRL-U	53	5	85	U	117	u
22	CTRL-V	54	6	86	V	118	v
23	CTRL-W	55	7	87	W	119	w
24	CTRL-X	56	8	88	X	120	x
25	CTRL-Y	57	9	89	Y	121	y
26	CTRL-Z	58	:	90	Z	122	z
27	ESC	59	;	91	[	123	{
28	CTRL-\	60	<	92	\	124	|
29	CTRL-]	61	=	93	]	125	}
30	CTRL-^	62	>	94	^	126	~
31	CTRL-_	63	?	95	_	127	DEL

2.3.2 Binary Coded Decimal

Binary coded decimal (BCD) is a way of representing decimal numbers in binary form. To convert a number from decimal to BCD, each decimal digit is simply turned into its equivalent four-bit binary number. To convert a BCD number back to decimal, it must be broken up into four-bit groups. Each is then converted back to its decimal equivalent. *Table 3* shows the decimal digits and their four-bit binary equivalents. For example, the decimal number 127_{10} equals the BCD number 000100100111_{BCD}. Similarly, the BCD number 1001100001110110_{BCD} equals the decimal number 9876_{10}.

Be aware that BCD numbers are not true binary numbers! If you plug a BCD number into a binary-to-decimal conversion app, you won't get the correct translation. For example, if you treat the BCD version of 127_{10} as a true binary number, you will get a decimal value of 295_{10} back when you convert it. Even though true binary numbers and BCD numbers look much the same, they are different. It's crucial that you know what you're working with before you attempt any conversion or interpretation.

2.3.3 Gray Code

Gray code is a modified binary code used by rotational or linear motion encoders to convert motion positions to a digital value. A common application of a rotational encoder is to provide angular position information for rotating shafts. These encoders are mechanical devices that output Gray code values as the rotational or linear motion occurs. Gray code was developed to reduce errors when transmitting data from mechanical encoders to devices like PLCs. When normal binary numbers change from one value to the next, more than one bit may flip at a time. For example, as you go from 0111_2 (7_{10}) to 1000_2 (8_{10}), all four bits flip. In a mechanical encoder, the bits may not change at precisely the same instant, potentially leading to erroneous results if a PLC reads them at the wrong moment.

Numbers in a Gray code sequence never have more than one bit flip at a time as numbers increase or decrease. *Table 4* shows a sixteen-value Gray code compared to a standard binary code. If a PLC is connected to a Gray code device, it generally includes a translation routine in its program to convert Gray code values back to "normal" numbers. Some special modules interface to Gray code devices and perform the translation automatically.

Table 3 Decimal to BCD Conversion Values

Decimal	BCD
0	0000
1	0001
2	0010
3	0011
4	0100
5	0101
6	0110
7	0111
8	1000
9	1001

Table 4 Gray Code and Standard Binary Values

Gray Code	Standard Binary
0000	0000
0001	0001
0011	0010
0010	0011
0110	0100
0111	0101
0101	0110
0100	0111
1100	1000
1101	1001
1111	1010
1110	1011
1010	1100
1011	1101
1001	1110
1000	1111

Additional Resources

Programmable Controllers, Thomas A. Hughes. Fourth Edition. 2004. Research Triangle Park, NC: International Society of Automation.

Programmable Logic Controllers, W. Bolton. Sixth Edition. 2015. Oxford, UK: Newnes/Elsvier.

The following websites offer resources for products and training:

Rockwell Automation, Inc., **ab.rockwellautomation.com**

AutomationDirect, **www.automationdirect.com**

2.0.0 Section Review

1. What is the value (weight) of the fourth column of a binary number?
    a. 2
    b. 8
    c. 16
    d. 18

2. Convert the number 01011101_2 to decimal.
    a. 68_{10}
    b. 84_{10}
    c. 93_{10}
    d. 186_{10}

3. Convert the number 39_{10} to binary.
    a. 100111_2
    b. 100100_2
    c. 111001_2
    d. 001001_2

4. Convert the number $7F_{16}$ to binary.
    a. 11111111_2
    b. 01111111_2
    c. 10001111_2
    d. 11110111_2

5. Characters that modify a device's behavior rather than representing text are known as _____.
    a. unprinted characters
    b. invisible codes
    c. device modifiers
    d. control codes

6. Convert the number 01110010_{BCD} to decimal.
    a. 72_{10}
    b. 73_{10}
    c. 114_{10}
    d. 115_{10}

SECTION THREE

3.0.0 PLC HARDWARE

Objective

Describe and explain the function of various PLC hardware components.
  a. Describe typical power supplies.
  b. Describe and explain the operation of processors.
  c. Describe and explain the operation of I/O and communications modules.

Performance Task

1. Given an instructor-provided PLC diagram, identify the basic components in a PLC system.

Trade Terms

Address: A unique identifier that enables a PLC program to refer to specific modules and wiring terminal positions.

Analog-to-digital converter (ADC): A circuit that converts an analog voltage or current signal into a numeric equivalent.

Data table: The section of a PLC's memory map that contains data used by the control program.

Digital-to-analog converter (DAC): A circuit that converts a number into an equivalent analog voltage or current signal.

Ethernet: Refers to a family of wired networking hardware standards with speeds ranging up to gigabits per second; a major networking standard since the 1980s.

Interrupt: An event that causes a PLC processor to stop whatever it's doing and perform a specific task, usually one that's time-critical.

Memory map: A diagram of the way in which a PLC's memory is divided up and assigned to different tasks.

Nonvolatile memory: Memory that does not require electrical power to maintain its contents. Flash memory is a typical example.

Real-time clock: A battery-backed timekeeping circuit embedded in a PLC that maintains the current time and date.

Scan: A PLC processor cycle in which inputs are read, program logic is executed, and outputs are updated.

Serial network: A broad umbrella term for various low-speed communications systems in which data travels one bit at a time between devices over a simple cabling system. Examples include RS-232, RS-422, and RS-485.

Volatile memory: Memory that requires uninterrupted electrical power to maintain its contents. Also known as RAM.

Wi-Fi: Refers to a particular family of wireless networking hardware and software standards designed for both general and special-purpose communications.

The hardware components of a PLC system include a power supply, a processor, input/output modules, and communications modules. Each of these components has a separate function that may be performed in a variety of ways and under various conditions. In the smallest PLCs, all of these components are contained in a single package that gets bolted to a panel or clamped to a rail in an equipment rack. With larger PLCs, each of these components can be a separate module. Modules are plugged together in an equipment rack making up the complete system. Very large PLCs may be made from racks of modules located in several different places and connected by a network.

3.1.0 Power Supplies and Grounds

There may be one or more power supplies associated with each PLC system. One power supply provides power to operate the processor and the circuitry internal to the attached modules. Additional power supplies may be required to provide power to operate the field devices connected to some modules.

The power supply that provides power to the processor and module circuitry can be internal to the processor module, mounted inside the rack that contains the processor and modules, or mounted external to the rack. For very small PLCs, the power supply is often internal. Regardless of location, the function of the power supply is to convert the incoming power, usually 120 VAC at 60 Hz, to a usable level for the PLC. Normally, the power supply is protected with a fuse or circuit breaker, and an indicator light shows power supply operation. *Figure 12* shows a typical power supply.

The voltages used by the electronic devices found in most modern control circuits are very

Figure 12 PLC power supply.

small (often 5 VDC or less). Static discharges from improper handling can destroy them. Static discharges or stray voltages from the process equipment, processes, or non-process activities such as welding can destroy electronic devices if they are not properly grounded and shielded. When installing or repairing PLCs and modules, make sure to read and understand their grounding, bonding, and shielding requirements before starting the work. Since different PLCs and their applications have different grounding and bonding needs, always refer to the associated manuals for guidance. When in doubt, ask the area engineer or maintenance supervisor for assistance.

3.2.0 Processor Modules

The processor module contains a microprocessor and one or more types of memory. It may also contain a real-time clock that maintains the current time and date. These are similar to their equivalents in a general-purpose personal computer. PLCs are designed with the industrial environment in mind. They are also designed to run with minimal attention, often tucked away in a cabinet or equipment rack.

PLCs don't have the same problem-resolving mechanisms as a personal computer. For example, if a notebook computer crashes, the user can reboot it and the problem is usually resolved. Most PLCs are designed to handle faults differently than a mainstream computer since they are responsible for controlling costly or hazardous processes.

Processor modules usually have front-panel lights or LCD displays to provide the user with status information about the PLC's operation. These indicators are very useful for troubleshooting. Also provided on many processors is a key switch used to change the processor mode from RUN to PROGRAM. Most processors also have a USB or serial port for connecting to a programming computer. *Figure 13* shows a typical processor module.

3.2.1 Processor Memory

In order to store and execute the control program, the processor must have sufficient memory. PLCs typically contain a mixture of volatile memory and nonvolatile memory. The program itself usually resides in nonvolatile memory so that it won't be lost if the power goes down. Volatile memory contains data and other temporary information that the PLC generates and uses as the control program runs.

A processor's memory size partly determines how big the final PLC system can become. It also influences how complex the control program can be. Very small PLCs have minimal memories designed to hold simple programs, while giant PLC processors contain fairly large memories. When purchasing a processor, consider the potential growth of the system and get one that has enough memory to handle a larger program.

The PLC's memory is organized into various areas, depending on how the processor uses the data. This organizational diagram of memory is called a memory map. Almost all PLCs have

Figure 13 PLC processor with front panel controls and ports.

different memory maps. In general, all programmable controllers must have memory allocated for the following six items:

- Executive program
- Processor work area
- Scratch pad
- Diagnostics
- Data table
- User program

The executive program, processor work area, scratch pad, and diagnostics memory spaces are all used by the processor for its own internal operations and are not accessible to the user. These spaces are similar to those occupied by the operating system of a personal computer.

The data table and user program make up the application memory space. This space stores the control program itself and any data that will be used by the processor to perform its various control functions. Each processor has a specific amount of application memory, which varies from processor to processor.

All data is stored in what is called the data table. The data table is divided into four areas: the input table, output table, internal storage bits, and storage registers.

The input table is an array of bits that reflect the status of the PLC's inputs, with each input having its own location in the data table. The data in the input table changes in response to changing input values coming from the I/O modules.

The output table is functionally similar to the input table, but its purpose is to control the operation of output devices. The processor controls the states of the bits in the output table and therefore controls the operation of the output modules.

The PLC stores various pieces of status information in the internal storage bit space. This information does not control I/O devices but is used by the program for making decisions.

Storage registers hold larger chunks of information (words). Words represent any quantity with a value that cannot be represented by a single 0 or 1 (a bit). In general, there are three types of storage registers: input registers, holding registers, and output registers. The total number varies, depending on the controller memory size and according to how the data table is configured. Values stored in the storage registers are in a binary or BCD format. Each register can generally be loaded, altered, or displayed using a connected computer or a front panel LCD, if present.

Input registers store numerical data received from devices such as thumbwheel switches, shaft encoders, and other devices that provide BCD or Gray code output. Analog signals also supply numerical data that must be stored in input registers. The current or voltage signal generated by various analog transmitters is converted by the analog interface into a number and stored in an input register.

Holding registers are required to store variable values that are program-generated by instructions (such as math, a timer, or a counter) or constant values that are entered via the programming computer or some other data entry method.

Output registers are used to provide storage for numerical or analog values that control various output devices. Typical devices that receive data from output registers are alphanumeric LED displays, chart recorders, analog meters, speed controllers, and control valves. Output registers are essentially holding registers that are used to control outputs.

3.2.2 Processor Scans

The processor module controls the system by executing the control program. During program execution, the processor reads all the inputs, evaluates and acts upon them according to the program logic, and then modifies any outputs affected. The process of reading the inputs, executing the program, and updating the outputs is known as a scan. The time required to make a single scan (called the scan time) can vary from a fraction of a millisecond to 100 milliseconds.

Scan time is a function of both the number of I/O modules that the processor is working with and the processor's basic speed. A tiny PLC may have a very fast scan time because it is working with no more than a dozen I/O connections. A PLC system that will handle tens of thousands of I/O connections needs a very powerful processor or the scan time will be unacceptably slow.

The scan is normally a continuous and sequential process of reading the status of inputs, executing the control logic, and updating the outputs. When an immediate response is needed, most programs have an interrupt capability that stops the normal flow of operation so that critical situations may be handled rapidly.

3.3.0 I/O and Communications Modules

All PLCs contain one or more modules that provide I/O capability to interact with the outside world. The modules may be built into the processor or they may be separate; functionally, it doesn't matter. Many PLCs also include a communications module that allows the processor to communicate with other PLCs or with a general-purpose network running in the building. Very small PLCs often lack this capability, but almost all medium and large PLCs have a communications module of some kind.

3.3.1 I/O Modules

The input/output system is the interface between the controlled devices and the control program running in the processor. Each module may either be an input module or an output module. The input module is designed to receive signals from the field devices. It conditions and isolates these signals before passing them on to the processor. An output module conditions and isolates a signal from the processor for use in activating or deactivating a field device. There are four general classes of I/O modules:

- Discrete
- Numerical data
- Analog
- Special

3.3.2 Discrete Modules

Discrete I/O devices are often compared to a switch that is either open or closed. Discrete devices have only two states: they are either OFF or ON. The discrete input signal may be AC or DC and of any magnitude or polarity. When the input is present, a voltage is supplied to the I/O module, where it is converted to a digital signal that is then passed to the processor. The specifications of the discrete I/O module must match the field device's requirements for everything to work correctly.

Some devices supply current to an input when they're active, while others draw current from the input instead. Similarly, some controlled devices expect an output to supply power or to provide a path to ground for power. These behaviors are called *sinking* or *sourcing*, respectively. The module must be able to operate in sink or source mode as required by the connected device. If it does not match the device's needs, an input module may not sense the input value reliably. Similarly, a mismatched output module may not be able to control the attached device. A block diagram of a typical AC/DC discrete input interface circuit is shown in *Figure 14*.

The power section of the input circuit converts an AC input voltage to a DC level, filters any line and/or switch noise spikes from the signal, and then determines whether the input signal is above a threshold voltage level. Any voltage above the threshold level is a valid signal and is passed to the logic portion of the input interface. The power and logic portions of the interface device are electrically isolated from the processor to prevent damage from noise and power surges.

Input modules often have LED status indicators that provide a visual indication of what the module is detecting. Connections are usually made with standard wiring terminals. *Figure 15* shows an example of several devices wired to an input module.

Figure 16 shows the block diagram of a typical AC/DC discrete output circuit. Like the input circuit, it consists of logic, isolation, and switching circuitry. In this case, when the logic signal from the processor is present, it is converted to a signal suitable to drive the output device. Output switching technologies include electromechanical relay contacts, SCRs, and TRIACs. Current and voltage ratings vary, so be sure that what you're connecting to the output module is appropriate for the particular switching technology. Most output modules include a status LED to show the condition of each output. *Figure 17* shows an example of several devices wired to an output module.

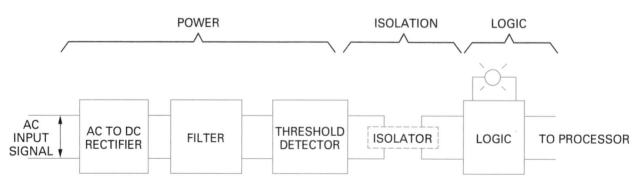

Figure 14 Discrete input circuit.

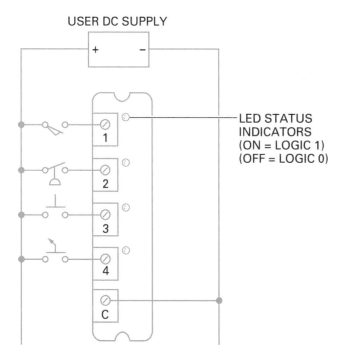

Figure 15 Input module connection diagram.

3.3.3 Numerical Data Interface Modules

The integration of the microprocessor into PLC architecture in the 1970s introduced new capabilities for arithmetic operation and data manipulation. This expanded processing capability led to a new class of input/output interfaces known as numerical data I/O. Numerical input interfaces allow measured quantities to be received from instruments and other devices that provide numerical data, while numerical output interfaces allow control of devices that require numerical data.

Numerical data I/O interfaces provide connections to multi-bit digital devices. The multi-bit interfaces are like the discrete I/O in that the processed signals are 0's and 1's. The difference is that with the discrete I/O only a single bit is involved. Multi-bit interfaces communicate in words, consisting of groups of bits. For example, a set of numeric thumbwheels communicates its position as groups of BCD digits (four bits). A numerical module can capture this information and make it available to the PLC.

3.3.4 Analog Modules

Analog input modules contain the circuitry necessary to accept analog voltage or current signals from field devices. The module then uses an analog-to-digital converter (ADC) to change these values into a number that represents the original signal. It then passes this number to the processor, which stores it in memory for later use.

Typically, analog input interfaces have high impedance inputs that don't significantly load the device they're connected to. The signal line from the analog field device generally uses shielded conductors. The shielded cable provides a better interface medium and helps to maintain good common-mode rejection of noise, such as that from powerline frequencies. The input stage of the analog module often provides filtering and isolation circuits to protect it from additional field noise.

The converted analog value is expressed as a range of numbers either in BCD or decimal format. The lowest value (typically 0) represents the bottom of the measuring scale and the maximum value represents the top. A typical analog input connection is illustrated in *Figure 18*.

An analog output module receives numerical data from the processor, which it converts into an equivalent voltage or current signal with a digital-to-analog converter (DAC). Again, there is usually isolation between the module and the field device that the analog signal goes to. Analog modules may require external power supplies to interface with field devices.

3.3.5 Special Modules

While discrete, numerical, and analog modules handle the majority of PLC interfacing needs, there are occasions when specialized devices may need to be connected. These situations are covered by different kinds of specialized modules. Examples include high-speed encoder/counter, servo control, stepper motor control, synchronized axis control, velocity control, temperature monitoring/control, and smart transmitters.

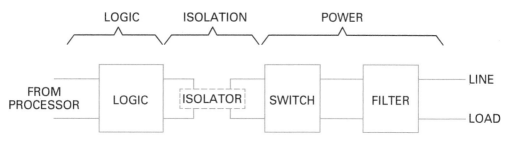

Figure 16 Discrete output circuit.

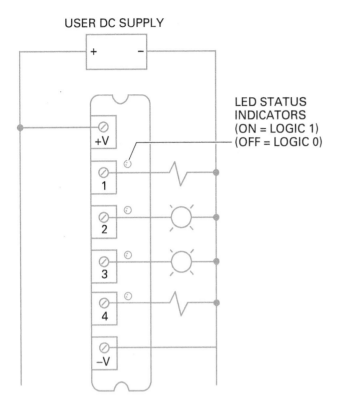

Figure 17 Output module connection diagram.

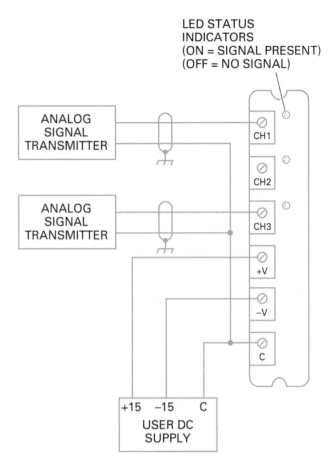

Figure 18 Analog input module connection diagram.

Of these, the most common are the encoder/counter, temperature monitoring/control, stepper motor control, and smart transmitter interface modules, which are briefly described as follows:

- *Encoder/counter modules* – These types of modules provide a high-speed counter external to the PLC processor. Normally they operate independently of the processor and keep track of high-speed input pulses while the processor is engaged with other program activities. The registered count value is periodically sent to the processor. Encoder/counters are often used in process control with turbine flowmeters that produce pulse outputs. Other types of encoder modules read data from devices that output information in a specific format, such as BCD or Gray code. Position encoders often supply data in these forms.
- *Temperature monitoring/control modules* – These modules condition inputs from thermocouples. The thermocouple signals are filtered, amplified, and converted to a numeric value by an ADC. The numerical values for each input are processed by an internal microprocessor as an input to a PID loop. The processor executes a PID algorithm on each input. The output of each loop is sent to the PLC as both a numerical value and as a time-proportioned output (TPO) signal. The PLC program can send either the numerical value to an analog output module or the TPO signal to a digital output module in order to close the loop.
- *Stepper motor control modules* – These output modules generate a pulse train that is compatible with stepper motor translators. The pulses normally represent distance, speed, and direction commands to the stepper motor. The desired motor position is dictated by the preset count of output pulses, a forward or reverse direction command, and an acceleration or deceleration command that is determined by the pulse rate. When the output module is initialized by a PLC program instruction, the module will send the output pulses as determined by the PLC program to the translator. The output module will generally not accept any commands from the PLC program until the move is completed as determined by status information returned from the translator. Some output modules may have a provision for an override command that resets the stepper motor to its current position.

- *Smart transmitter interface modules* – These modules allow a PLC to access all of the analog and digital information generated by HART-compliant field devices. The digital process values can be used to qualify analog device inputs, enabling PLC-based programs to compensate more precisely and control a process more accurately. The HART communication protocol, developed by Rosemount™, is a major industry standard field communications protocol for instrumentation networks.

3.3.6 Communications Modules

PLCs need to communicate for a variety of reasons. Programming and debugging requires some type of computer to be connected to the PLC. PLCs often have to communicate with other PLCs, remotely located racks of modules, or other electronic devices. There are a number of ways that these goals may be accomplished, but they are always addressed by communications modules. In many cases, the communication module is part of the processor, but with larger PLC systems external modules may be added as well.

All PLCs have at least one means by which a programming computer can be attached for software development and debugging. In most recent PLCs, this is accomplished through a USB port located on the front of the processor module. This method is simple, but it requires the programmer to work within a few feet of the PLC location, which is not always convenient.

Connecting PLCs to a network makes communication especially simple and convenient. Most industrial settings already have networks in place for standard office and information technology needs. PLCs can join these networks and take advantage of their reach. Ethernet and Wi-Fi are currently the most common networking technologies, providing wired and wireless communications, respectively. Many PLC processors have a built-in Ethernet port.

Being connected to a building network offers significant advantages, particularly when PLCs have to be programmed or debugged. Instead of having to go to the PLC, the programmer can connect to it over the network from any convenient location. It's even possible for a PLC technician to access the PLC from home if the facility's network is connected to the Internet. Similarly, PLCs connected to a facility's network can communicate with any other device on the network as long as they use a common communications protocol. This enables many types of PLC and device interactions.

Many PLCs must also communicate over a serial network, such as RS-232, RS-422, and RS-485. All of these are relatively low speed compared with Ethernet and Wi-Fi, but they have the advantages of being simple, inexpensive, and well-established in industry. Many older PLC systems use serial communications almost exclusively, even for programming and debugging tasks. Serial ports are often built in to the processor module as well. *Figure 19* shows a dedicated serial communications module.

PLC manufacturers have introduced many proprietary forms of serial communications over the years. Some of these use standard serial ports and cabling, while others require their own specialized hardware. Some are just communications protocols designed to run over a variety of networking technologies, while others are tied to specific hardware. A few examples that technicians may encounter are Modbus®, DF1, DeviceNet™, and PROFIBUS®. There are many others.

3.3.7 Module Addressing

If you want to tell your friends where you live, you give them your house number, the street, the city, and the state. Computer systems also need

Figure 19 A serial communications module with multiple ports.

very specific instructions as to where a signal is coming from and where it needs to go. Each PLC input or output must have an address attached to it. PLC programs usually refer to these addresses when they access different modules.

Address schemes differ by manufacturer, but most will be either four or five digits. The following is an example of an Allen-Bradley address:

I:013/12

The I indicates that the address is associated with an input. (An O would designate an output.) The first two digits to the right of the colon identify the rack number. In this case, the address is an input located in rack number 01. The third digit to the right of the colon is for the module group (a slot within the rack). This particular input is in slot 3. The final two digits are for the wiring terminal number associated with the signal; in this case 12.

Additional Resources

Programmable Controllers, Thomas A. Hughes. Fourth Edition. 2004. Research Triangle Park, NC: International Society of Automation.

Programmable Logic Controllers, W. Bolton. Sixth Edition. 2015. Oxford, UK: Newnes/Elsvier.

The following websites offer resources for products and training:

Rockwell Automation, Inc., **ab.rockwellautomation.com**

AutomationDirect, **www.automationdirect.com**

3.0.0 Section Review

1. The power supply in a small PLC is often _____.
   a. underpowered
   b. remote
   c. external
   d. internal

2. A PLC's control program is usually stored in _____.
   a. volatile memory
   b. nonvolatile memory
   c. hard disk memory
   d. core memory

3. A PLC processor cycle is called a(n) _____.
   a. instruction fetch
   b. IPS
   c. scan
   d. I/O cycle

4. Discrete I/O module inputs and outputs are usually protected by _____.
   a. isolation
   b. circuit breakers
   c. opto-DIACs
   d. resistors

5. To generate its output signals, an analog output module uses a(n) _____.
   a. DAC
   b. ADC
   c. optoisolator
   d. TRIAC

6. RS-485 is an example of a _____.
   a. processor type
   b. wireless network
   c. serial network
   d. programming language

SECTION FOUR

4.0.0 PLC PROGRAMMING

Objective

Describe PLC programming concepts.
a. Identify various programming languages used to program PLCs.
b. Explain how ladder diagramming is used and identify the six related categories of instructions.
c. State typical guidelines for PLC programming and installation.

Performance Tasks

2. Given an instructor-provided ladder diagram program, point out commonly used symbols and their meaning.
3. Implement a simple logic circuit using an instructor-provided PLC platform or simulator.

Trade Terms

Language: In the context of computers and PLCs, a language is a set of instructions and an associated grammar that governs how they are used to develop programs that control the processor.

Simulator: A piece of software that mimics the operation of a different computer system. Simulators allow software development and testing without the need to have the actual hardware present.

A PLC system does nothing without a control program to process its inputs, make decisions, and alter its outputs. PLC programming is a unique discipline, similar in some ways to regular computer programming. However, it is highly specialized due to its industrial aspects, and requires more knowledge of the hardware than does ordinary computer programming.

PLC technicians may or may not write whole control programs from the ground up, but they often make minor alternations to existing programs in order to accommodate changes to the system. This section describes programming languages and provides a basic overview of PLC programming.

4.1.0 PLC Programming Languages

Over the years, PLC programming has developed in a number of different directions, driven by various industry forces. Many different programming systems have been introduced, some of them specific to certain brands of PLCs. Others are fairly standardized across the industry and work on many different brands. Each of the following is a PLC language that is standardized and well-supported:

- Ladder diagram (LD)
- Function block diagram (FBD)
- Sequential function chart (SFC)
- Structured text (ST)
- Instruction list (IL)

Each language has its advantages and disadvantages. Some are better supported than others by particular manufacturers. Programmers tend to select the language that best fits a particular need most productively. The following sections briefly summarize each type.

4.1.1 Ladder Diagram (LD)

LD programming, also known as ladder logic, is the most widely used and oldest form of PLC programming. This is because it is most like the conventional relay ladder schematics that were used to diagram the relay logic systems that the earliest PLCs replaced. LD programs are created in a graphical environment by successively entering rungs to a picture of the ladder. Each rung is comprised of input and output symbols that are very similar to conventional industrial electrical components. LD programming is explained in greater detail later in this section.

4.1.2 Function Block Diagram (FBD)

FBD programming is a lot like creating an electrical schematic. It's done graphically, on a computer screen, and consists of blocks that perform certain PLC functions, such as logic, timing, evaluation, and control. Blocks are dropped onto the workspace and wired together by connecting them with lines that represent links. In some ways, FBD programming is similar to LD programming, but with more flexibility and sophistication.

4.1.3 Sequential Function Chart (SFC)

SFC programming is similar to FBD. It is graphical as well, and done by connecting blocks with lines. However, SFC programs are sequences of operations in which one operation can't happen until another completes. Many operations may be going on in parallel. In many ways, an SFC program is much like a flowchart—a diagram of the operations and decisions that must take place for specific things to happen.

4.1.4 Structured Text (ST)

ST is similar to many mainstream computer programming languages. ST programs resemble English and can be understood to some degree just by reading them. They're written using a text editor and are lines of code that specify the operations and decisions that define the program's behavior. Technicians who have had a conventional computer programming background often prefer ST since it is much more familiar than the electrically oriented programming tools like LD and FBD. ST draws much of its style from the computer languages Pascal, BASIC, and C.

4.1.5 Instruction List (IL)

Like ST, IL programming is text-based and done with a text editor. Unlike ST, IL is essentially only readable to experienced programmers. Each program line is a short, abbreviated instruction called a *mnemonic*. Most are just a few letters long and look like LD, ST, and JMP. Each stands for a very basic operation that the PLC processor can perform. By stringing together many of these primitive instructions, sophisticated programs can be developed. Writing IL programs is similar to assembly language programming, a method used for programming general-purpose computers that was much more common several decades ago than it is now. It requires a very specific kind of mindset to be a good low-level programmer.

4.1.6 Development Tools

While the languages used in PLC programming are reasonably well standardized, each manufacturer has its own development tools that run on a standard personal computer. These vary in their sophistication and the steepness of their learning curves. All offer the same basic features, including a means of creating the program. This will be some kind of text editor for an ST or IL program, and a graphical editor for LD, FBD, and SFC. There will also be a tool that checks the program for errors and prepares it for uploading into the PLC. Finally, there will be tools for interacting with the PLC, performing debugging tasks, and examining PLC memory.

Besides these basic tasks, some manufacturers also offer a PLC simulator that mimics the basic functionality of the PLC on a regular computer. Simulators can be used for program testing without the need for an actual PLC. This can be handy if a spare PLC isn't available or if the system is controlling a live process and can't be used for testing purposes. PLC simulators are also often used in programming classes since they allow students to learn programming skills without having to invest in the expense of actual PLC hardware.

4.2.0 Ladder Diagram Programming

LD programming has been used for many decades and is one of the most popular ways of programming a PLC, particularly for those with a background in plant electrical systems. LD programs use a symbolic instruction set in which the symbols are connected as rungs in the ladder diagram. The symbols that are used to represent the various operations have been selected to closely approximate common electrical component symbols.

The six general categories of instructions available for use in LD programming are as follows:

- Relay
- Timer/counter
- Arithmetic
- Data manipulation
- Data transfer
- Program control

Each of these instructions is entered onto a rung within the ladder. When the input to a rung is present, it is said to be TRUE. When a given rung's inputs are TRUE, this causes the output to become energized (also called ON or TRUE). When logic continuity exists across a rung, the rung is TRUE and the output is energized.

4.2.1 Relay

Relay instructions are comprised of input contacts and output coils. These instructions are the most fundamental of ladder logic instructions and allow the PLC to communicate with external devices.

Inputs from a PLC's input module are shown as contacts on a ladder rung. These contacts can be either normally open (NO) or normally closed (NC) in their deactivated state (also called OFF or FALSE). Examples of possible inputs include switches, pushbuttons, limit switches, and output coil contacts. By convention, input contacts are entered on the left side of a ladder rung and output relay coils are entered on the right. Typical symbols for input contacts are shown in *Figure 20*.

Outputs from a PLC's output module are shown as a coil on a ladder rung. These coils can be of two different types: standard or latch. A standard coil will energize when the rung is TRUE and de-energize when a rung is FALSE. A latch coil is shown like a standard coil with the letter L in the center of the coil. When the rung is TRUE, a latch coil will energize. When the unlatch coil (the coil with letter U in the center) with the same address is energized, the latch coil will unlatch. Typical symbols for various output coils are shown in *Figure 21*.

Figure 22 illustrates how the contacts and coils are used to represent logic flow in a ladder diagram. Rung 0 shows a Normally Open contact that is deactivated, resulting in a de-energized coil. Rung 1 shows a Normally Open contact that is activated, resulting in an energized coil. Rungs 2 and 3 show the logic states of Normally Closed contacts that result in the exact opposites of the coil logic states in rungs 0 and 1.

Figure 23 shows the operation of a latching/unlatching coil and its input contacts. In rung 0, output coil 3 will latch on when input contact 1 is closed. If contact 1 then opens, output 3 will remain latched on until unlatched by unlatch coil 3. Unlatch coil 3 will be energized when input contact 2 is closed.

Each of these input contacts could be a pushbutton connected to the PLC's input module. The output coils of the ladder control some point on the PLC's output module. When the ladder's coil is energized, so is the device connected to the PLC's output module. Some PLCs only allow internal outputs to be latched.

Observe that in *Figure 23*, the inputs are labeled I:1 and I:2, while the output is labeled O:3. This is

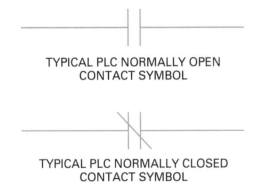

Figure 20 Input contact symbols.

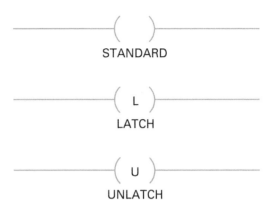

Figure 21 Output coil symbols.

a simplified version of the Allen Bradley addressing scheme discussed earlier. This system will be used in the examples throughout the remainder of this module. Other addressing schemes are used by different manufacturers, and some are more difficult to interpret.

4.2.2 Timer and Counter

Timers and counters are output instructions that provide the same functions as hardware timers and counters. They are used to activate or deactivate a device after an expired interval or count. The timer and counter instructions are generally considered internal outputs. Like the relay-type instructions, timer and counter instructions are fundamental to the ladder diagram instruction set.

The operations of the software timer and counter are similar in that they are both counters. A timer counts the number of times that a fixed interval of time (such as 0.1 second or 1.0 second) elapses up to a preset value. For example, to time an interval of 3 seconds, a timer counts three one-second intervals. A counter simply counts the occurrences of an event. Both the timer and counter instructions require a register location in memory to store the elapsed count and another register to store a preset value. The preset value will determine the number of event occurrences or time intervals that are to be counted.

The timer on delay (TON) output instruction is programmed to provide time-delayed action or to measure the duration over which some event is occurring (*Figure 24*, Rung 3). If the rung path for the timer is TRUE, the timer begins counting time intervals until the accumulated time equals the preset value. If the input remains TRUE when the accumulated time equals the preset value, the timer done (DN) output is energized (or de-energized), and the timed-out contact associated

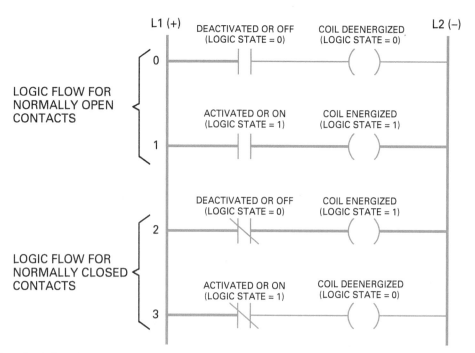

Figure 22 Logic flow for NC and NO contacts.

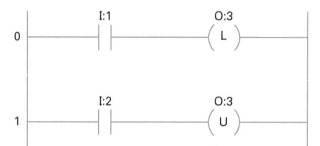

Figure 23 Latching/unlatching coil operation.

with the output is closed (or opened), as shown in rung 4. If the rung logic for the TON goes FALSE at any time, the accumulated value is reset to 0.

The timer off delay (TOF) output instruction is programmed to provide time-delayed action (Rung 5). If the rung path is FALSE, the timer begins counting time intervals until the accumulated time equals the programmed preset value. When the accumulated time equals the preset time, the timer done (DN) output is de-energized (or energized), and the timed-out contact (Rung 6) associated with the output is opened (or closed). If the input logic to the TOF is restored to TRUE at any time, the accumulated value is set to 0, and the DN output is restored to its original condition.

Timers may also be retentive. A retentive timer on (RTO) output instruction is one that stores its timed value until it has been reset by a reset instruction. The RTO (Rung 7) will retain its accumulated count, even if the input logic or power is lost. If the input rung is TRUE and remains TRUE, the timer will count time intervals until the accumulated value equals the preset value. If the input becomes FALSE before the timer has timed out, the accumulated value is retained. When the input is restored to TRUE, the timer will resume the count until the accumulated value equals the preset value and the timer done (DN) bit is set. The DN bit can be a NO or NC instruction as desired for use in the rest of the program. To clear the RTO accumulated value, a reset (RES) instruction for the timer must be set by a NO reset input (Rung 8).

The count up (CTU) output instruction will increment (increase by 1) each time the counted event occurs (*Figure 25*, Rung 0). The CTU increments its accumulated value each time the up-count event makes a TRUE-to-FALSE transition. When the accumulated value reaches the preset value, the counter done (DN) output is closed, and the DN contact associated with the referenced output is closed (or opened).

The count down (CTD) output instruction will decrement (decrease by 1) each time the counted event occurs (Rung 1). In normal use, the down-counter is usually used in conjunction with the up-counter to form an up/down-counter.

As an example, while the CTU counts the number of filled bottles that pass a certain point, a CTD with the same reference address would subtract one from the register each time an empty or improperly filled bottle goes by. Counters require a reset instruction to clear their count register. In this example, a CTU is used to determine the completion of a case of filled bottles. The counter done (DN) contacts signal the completion of the case when the register reaches the preset value so that the next case can be moved to the fill point.

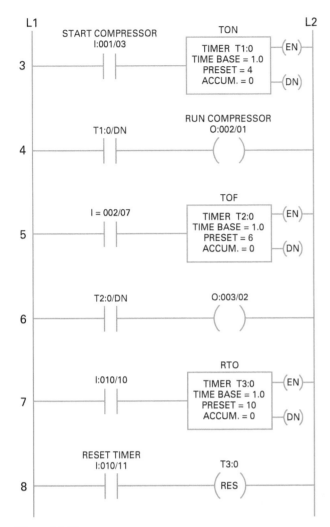

Figure 24 Timers.

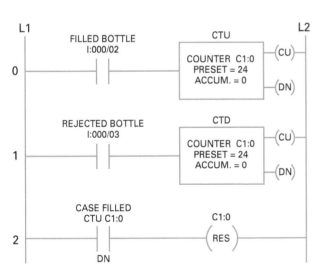

Figure 25 Counter application.

The DN bit is also used to reset the counter for the next case.

A momentary FALSE-to-TRUE transition is required for both counters to reset. Unlike timers, an up-counter will continue to increment the register value after the preset has been reached unless the counter is reset. If the accumulated value exceeds the maximum range of the counter, an overflow (OV) bit will be set. This overflow signal can be used to link multiple counters together for applications that require counts greater than the maximum value of the counter. Chaining several counters together is called *cascading*.

4.2.3 Arithmetic

Arithmetic operations include the four basic operations of addition, subtraction, multiplication, and division. These instructions use the contents of two registers to perform the desired function and then store the result in another register. *Figure 26* shows the four arithmetic operations.

The ADD instruction performs an addition on two values stored in the referenced memory locations. The SUB instruction performs a subtraction on two registers' contents. The MUL instruction performs the multiplication operation. It uses two words in a register to hold the result of the

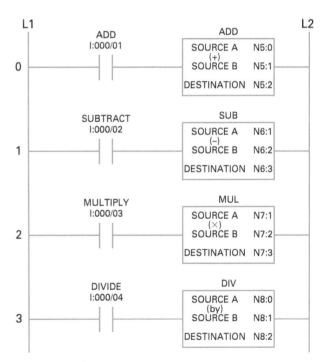

Figure 26 Arithmetic operations.

operation since a multiplication can produce a much larger number than the original two. The DIV instruction performs a division on two registers' contents. The result of the division is held in two words of a third register. The first word generally holds the integer portion of the operation, while the second word holds the decimal fraction component.

4.2.4 Data Comparison

The data comparison instructions are an enhancement of the basic ladder diagram instruction set. Whereas the relay-type instructions were limited to the control of internal and external outputs based on the status of specific bit addresses, the data comparison instructions allow multi-bit operations. In general, the comparison of data using ladder diagram instructions involves simple register operations to compare the contents of two registers.

In the ladder language, there are three basic data comparison instructions: equal-to (EQU), greater-than (GRT), and less-than (LES). Based on the result of a greater-than, less-than, or equal-to comparison, an output can be turned ON or OFF, or some other operation can be performed. The comparison can be performed on either addresses or values. *Figure 27* shows typical data comparison instructions.

4.2.5 Data Transfer

Data transfer instructions involve the transfer of the contents of one register to another. Data transfer instructions can address any location in the memory data table, with the exception of areas not accessible to user programs. Pre-stored values can be automatically retrieved and placed in any new location. The common instructions used are bit distribute (BTD), move (MOV), and masked move (MVM). Similar data transfer instructions are used by most PLC manufacturers. *Figure 28* and *Figure 29* show typical data transfer operations.

The BTD output instruction is used to move up to sixteen bits of data within or between words. The source of the data is not affected. The destination is overwritten by the data bits being moved. If the length of the moved bit field exceeds the bit field of the destination word, the excess bits are lost. When the BTD instruction is TRUE, the desired bit field is moved from the source word to the destination word during a PLC scan. If bits are to be moved within the source word, as shown in *Figure 28*, the source and destination word address is the same.

The MOV output instruction is used to copy the data in a source register to a destination. When the MOV instruction is TRUE, the contents of the source is transferred to the destination during a scan cycle. The source may be a program constant or a data address that is used by the instruction to read a copy of the value. This copy is used by the instruction to overwrite any data stored at the destination.

The MVM output instruction is used to copy the contents of the source to the destination while allowing portions of the source data to be masked (hidden) from the transfer. The source data is not changed. This instruction can be used to copy I/O values, binary values, or integer values.

For instance, bit data such as status or control bits can be extracted from an address that contains bit and word data. The source can be a program constant or data address. The mask can be an address or hexadecimal value that specifies which bits to hide. The mask bits must be set to 1 by the programmer to allow the desired data bits to pass to the destination. Mask bits set to 0 hide the data in those locations. The moved data overwrites the destination data. Note that in *Figure 29*, the mask (FF00) used for the source has the higher eight bits set to 1 (the two Fs) to allow transfer while the lower eight bits are hidden (the two 0s).

4.2.6 Program Control

Program control operations are accomplished using a series of conditional and unconditional branches and return instructions. These instructions provide a means of executing sections of the control logic if certain conditions are met.

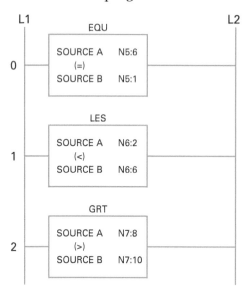

Figure 27 Examples of data comparison instructions.

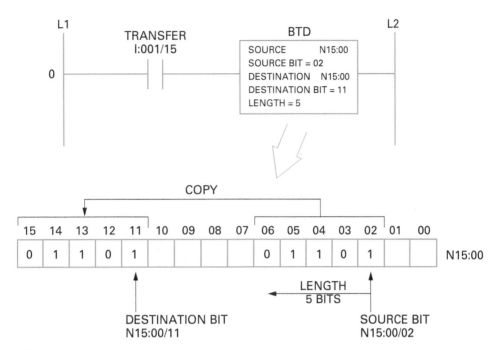

Figure 28 BTD transfer operation.

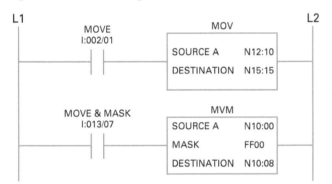

Figure 29 MOV and MVM transfer operations.

The master control relay (MCR) output instruction is used in pairs (conditional and unconditional) to fence and activate or de-activate the execution of a group of ladder rungs (*Figure 30*). When the MCR rung condition is TRUE, the conditional MCR output is activated, and all rung outputs within the zone can be controlled by their respective input conditions. If the MCR output is turned OFF, all non-retentive outputs within the zone will be de-energized and reset. With the MCR output OFF, PLC scan time is reduced by the FALSE state of the non-retentive outputs, even though they are scanned.

The zone control (ZCL) instruction is similar to the MCR instruction. It determines if a group of ladder rungs will be evaluated or not. If the referenced ZCL output is activated, the outputs within the zone are controlled by their respective rung input conditions. If the ZCL output is turned OFF, the outputs within the zone will be held in their last state.

The jump (JMP) instruction is used to skip portions of the ladder logic. The JMP allows the normal sequential program execution to be altered if certain conditions exist. If the rung condition is TRUE, the JMP coil reference address tells the processor to jump forward and execute the rung labeled with the same reference address as the JMP coil. The order of execution can be altered to execute a rung that needs immediate attention.

The jump to subroutine (JSR) instruction, when TRUE, causes the processor to execute a subroutine file outside the main program during a scan and then return to the main program. If required, data passed to and received from the subroutine is defined. An optional subroutine (SBR) instruction can be used to store incoming data. It is used only if data must be passed to and from the subroutine. If used, it must be the first instruction on the first rung of the ladder for the subroutine program. The subroutine is ended by a return (RET) instruction. If required, the RET instruction causes data that is returned to the JSR instruction in the main program to be stored.

The LBL instruction is used to identify a ladder rung that is the target destination of a JMP or JSR instruction. The LBL reference number must match that of the JMP or JSR instruction with which it is used. The LBL instruction does not contribute to logic continuity and, for all practical purposes, is always logically TRUE. It is placed as the first condition instruction in the rung. An LBL instruction referenced by a unique address can be defined only once in a program.

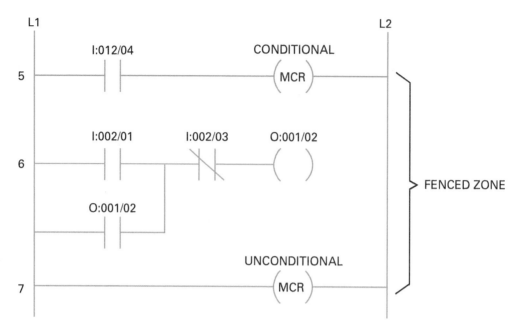

Figure 30 Typical MCR instructions.

4.3.0 Guidelines for Programming and Installation

While each manufacturer will recommend specific methods of programming and installing a PLC, some or all of the following guidelines may be applicable, depending on the situation. Due to the number of different PLC development tools available, detailed programming procedures, including solutions to the many types of control problems that can be encountered, are beyond the scope of this module.

4.3.1 Programming

Use the following general guidelines to program a PLC:

- *Determine the control task* – This task is very important and should be done by someone who is very familiar with the controlled process. In many cases, the programmer will consult with an operations person or process engineer. This person provides the "control narrative" used to create the program.
- *Determine the control strategy* – Based on the controlled task, this is the sequence of detailed actions that will be necessary in the PLC program to produce the desired result from start to finish.
- *Program implementation* – After the control strategy has been thoroughly reviewed, implement the general programming guidelines for system modernizations or new systems provided in *Table 5*.

4.3.2 Installation

Use the following general guidelines to install a PLC:

- Make sure that the PLC selected has 25- to 50-percent extra capacity for future expansion.
- Select an enclosure as recommended by the manufacturer so that the PLC is easily accessible for wiring, testing, troubleshooting, and local programming.

Table 5 Programming Guidelines

System Modernizations	New Systems
Understand the process and/or machine(s) function	Understand the system functions that are required
Review existing machine logic and optimize. Create an updated ladder diagram and optimize	Recheck control strategy
Assign I/O and internal addresses to inputs and outputs	Flowchart the operation of the process
Translate ladder diagram into PLC coding	Create ladder diagrams (or other logic symbology)
If available, run simulator software to evaluate program operation	Assign I/O and internal addresses to inputs and outputs
	Translate ladder diagram or other symbology into PLC coding
	If available, run simulator software to evaluate program operation

- The location and enclosure must be tailored to mitigate the effects of the surrounding environmental conditions, such as temperature, humidity, electrical noise, and vibration.
- Group I/O modules in base units according to their type. If possible, physically separate input modules from output modules.
- Incoming and output power wiring, external input power devices, power supplies, output power starters, contactors, relays, and other electromechanical devices should be located at the top of the enclosure to minimize wiring runs and electrical noise.
- Make sure that power wiring in ducts or raceways is separated from all low-level I/O wiring. Make sure that both the PLC I/O wiring and electrical wiring materials and installations conform to the *National Electrical Code®* (*NEC®*) and local codes as well as to the manufacturer's recommendations.
- Make sure to ground the system components in accordance with the manufacturer's recommendations as well as the *NEC®* and local codes.
- AC power for the PLC and field devices should be appropriately fused and should come from a common source at the PLC location. If necessary, install line filters to remove electrical noise.
- Personnel safety devices should be external to the PLC and hardwired using reliable electromechanical devices to remove power from the controller and/or inputs and outputs in the event of a personnel emergency or PLC failure. Master control relay circuits can be used to remove all power. Safety control relay circuits can be used to disable only the inputs or the outputs without shutting down the PLC.
- Check that all primary and PLC power supply wiring is correctly connected to the PLC.

4.3.3 I/O Wiring

Use the following general guidelines for I/O wiring:

- Remove and lock out all controller and I/O power.
- Check that all I/O modules are in the correct locations in accordance with the I/O address assignment document.
- Loosen all terminal screws on each I/O module.
- Locate the wire bundle corresponding to each module. Identify each of the wires.
- Start wiring each module from the bottom up, bending the wires at a right angle at the appropriate terminal.
- Cut the wires so that they extend about ¼" beyond the terminal. Then strip approximately ⅜" of insulation from each of the wires.
- Insert each wire under the appropriate terminal pressure plate, and tighten the screw.
- If using shielded wire, follow the manufacturer's recommendations for terminating the shield. In most cases, ground only one end of the shield, usually at the PLC end, to prevent ground loops.
- After all wires are connected, pull gently on the wires to make sure that they are securely clamped under their terminals.
- If necessary, install any special leakage suppression for input devices that have known current leakage problems or inductive spike protection for outputs to inductive load devices like motors.

Use the following general guidelines to make static input wiring checks:

- Place the PLC in a mode that will inhibit automatic operation.
- Apply power to the PLC. Verify that any system indicators show proper operation.
- Activate each emergency stop, and check that the appropriate devices are de-energized.
- Reestablish power to the PLC.
- Manually activate each input device, and check that the proper input is registered at the PLC.
- Take precautions to avoid injury or damage when activating input devices connected in series with loads that are external to the PLC.

Use the following general guidelines to make static output wiring checks:

- Remove power from the PLC and the I/O modules.
- Locally disconnect all output devices that could result in mechanical motion.
- Apply power to the PLC. Verify that any system indicators show proper operation.

- Activate each emergency stop, and check that the appropriate devices are de-energized.
- Reestablish power to the PLC.
- Use the processor front panel controls or an attached computer to manually force each output to ON by setting the corresponding terminal address to 1. Verify that the proper output indicator lights and that the output device energizes (except for the disconnected ones).
- Once all outputs have been checked, de-activate all outputs and reconnect each disconnected device, one at a time, and repeat the forcing function for that device to make sure that it energizes. Take precautions to make sure that energizing the device will not cause injury or damage when the load operates independently.

4.3.4 Dynamic System Checkout

Use the following general guidelines to make dynamic system checks:

> NOTE: During dynamic checkout, it is advisable to bring large systems up one section at a time. Usually, large systems have remote subsystems that can be brought up one at a time for checkout before the entire system is brought up.

- Make sure that all corrections from the previous checks have been incorporated into the PLC program.
- Load the program into the PLC memory.
- Switch the PLC to a test or emulation mode that will allow running and debugging of the program with the outputs disabled.
- Use a single scan mode or the emulation mode to check each rung of the ladder diagram for proper operation.
- Correct the program and program documentation to reflect all changes.
- Place the PLC in the run mode, and verify that the subsystem or system operation is correct, as applicable.

Additional Resources

Programmable Controllers, Thomas A. Hughes. Fourth Edition. 2004. Research Triangle Park, NC: International Society of Automation.

Programmable Logic Controllers, W. Bolton. Sixth Edition. 2015. Oxford, UK: Newnes/Elsvier.

The following websites offer resources for products and training:

Rockwell Automation, Inc., **ab.rockwellautomation.com**

AutomationDirect, **www.automationdirect.com**

4.0.0 Section Review

1. Which PLC programming language is the most common, and is also the oldest?
   a. Function block diagram (FBD)
   b. Instruction list (IL)
   c. Structured text (ST)
   d. Ladder diagram (LD)

2. Which PLC programming language is similar to English and is fairly easy for most technicians to understand?
   a. Function block diagram (FBD)
   b. Instruction list (IL)
   c. Structured text (ST)
   d. Ladder diagram (LD)

3. PLC programs can be tested using a special program that runs on a personal computer called a _____.
   a. processor
   b. simulator
   c. replicator
   d. hard PLC

4. In LD programming, relay instructions consist of input contacts and output _____.
   a. coils
   b. isolators
   c. TRIACs
   d. jumps

5. In LD programming, timer instructions decide when the timer has reached the target time by checking the _____.
   a. timer complete contact
   b. preset register
   c. timer done contact
   d. real-time clock

6. When planning a PLC system, it's wise to pick components with extra _____.
   a. emergency stops
   b. air flow
   c. capacity
   d. timers

Summary

PLCs have become one of the most popular ways of controlling industrial processes. Their flexibility, both in hardware and software, makes them an ideal tool to handle any control situation that requires intelligence and the ability to make changes. PLC hardware consists of a power supply, a processor, and various modules that provide the necessary I/O and communications features required by the application.

Programming a PLC can be done in one of five different industry-standard languages. Ladder Diagram (LD) programming is the oldest and most common. It consists of symbols that resemble common electrical functions that are connected together in ways similar to that in which relay logic was wired in the past.

Proper planning is the key to a successful installation. Understanding PLCs and their components, power supplies, wiring, and languages is absolutely necessary for being a successful instrumentation technician.

12406-16 Programmable Logic Controllers

Review Questions

1. What is the term used to refer to how an entire PLC system is put together?
   a. Development environment
   b. Memory map
   c. Network
   d. Architecture

2. A PLC system consists of a power supply, processor, and various _____.
   a. ports
   b. simulators
   c. modules
   d. terminals

3. PLC inputs are often protected from hostile voltages by _____.
   a. optocouplers
   b. TRIACs
   c. SCRs
   d. relays

4. Convert 105_{10} to binary.
   a. 1001011_2
   b. 1101001_2
   c. 0011101_2
   d. 1001101_2

5. Hexadecimal numbers are based on powers of _____.
   a. 2
   b. 8
   c. 10
   d. 16

6. The most common text encoding system used in the PLC world is _____.
   a. ITU-T
   b. IANA
   c. ASCII
   d. EBCDIC

7. Gray code is often used with mechanical _____.
   a. encoders
   b. rotors
   c. keyboards
   d. HMIs

8. To prevent damage from environmental electrical discharges, PLCs must be properly _____.
   a. isolated
   b. activated
   c. grounded
   d. terminated

9. The way a PLC organizes its memory is called a memory _____.
   a. allocation
   b. map
   c. chart
   d. workspace

10. A location in PLC memory used to store a word of data is called a _____.
    a. register
    b. bit
    c. timer
    d. scratch pad

11. A set of thermocouples would most likely be connected to a PLC through a(n) _____.
    a. discrete module
    b. analog module
    c. numerical module
    d. special module

12. Which PLC programming language is much like an electrical schematic with blocks connected by lines representing links?
    a. Function block diagram (FBD)
    b. Structured text (ST)
    c. Instruction list (IL)
    d. Ladder diagram (LD)

13. Which text-based PLC programming language uses short, abbreviated instructions that represent basic PLC operations?
    a. Functional block diagram (FBD)
    b. Structured text (ST)
    c. Instruction list (IL)
    d. Ladder diagram (LD)

14. In LD programming, an output that stays energized until a second coil is made TRUE is called a(n) _____.
    a. locked relay
    b. latch relay
    c. unlockable relay
    d. sealed relay

15. In LD programming, which instructions are used to keep track of event occurrence?
    a. Relay
    b. Arithmetic
    c. Data transfer
    d. Timer/counter

Trade Terms Introduced in This Module

Address: A unique identifier that enables a PLC program to refer to specific modules and wiring terminal positions.

Analog-to-digital converter (ADC): A circuit that converts an analog voltage or current signal into a numeric equivalent.

Control code: A character that when received causes a device to modify its behavior in some way.

Data table: The section of a PLC's memory map that contains data used by the control program.

Digital-to-analog converter (DAC): A circuit that converts a number into an equivalent analog voltage or current signal.

Encoded: The process of representing one kind of information as another. For example, a character encoding system represents each character with an equivalent number.

Ethernet: Refers to a family of wired networking hardware standards with speeds ranging up to gigabits per second; a major networking standard since the 1980s.

Human-machine interface (HMI): Any device that allows a human to control and monitor the status of a computer or PLC.

Input/output (I/O): The component of a computer or PLC that allows it to interact with the outside world.

Interrupt: An event that causes a PLC processor to stop whatever it's doing and perform a specific task, usually one that's time-critical.

Isolated: Connected in such a way that a control signal does not directly interact at the electrical level with a device.

Language: In the context of computers and PLCs, a language is a set of instructions and an associated grammar that governs how they are used to develop programs that control the processor.

Memory map: A diagram of the way in which a PLC's memory is divided up and assigned to different tasks.

Modules: Components in a PLC system that provide particular features, such as I/O or communications.

Motion encoder: An electromechanical device that converts a rotary or linear motion into a number representing the current position.

Nonvolatile memory: Memory that does not require electrical power to maintain its contents. Flash memory is a typical example.

Real-time clock: A battery-backed timekeeping circuit embedded in a PLC that maintains the current time and date.

Ruggedized: Designed to work reliably in a potentially harsh environment, such as one with unusual temperatures, electrical noise, or physical risk.

Scan: A PLC processor cycle in which inputs are read, program logic is executed, and outputs are updated.

Serial network: A broad umbrella term for various low-speed communications systems in which data travels one bit at a time between devices over a simple cabling system. Examples include RS-232, RS-422, and RS-485.

Simulator: A piece of software that mimics the operation of a different computer system. Simulators allow software development and testing without the need to have the actual hardware present.

Universal serial bus (USB): A communications system built into many computers, PLCs, and portable electronic devices to allow simple, high-speed cabled connections.

Volatile memory: Memory that requires uninterrupted electrical power to maintain its contents. Also known as RAM.

Wi-Fi: Refers to a particular family of wireless networking hardware and software standards designed for both general and special-purpose communications.

Additional Resources

This module presents thorough resource for task training. The following resource material is recommended for further study.

Programmable Controllers, Thomas A. Hughes. Fourth Edition. 2004. Research Triangle Park, NC: International Society of Automation.

Programmable Logic Controllers, W. Bolton. Sixth Edition. 2015. Oxford, UK: Newnes/Elsvier.

The following websites offer resources for products and training:

Rockwell Automation, Inc., **ab.rockwellautomation.com**

AutomationDirect, **www.automationdirect.com**

Wi-Fi® is a registered trademark of Wi-Fi Alliance, **wi-fi.org**.

HART®, *Wireless*HART®, and HART-IP™ are registered trademarks of the FieldComm Group™, **www.fieldcommgroup.org**

Figure Credits

emel82/Shutterstock.com, Module Opener

Rockwell Automation, Inc., Figures 1–3

Unitronics, Inc., Figure 4

Automationdirect.com, Figures 12, 13, 19

Section Review Answer Key

Answer	Section Reference	Objective
Section One		
1. c	1.0.0	1a
2. d	1.1.0	1a
3. c	1.1.1	1a
4. a	1.1.1	1a
5. b	1.2.3	1b
Section Two		
1. b	2.1.0	2a
2. c*	2.1.1	2a
3. a*	2.1.2	
4. b*	2.2.1; Table 1	2b
5. d	2.3.1	2c
6. a*	2.3.2; Table 3	2c
Section Three		
1. d	3.1.0	3a
2. b	3.2.1	3b
3. c	3.2.2	3b
4. a	3.3.2	3c
5. a	3.3.4	3c
6. c	3.3.6	3c
Section Four		
1. d	4.1.1	4a
2. c	4.1.4	4a
3. b	4.1.6	4a
4. a	4.2.1	4b
5. b	4.2.2	4b
6. c	4.3.2	4c

*Calculations for these answers are provided on the following page(s).

Section Review Calculations

Section 2.0.0

Question 2

Convert the number 01011101_2 to decimal.

$01011101_2 =$
$(0 \times 2^7) + (1 \times 2^6) + (0 \times 2^5) + (1 \times 2^4) + (1 \times 2^3) + (1 \times 2^2) + (0 \times 2^1) + (1 \times 2^0) =$
$0 + (1 \times 64) + 0 + (1 \times 16) + (1 \times 8) + (1 \times 4) + 0 + (1 \times 1)$
$0 + 64 + 0 + 16 + 8 + 4 + 0 + 1 = 93_{10}$

01011101_2, converted to decimal, is equal to **93_{10}**.

Question 3

Using the Subtractive Method:

32 is the largest power of two that can be used.

Powers of 2	Subtract	Record Binary Digits
$2^5 = 32$	$39 - 32 = 7$	write down a **1**
$2^4 = 16$	Can't subtract 16	write down a **0**
$2^3 = 8$	Can't subtract 8	write down a **0**
$2^2 = 4$	$7 - 4 = 3$	write down a **1**
$2^1 = 2$	$3 - 2 = 1$	write down a **1**
$2^0 = 1$	$1 - 1 = 0$	write down a **1**

Arrange recorded binary digits from *first to last*.

$39_{10} = 100111_2$

39_{10}, converted to binary, is equal to **100111_2**.

Using the Division Method:

Divide by 2	Record Remainders
$39 \div 2 = 19$, *R1*	write down a **1**
$19 \div 2 = 9$, *R1*	write down a **1**
$9 \div 2 = 4$, *R1*	write down a **1**
$4 \div 2 = 2$, *R0*	write down a **0**
$2 \div 2 = 1$, *R0*	write down a **0**
$1 \div 2 = 0$, *R1*	write down a **1**

Arrange the recorded remainders from *last to first*.

$39_{10} = 100111_2$

39_{10}, converted to binary, is equal to **100111_2**.

Question 4

Convert each hexadecimal digit to its four-bit binary equivalent:

$7_{16} = 0111_2$
$F_{16} = 1111_2$
$7F_{16} = 01111111_2$

$7F_{16}$, converted to binary, is equal to **01111111$_2$**.

Question 6

Convert each four-bit binary coded decimal to its decimal equivalent:

$0111_{BCD} = 7_{10}$
$0010_{BCD} = 2_{10}$
$01110010_{BCD} = 72_{10}$

01110010_{BCD}, converted to decimal, is equal to **72$_{10}$**.

NCCER CURRICULA — USER UPDATE

NCCER makes every effort to keep its textbooks up-to-date and free of technical errors. We appreciate your help in this process. If you find an error, a typographical mistake, or an inaccuracy in NCCER's curricula, please fill out this form (or a photocopy), or complete the online form at **www.nccer.org/olf**. Be sure to include the exact module ID number, page number, a detailed description, and your recommended correction. Your input will be brought to the attention of the Authoring Team. Thank you for your assistance.

Instructors – If you have an idea for improving this textbook, or have found that additional materials were necessary to teach this module effectively, please let us know so that we may present your suggestions to the Authoring Team.

NCCER Product Development and Revision
13614 Progress Blvd., Alachua, FL 32615

Email: curriculum@nccer.org
Online: www.nccer.org/olf

❏ Trainee Guide   ❏ Lesson Plans   ❏ Exam   ❏ PowerPoints   Other _____

Craft / Level: _____   Copyright Date: _____

Module ID Number / Title: _____

Section Number(s): _____

Description: _____

_____

_____

_____

_____

Recommended Correction: _____

_____

_____

_____

_____

Your Name: _____

Address: _____

_____

Email: _____   Phone: _____

12407-16
Distributed Control Systems

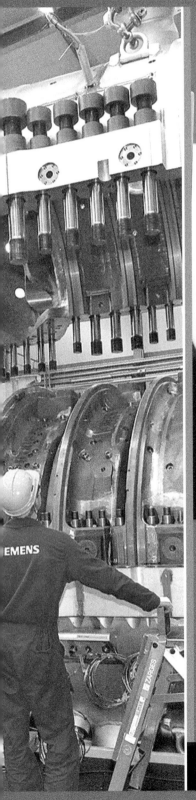

Overview

Thanks to embedded computer technology, the world of instrumentation has changed rapidly in the last two decades. One of the most significant changes is the way in which control systems are put together. While there are several different methods used, one of the most popular, particularly with large processes, is the distributed control system. This module explores DCSs, their components, operation, and security, and compares them to other modern control technologies.

Module Six

Trainees with successful module completions may be eligible for credentialing through the NCCER Registry. To learn more, go to **www.nccer.org** or contact us at 1.888.622.3720. Our website has information on the latest product releases and training, as well as online versions of our *Cornerstone* magazine and Pearson's product catalog.

Your feedback is welcome. You may email your comments to **curriculum@nccer.org**, send general comments and inquiries to **info@nccer.org**, or fill in the User Update form at the back of this module.

This information is general in nature and intended for training purposes only. Actual performance of activities described in this manual requires compliance with all applicable operating, service, maintenance, and safety procedures under the direction of qualified personnel. References in this manual to patented or proprietary devices do not constitute a recommendation of their use.

Copyright © 2016 by NCCER, Alachua, FL 32615, and published by Pearson Education, Inc., New York, NY 10013. All rights reserved. Printed in the United States of America. This publication is protected by Copyright, and permission should be obtained from NCCER prior to any prohibited reproduction, storage in a retrieval system, or transmission in any form or by any means, electronic, mechanical, photocopying, recording, or likewise. To obtain permission(s) to use material from this work, please submit a written request to NCCER Product Development, 13614 Progress Blvd., Alachua, FL 32615.

From *Instrumentation Level Four, Trainee Guide*, Third Edition. NCCER.
Copyright © 2016 by NCCER. Published by Pearson Education. All rights reserved.

12407-16
DISTRIBUTED CONTROL SYSTEMS

Objectives

When you have completed this module, you will be able to do the following:

1. Define a distributed control system and describe its evolution and relationship to other kinds of control systems.
   a. Define a distributed control system.
   b. Describe the evolution of DCS technology.
   c. Compare a DCS to other types of control systems.
2. Identify and describe components and systems related to DCSs.
   a. Describe the hardware components of a typical DCS.
   b. Describe servers and workstations used with DCSs.
   c. Describe DCS fieldbuses, networks, and communications protocols.
   d. Describe human-machine interfaces used with DCSs.
3. Describe common considerations for the maintenance of DCS technology.
   a. Describe various considerations for preventive and/or periodic instrument maintenance.
   b. Describe considerations and approaches to the calibration and repair of instrumentation.
   c. Explain the importance of expertise in the servicing of instrumentation and how information can be obtained.
   d. Identify security issues associated with a DCS and explain how they can be addressed.

Performance Task

Under the supervision of the instructor, you should be able to do the following:

1. Develop a diagram of the basic system architecture of a DCS, including the components and information flow.

Trade Terms

Algorithm	Fieldbus	Partitioning	Supervisory Control
Backbone	Field device	Patch	and Data Acquisition
Blacklist	Firewall	Permissive	(SCADA)
Cloud-based storage	First-out logic	Protocol	Third-party solutions
Drop	Functional	RAID array	Transmission Control
Eavesdropping	Gateway	Real time	Protocol/Internet
Encryption	Hierarchical	Reformed	Protocol (TCP/IP)
Enterprise-level	Historian	Router	Vulnerability
Fail safe	Malware	Server	Whitelist
Fiber optic network	Network switch		

Industry Recognized Credentials

If you are training through an NCCER-accredited sponsor, you may be eligible for credentials from NCCER's Registry. The ID number for this module is 12407-16. Note that this module may have been used in other NCCER curricula and may apply to other level completions. Contact NCCER's Registry at 888.622.3720 or go to **www.nccer.org** for more information.

Contents

- **1.0.0 DCS Basics** 1
  - 1.1.0 Distributed Control 1
    - 1.1.1 DCS Overview 1
  - 1.2.0 DCS Evolution 2
  - 1.3.0 DCS vs. Other Control Systems 3
    - 1.3.1 DCS vs. PLC 4
    - 1.3.2 DCS vs. SCADA 4
    - 1.3.3 Safety Instrumented Systems 4
- **2.0.0 Components of a DCS System** 6
  - 2.1.0 DCS Hardware 7
    - 2.1.1 Field Devices 7
    - 2.1.2 Controllers 7
    - 2.1.3 Controller I/O 8
    - 2.1.4 Controller Applications 8
    - 2.1.5 Redundancy 9
  - 2.2.0 Servers and Workstations 9
    - 2.2.1 The Function of a Server 9
    - 2.2.2 Software Server 11
    - 2.2.3 Database Server 12
    - 2.2.4 Other Servers 12
    - 2.2.5 The Function of a Workstation 13
    - 2.2.6 Engineering Workstations (EWs) 13
    - 2.2.7 Operator Workstations (OWs) 13
    - 2.2.8 Server and Workstation Operating Systems 13
  - 2.3.0 DCS Communications 13
    - 2.3.1 Basic Networking 15
    - 2.3.2 Fieldbuses 15
    - 2.3.3 Modbus 16
    - 2.3.4 PROFIBUS 16
    - 2.3.5 FOUNDATION Fieldbus 16
    - 2.3.6 Higher-Level Networks 17
    - 2.3.7 Ethernet Networks 18
    - 2.3.8 Industrial Ethernet 19
  - 2.4.0 Human-Machine Interfaces 19
    - 2.4.1 Operator Workstation Graphics 20
    - 2.4.2 Process Graphics 20
    - 2.4.3 Analog Control Graphics 21
    - 2.4.4 Discrete Control Graphics 22
    - 2.4.5 Informational Screens and Navigation 23
    - 2.4.6 The Alarming System 23
    - 2.4.7 Detailed Point Displays 24
    - 2.4.8 Trends 25
    - 2.4.9 Historical Trends 25
- **3.0.0 Maintaining a DCS** 27
  - 3.1.0 The Importance of Maintenance 27
    - 3.1.1 Preventative Maintenance Scheduling 27
    - 3.1.2 History Files 28
    - 3.1.3 Maintenance 28

3.2.0	Calibration and Repair	28
3.2.1	Calibration	28
3.2.2	Repair	28
3.2.3	Troubleshooting a DCS	29
3.2.4	Field Device Failures	29
3.2.5	DCS Component Failures	30
3.2.6	Power Supply Failures	30
3.2.7	Controller Failures	30
3.2.8	Network Failures	30
3.3.0	Acquiring Expertise	31
3.3.1	Equipment Knowledge	31
3.3.2	Repair Tool Skills	31
3.4.0	DCS Security	31
3.4.1	Access Control	32
3.4.2	Attacks from Without	32
3.4.3	Attacks from Within	32

Figures and Tables

Figure 1	A typical DCS showing the major components and the levels that make up its hierarchical structure	2
Figure 2	Operator workstations in a control room	3
Figure 3	PLCs are often used for machine control	4
Figure 4	A SCADA station allows plant personnel to interact with the process	4
Figure 5	Controller and I/O module	7
Figure 6	Basic process control channel of a DCS	8
Figure 7	Digital logic drawing	10
Figure 8	Analog logic drawing	11
Figure 9	RAID arrays improve server performance and reliability	11
Figure 10	DCS operator workstation and typical screen	14
Figure 11	The OSI Model for networks	15
Figure 12	DCS using both Modbus and PROFIBUS fieldbuses	17
Figure 13	FOUNDATION fieldbus network	18
Figure 14	Fiber optic network cables	18
Figure 15	Ethernet switch	19
Figure 16	A wireless hot spot	19
Figure 17	Industrial Ethernet connectors	20
Figure 18	Operator workstation process graphic	21
Figure 19	Analog control graphic	22
Figure 20	Alarm screen	23
Figure 21	Detailed point screen	24
Figure 22	Trend display	25
Figure 23	Large corporate networks can be very complex	31
Figure 24	Physical access control is the first step toward network security	32
Table 1	Examples of module I/O	9

SECTION ONE

1.0.0 DCS BASICS

Objective

Define a distributed control system and describe its evolution and relationship to other kinds of control systems.

a. Define a distributed control system.
b. Describe the evolution of DCS technology.
c. Compare a DCS to other types of control systems.

Trade Terms

Drop: A controller, server (see *server*), workstation, or other DCS component connected to a network.

Fail safe: A system that puts itself into a safe or minimally hazardous condition in the event of a failure.

Fiber optic network: A networking technology in which data is transmitted through fiber optic cables as pulses of light.

Fieldbus: A general term for an industrial process network and protocols used to connect instruments, sensors, and controllers together using simple cabling systems.

Hierarchical: Arranged in layers, with higher layers being more sophisticated and performing more supervisory functions than lower ones.

Protocol: A set of rules for communications between two devices. Using a common protocol enables devices to communicate, even if they are made by different manufacturers.

Server: A networked computer that provides information or services to other "client" machines.

Supervisory Control and Data Acquisition (SCADA): An industrial control system that emphasizes data collection and display. Human operators make and execute high-level control decisions over the process through the system.

Third-party solutions: Products developed by an outside company to solve a problem or meet a need associated with a particular manufacturer's technology.

Transmission Control Protocol/Internet Protocol (TCP/IP): A family of communications protocols used to connect many different types of networked computers together. It is the protocol basis of the global Internet.

The key to success in any industrial process is control. Lack of control can result in wasted materials, substandard products, and unsafe working conditions. Industrial control takes many different forms, some of which you may have studied in other modules. One very popular approach to control that is particularly common with large-scale processes is the distributed control system (DCS).

DCS evolved over several decades from basic digital systems designed to replace traditional pneumatic and electronic control systems. Today, DCS is powerful, sophisticated, and highly capable. In many ways, it resembles other digital control systems, such as programmable logic controllers (PLCs). But DCS has its own unique set of features and its own target industries. This section presents DCS basics and compares the DCS to other modern control systems.

1.1.0 Distributed Control

The word *distributed* is the key to understanding DCS. Unlike many historical and present-day control systems, DCS does not rely on control from a single, centralized location. Instead, there are many different control points, with most of them located close to the elements that they're managing. The concept behind DCS is that control is spread out in layers that have a hierarchical relationship to each other. Because DCS doesn't rely on a single, centralized control, the whole system won't go down if a single point fails.

1.1.1 DCS Overview

Broadly speaking, a DCS consists of five major components:

- Sensors and actuators
- Controllers
- Fieldbus networks
- Building networks
- Servers, workstations, and operator workstations

Figure 1 shows a typical DCS arrangement. Sensors and actuators operate at the lowest level. These interface directly with the process, providing feedback about its activities and a means by which to control it. The controller interacts with the sensors and actuators, making decisions on the basis of the data supplied by the sensors. It controls the actuators to modify the process. Large DCSs contain many sensors, actuators, and controllers. Each group provides a local cluster of control.

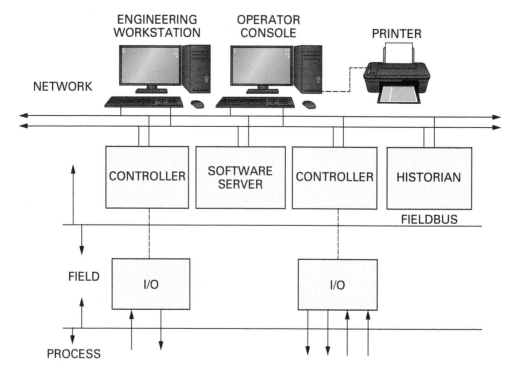

Figure 1 A typical DCS showing the major components and the levels that make up its hierarchical structure.

Fieldbus networks link sensors, actuators, and controllers together using a simple cabling system and a common communications protocol. They streamline wiring and eliminate a lot of unnecessary connectors and complexity. A typical plant environment has many different fieldbus networks, each associated with a particular area of control.

Generally, controllers are connected to the building network, which typically is some form of Ethernet, a fiber optic network, or wireless (Wi-Fi). This network is the high-speed channel that unites all of the higher-level pieces that make up the DCS. Servers and workstations, as well as operator workstations, also connect to the building network. A connection to the network is known as a drop.

Servers and workstations are usually PC-type computers, although smartphones and tablet devices are more commonly being used as workstations. Servers store specific kinds of data and host applications related to the operation of the DCS. Workstations allow plant personnel and engineers to interact with the DCS in various ways, either for ordinary process control or for programming and configuring some component in the DCS.

Operator workstations (OWs), also known as human-machine interfaces (HMIs), provide the means for plant personnel to monitor and control different parts of the process (*Figure 2*). These may be located on the plant floor near the process itself or they may be located in offices or control rooms. OWs vary from simple control panels to sophisticated graphical displays with touchscreens. Smartphone and tablet apps can also function as OWs in some systems.

Because no two DCSs are alike, the information presented in this module is generalized. Each manufacturer implements a DCS differently. Typical plants may contain several different brands of DCS, tied together with special third-party solutions designed to allow them to communicate. DCSs are often organic in that they grow over time and acquire new functions and duties that weren't envisioned when the system was first put together.

1.2.0 DCS Evolution

Today's industrial plants require the control of tremendous amounts of equipment and the ability to relay vast quantities of data back to operations personnel and engineers. This would not be possible without a DCS. The modern DCS is an evolution of earlier control systems—pneumatic and electronic, primarily—whose function was to control multiple processes and provide data to operators. These early control systems were generally referred to as *analog control systems*. Analog systems were very effective at controlling plant processes. Until the 1990s, a DCS could not match their speed in some applications.

Early DCSs were based on the small computers of the late 1960s, known as minicomputers. While small and inexpensive compared to the

Figure 2 Operator workstations in a control room.

large computers of the era, compared to computers today they were anything but small and inexpensive. Each minicomputer controlled specific aspects of a process and gathered data that could be analyzed by larger computer systems. Communications between minicomputers was limited, so these early systems were distributed only to a basic degree.

When the microprocessor was developed in the 1970s, engineers began to experiment with using digital control in localized parts of industrial processes. These early digital controllers were connected by simple networks to larger computer systems that processed data and provided basic user interfaces to the process. Systems grew in sophistication through the 1980s. Networking was rapidly becoming powerful and useful, as was the desktop PC. In some places, networked PCs even began to replace traditional large computers, since they were almost as powerful in many cases.

By the late 1990s, most DCS manufacturers were tapping into the mature commercial IT infrastructure as the basis for much of their technology. Microsoft® operating systems and other software products provided the majority of the server and workstation-level functionality, displacing earlier operating systems and databases. General-purpose networks like Ethernet and communications protocols like Transmission Control Protocol/Internet Protocol (TCP/IP) were handling higher-level network traffic. With the Internet becoming widely available to the general public, companies started using it as a means of communicating between physically distant facilities.

Today, DCS systems are powerful and sophisticated. Industry standards are well established and software continues to be developed, adding new features on a regular basis. Wireless networking, high-quality display technology, web-based interfaces, and portable electronic devices have greatly expanded the OW side of DCS. And with virtually all facilities having on-site networks and the IT personnel to manage them, getting connected is no longer the challenge that it once was.

1.3.0 DCS vs. Other Control Systems

At this point, you may be wondering how DCS differs from other kinds of intelligent process control and automation systems. You've probably noticed that DCS technology sounds a lot like PLC technology. Depending on your background, you might think that it resembles Supervisory Control and Data Acquisition (SCADA) systems as well. So, what's the difference between these three types of control systems? Or is there a difference?

At one time, the differences between the three technologies was much more significant than it is now. Essentially, each technology evolved independently to fill a particular need in industry. While they used similar tools (computer systems, microprocessors, and networks), each system grew to accommodate their particular challenges in different ways.

1.3.1 DCS vs. PLC

PLC technology was developed to replace relay-based logic used in machine control. The original PLCs didn't add any new functionality—they just got rid of the relays. One thing that early PLCs didn't handle was analog control situations. The first PLCs offered discrete control only, which is what the original relay logic did. Later, as microprocessors became more powerful, manufacturers started making PLCs that could do more things, such as interacting with analog information.

The first PLCs were standalone devices, associated with individual machines rather than processes as a whole. As general-purpose computer networks started to appear, PLC manufacturers added networking capability and more sophisticated HMIs. In some facilities, networks of PLCs are the norm rather than the exception.

Today, there isn't nearly as much difference between DCS and PLC technology as there once was. Both use many of the same technologies, particularly in fieldbuses and networks. The real difference is more in their areas of emphasis. PLCs tend to focus on machine control (*Figure 3*), while DCSs focus more on process control. PLCs also can be standalone technologies, associated with just one machine and not connected to anything else. By its very definition, a DCS isn't standalone. In many cases, however, control systems will contain both DCS and PLC elements working together. It's fair to say that DCSs and PLCs have converged, with significant overlap.

1.3.2 DCS vs. SCADA

SCADA was another computer-based control system that began to emerge in the 1960s. Like early DCSs, it used minicomputers to interact with various sensors. SCADA systems did not focus on automatic, feedback-based control. Instead, SCADA systems collected data from sensors using centralized computers that organized, logged, and displayed it. This was the data acquisition aspect of the system. The collected data often got pushed up to large corporate IT systems that used it to prepare reports designed to help facilitate executive-level decisions.

SCADA systems were also often used to link physically distant facilities, such as pipeline control and power plants. SCADA didn't emphasize automatic control. Instead, it provided convenient HMIs for humans to use in making control decisions and adjustments based on the data it supplied (*Figure 4*).

Today, SCADA, PLCs, and DCSs often work together. PLCs and DCSs perform automatic control functions, while SCADA systems often collect and display results or allow high-level process adjustment decisions to be entered by human supervisors. In a large facility, it's likely that you could encounter all three systems performing different tasks in one giant system. Much of the technology is the same; the differences lie in the aspect that is being emphasized.

1.3.3 Safety Instrumented Systems

A safety instrumented system (SIS) is a parallel system that is sometimes found alongside any of the industrial control systems already discussed. An SIS exists for the purpose of safety in situations where a lack of safety could be catastrophic.

In many industrial processes, a control system failure could result in a bad product or a loss of revenue, but the results wouldn't actually be dangerous. However, there are some industrial processes, known as *critical processes*, that have almost no margin for failure. A malfunctioning control system could result in an explosion, fire, dangerous chemical reaction, or containment failure. For processes of this type, an extra layer of safety is necessary. The SIS meets this need.

Figure 3 PLCs are often used for machine control.

Figure 4 A SCADA station allows plant personnel to interact with the process.

The SIS is an independent control system that runs in parallel with other control systems. Its sole purpose is to guarantee that the process never enters a dangerous condition. Should a failure occur, the SIS ensures that the process goes into a fail safe condition. It also works to ensure that dangerous conditions don't develop in the first place.

For example, a burner management system (BMS) ensures the safety of burners that heat boilers. The BMS monitors combustion, fuel and air delivery, and burner performance. Should an unsafe condition develop, it shuts everything down in an appropriate manner. The 2015 *NFPA 85, Boiler and Combustion System Hazards Code* defines the rules for a BMS. As with all SIS elements, a BMS is separate from other parts of the DCS, so a failure in the main control system won't compromise the safety of the system. The Triconex® system by Schneider Electric™ is one example of an SIS family of components that includes BMS capabilities.

Various standards bodies define the parameters of an SIS. *IEC 61511* and *IEC 61508* are two examples connected with the process industry. The IEC has similar standards for other industries, such as the nuclear power industry.

Additional Resources

The following websites offer resources for products and training:

ABB, a global leader in power and automation technologies, **www.abb.com**

Emerson Process Management, **www.emersonprocess.com**

Fieldbus Foundation, part of the FieldComm Group™, **www.fieldcommgroup.org**

Honeywell International, **www.honeywellprocess.com**

PROFIBUS®, the world leaders in industrial networking, **www.profibus.com**

1.0.0 Section Review

1. A modern DCS provides control over a large process through the use of _____.
    a. a centralized controller
    b. lots of connected components
    c. a distant controller on the Internet
    d. several minicomputers

2. Sensors, actuators, and controllers communicate using _____.
    a. the Internet
    b. the building network
    c. personal computers
    d. a fieldbus network

3. The 1970s-era invention that made the modern DCS possible is _____.
    a. server technology
    b. minicomputers
    c. fieldbuses
    d. the microprocessor

4. Which technology emphasizes data collection and display with the purpose of helping humans make control decisions?
    a. DCS
    b. PLC
    c. SCADA
    d. HMI

SECTION TWO

2.0.0 COMPONENTS OF A DCS SYSTEM

Objective

Identify and describe components and systems related to DCSs.
a. Describe the hardware components of a typical DCS.
b. Describe servers and workstations used with DCSs.
c. Describe DCS fieldbuses, networks, and communications protocols.
d. Describe human-machine interfaces used with DCSs.

Performance Task

1. Develop a diagram of the basic system architecture of a DCS, including the components and information flow.

Trade Terms

Algorithm: In the context of computer programming, this refers to code that performs a specific task. In analog control systems, it refers to symbols for components that perform various analog functions, such as PIDs and summers.

Backbone: A high-speed network channel designed to connect multiple networks together, usually across a facility, campus, or geographically distant area. Backbones often use fiber optic technology.

Cloud-based storage: Remote file storage accessed through the Internet and provided by a service company that maintains one or more servers in various locations.

Encryption: A security measure against data interception in which the data is turned into an unreadable form using cipher techniques prior to being transmitted. It is returned to its original form upon receipt.

Enterprise-level: Technology designed to provide high-level information services for company executives in order to facilitate large-scale planning and decision-making.

Field device: A general term for a sensor or actuator that interacts directly with a process. Field devices are at the bottom of the DCS hierarchy.

Firewall: A network device that divides a network into two sections and prevents unauthorized accesses or protocols from crossing between the sections.

First-out logic: The first element (possibly of several) that caused a process trip.

Functional: Diagram of a digital logic system expressed as connected logic components.

Gateway: A network device that translates from one type of physical layer and/or protocol to another. Gateways allow normally incompatible systems to communicate.

Historian: A program that keeps a record of process data for later analysis and to reveal trends. Historians can run on dedicated computers or on a server along with other programs.

Network switch: A kind of network hub to which many devices connect. Switches perform intelligent traffic management, channeling data between their ports based on hardware addresses.

Partitioning: Dividing process I/O across multiple drops or multiple I/O cards in order to prevent a failure from causing a process shutdown.

Patch: A fix for defects discovered in software after its release. Many patches are security-oriented.

Permissive: A condition in a control system that must be met before an action can occur. For example, a group of series-wired sensor switches would be permissives in an electrical control system.

RAID array: Stands for Redundant Array of Independent Disks; a group of hard drives acting as a unit in order to improve performance and/or provide error correction and redundancy.

Real time: A computer system that responds with minimal and predictable delays to events as they happen. Critical control systems must be real time.

Router: A network traffic manager that connects multiple networks together. Routers examine data at the protocol levels to decide where it should be forwarded.

The previous section described DCSs and explained how they differ from other modern control systems. A basic overview of the pieces that make up a DCS was provided. This section describes those pieces in greater detail. Since the whole philosophy of distributed control lies in delegating control to many smaller devices, it's crucial to understand how the different parts interact with each other. It's also important to understand the overall system hierarchy and how it is viewed by plant personnel, supervisors, management, and executives.

2.1.0 DCS Hardware

At the most fundamental level, a DCS consists of many different pieces of hardware that work together to form the complete system. Many of these pieces may already be familiar; they may even be found in non-DCS environments. This section explains how these pieces of hardware can work together to provide distributed control.

2.1.1 Field Devices

A field device includes sensors that monitor temperature, pressure, level, and flow. More exotic sensors measure chemical composition or detect the presence of gases. Sensors often connect to transmitters that contain transducers, signal conditioners, and amplifiers to prepare the signal before sending it on. Actuator field devices include relays and solid-state elements that control pumps, heating elements, valves, motors, and positioning actuators.

DCSs can interact with traditional field devices that don't include much intelligence. But in the modern industrial environment, it's more common for a field device to be smart. This term indicates that the device includes an embedded microprocessor to manage it and communicate with the outside world over a fieldbus using one of several industry standard control protocols. Smart field devices offer the significant advantage that they can provide much more information to the DCS than ordinary field devices. They also make it possible to perform tasks such as diagnostics, testing, and calibration more easily and, in some cases, remotely.

2.1.2 Controllers

In many ways, a controller (*Figure 5*) is similar to a PLC. It scans its inputs and sends the data obtained to application programs loaded into its processor's memory. The processor, in turn, scans the logic, turns discrete outputs on or off, and adjusts analog output signals to control the plant processes. Analog signals are continually adjusted based on the changing inputs.

Figure 6 shows a basic process control channel that's part of a DCS. Notice that the DCS controller fits between the transmitter and the final control element. It provides both indication (display) and control functions. Internally, the DCS controller runs the appropriate control algorithm (such as PID, for example) to manage the final element.

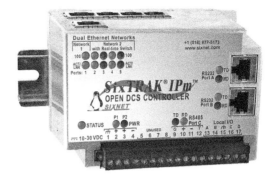

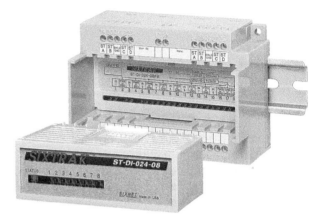

Figure 5 Controller and I/O module.

Another similarity to PLCs is that the I/O is mapped to the processor, which needs to know what type of I/O card resides in each location. This is often a slot with a hexadecimal address. Controller capabilities extend beyond that of an ordinary PLC. The following list explains the differences:

- Controllers are in run mode all the time. There is no way to globally turn off controller outputs other than by taking the controller off line.
- Networking capability is usually built into the controller module; some PLCs implement networking through an external card or module.
- Individual points in a controller can be removed from scan. This includes intermediate processing points that do not directly attach to field I/O. The processor can still monitor the raw value, but the point will simply hold its last value until returned to scan.
- Controllers are programmed and monitored directly across the network. Programming changes can be made during run time. PLCs usually must be taken out of run mode for programming.
- The entire DCS database resides on a server; the controller only gets the piece it needs.

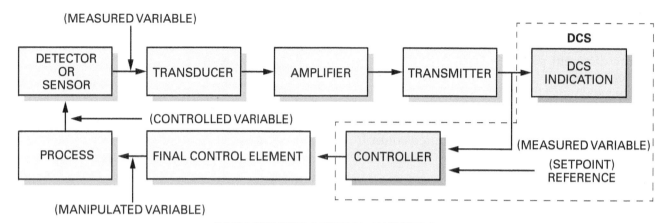

Figure 6 Basic process control channel of a DCS.

- Controllers almost always operate in redundant pairs, and care must be taken to prevent mismatches.
- Alarm information is stored as part of controller point data, and the data is immediately sent to the operator when limits are exceeded.
- All application logic and commentary is stored in the controller, so it can be seen by the user regardless of the interface being used.

2.1.3 Controller I/O

To be able to work efficiently with a DCS, a technician must have a complete understanding of how the process points relate to I/O and the various field devices. Most process measurements involve temperature, pressure, flow, speed, voltage, or current. Transmitters often output analog 4–20 mA signals. Sensors with discrete outputs typically close or open relay or solid-state contacts. Actuators with proportional control capability can accept 4–20 mA signals. Those that need discrete control also interact by switch contacts of some type.

When selecting the I/O for the DCS, the project engineer develops an instrument list of all the field devices used in the project. Devices to be connected to the DCS are identified. The next step is to determine which devices to send to which controller drops. For example, if filtered water pump "A" is part of a water treatment system and Drop 7 has the control application for that pump, all the I/O associated with that pump will go to Drop 7. This might include many different types of I/O, such as a pump temperature sensor (a thermocouple), a pump starter control (a relay), and a pump running sensor (a dry contact).

It is important to recognize that controllers must control their parts of the plant independently, without operator intervention in many cases. An exception to this would be critical pairs of equipment that could cause a plant trip if the drop they were in failed. In that case, it may be necessary to place the I/O in two different drops. This is called partitioning. It also common to partition within a drop and send the critical pair signals to two different I/O cards.

Table 1 shows some typical inputs and outputs and the modules to which they are connected. As you can see, the field devices are wired to compatible modules. DCS modules tend to be more specialized and expensive than PLC modules. As PLC and DCS technologies evolve, however, the modules are becoming more versatile, allowing a plant to maintain fewer spares.

2.1.4 Controller Applications

The programs that run in a controller are technically software, not hardware. But without them, the controller's hardware does nothing. Normally, there are at least two pieces of software running in the controller: the operating system and the application program.

Controller operating systems are similar to the ones that run on a regular personal computer. But in most cases they are more specialized and also require less memory and processor power. More significantly, controller operating systems are almost always real time. This means that they can respond to events, such as changes in I/O values, rapidly and in a predictable amount of time. Ordinary personal computer operating

Table 1 Examples of module I/O.

Inputs and Outputs	Channel	Monitored Function	Sensor Type
125VDC 16 Point Digital Input Card	Chan 2	Bunker D High Level	Level Switch
125VDC 16 Point Digital Input Card	Chan 4	Spray Wash On	Pressure Switch
32 Point Digital Output Card (Form C)	Chan 6	Spray Wash Start	Dry Contact Output
16 Point Analog Input Card (4–20 mA)	Chan 2	Spray Wash Pressure	Pressure Transmitter
16 Point Analog Input Card (4–20 mA)	Chan 3	Bunker D Heater Temp	Temperature Transmitter
8 Point Analog Input Card (Type J TC)	Chan 6	Bunker D Coal Temp	Type J Thermocouple
8 Point Analog Input Card (100 Plt. RTD)	Chan 8	Wash Motor Stator Temp	100 Ohm Platinum RTD
8 Point Analog Output Card (4–20 mA)	Chan 1	Spray Wash Pressure Valve	Control Valve I/P

systems are not real time, so there can be significant delays in responding to events. Obviously, in a process environment, delay is rarely acceptable.

The application program encompasses all the discrete and analog logic. This includes loops for monitoring as well as controlling the process. All control logic is either discrete or analog. Regardless of the methodology used, the end result must be to turn something on or off, or to position a final control element, such as a valve. At one time, discrete logic was separate from analog logic in the DCS. Today, it can be integrated on the same diagram.

Historically, control system logic was written on a diagrams called a functional. The symbols used to represent digital logic included AND and OR gates, as well as other common gates. Symbols for analog control were called algorithms, and included summing devices (summers), function generators, controller PIDs, high signal selects, and other devices. Once the diagrams were drawn, electronic or pneumatic modules were selected to perform the control and connected together. *Figure 7* and *Figure 8* show examples of these drawings.

In many DCSs, the applications engineer can create controller programs graphically so they look very much like traditional diagrams. This makes programming quick and efficient. Each manufacturer has its own approach, however, so what you encounter in the workplace will vary.

2.1.5 Redundancy

When designing a process, engineers pay particular attention to critical components that can shut down the process if they fail. The term *single point of failure* is often used to define such components. To avoid the consequences of single points of failure, process designers sometimes build redundancy into the process, which simply means that they duplicate critical components. How much redundancy gets designed in depends on the potential impact of a given failure. Components that are often duplicated include controllers, I/O cards, and power supplies. Servers and communications channels (fieldbuses and networks) may also be redundant.

2.2.0 Servers and Workstations

Two very important classes of equipment found in DCSs are servers and workstations. These are computers or computer-like devices connected to a network. A typical DCS contains a number of servers and workstations, each of which performs tasks specific to some part of process control.

2.2.1 The Function of a Server

Servers, as their name implies, provide services of some type to other devices (clients) on the network. For example, a file server stores and distributes files. Client computers access the files stored on the file server. A database server runs a high-end database application that manages the tables, indexes, and report forms associated with a database. Client computers send requests to the database server, which it addresses in appropriate ways. A web server stores static web pages and generates dynamic-content pages as required. These are delivered in response to requests from remote browsers connected to the Internet.

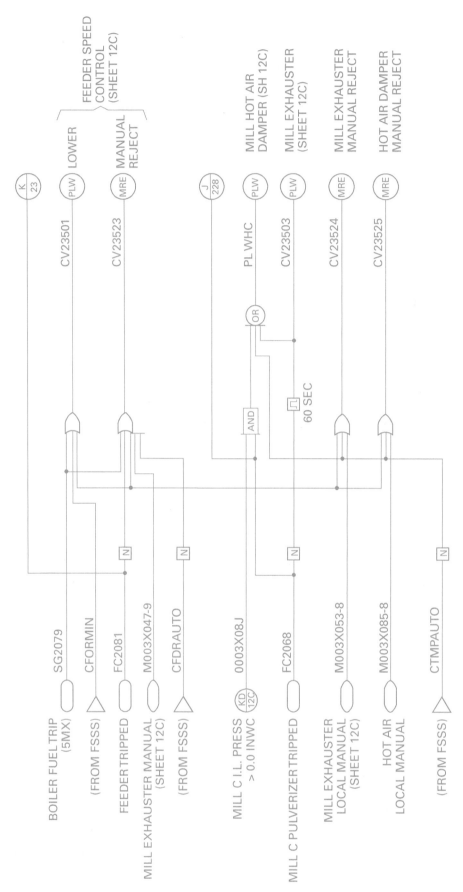

Figure 7 Digital logic drawing.

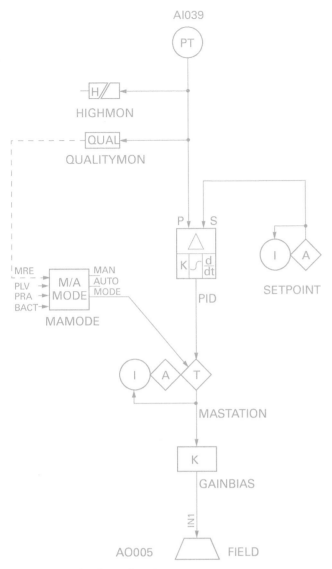

Figure 8 Analog logic drawing.

Figure 9 RAID arrays improve server performance and reliability.

Servers are usually PC-type machines. While any computer can be set up to act as a server, PCs configured and sold as servers usually contain high-reliability components to reduce the risk of a failure that could disable essential system functions. Examples of these components include error-correcting memory and redundant power supplies.

Servers often have larger and faster hard drives, particularly if they are used as file servers. File servers and other types of servers that must maintain data integrity typically contain RAID arrays (*Figure 9*). A RAID array is a group of hard drives that keeps multiple copies of the same data, along with error-correcting information. In the event of a drive failure, the remaining drives will keep services running and can be used to reconstruct the failed drive's contents when it's replaced. RAID arrays also can offer higher disk performance since data can flow from several drives at once.

Many servers include some type of backup system so critical information can be stored in a separate location. In the event of a server failure, fire, natural disaster, or theft, the backup can be used to get a replacement system up and running quickly. Backup systems can include external hard drives, magnetic tapes, optical media (DVDs), and cloud-based storage. How often backups are made depends on the server's importance and the potential costs and consequences of downtime.

2.2.2 Software Server

In a DCS, the main job of the software server is to control and store the system software files. It holds the master database and plant graphic files. No other computer can use the programs that modify the database and graphics without use of the software server. This maintains integrity of these files. The software server also may include a security server that restricts access to the system or limits the actions that different company personnel may perform. Software servers may be one or more physical computers. A single machine may run the various server programs, or they may be distributed over multiple machines.

For convenience, a software server may have a monitor and keyboard attached to enable engineers to use the application programs loaded in its memory. In this capacity, it can be used to load software onto operator workstations and controllers. This task may be necessary in the event of a failure elsewhere on the network. Other computers may access these same programs across the network as well. Because of this, controls may be in place to prevent two or more computers from modifying the same database or controller logic simultaneously.

The following list identifies some of the major functions of the software server:

- Maintains the DCS database in the proper location.
- Automatically updates the database with point and controls changes.
- Provides a portion of the database to OWs/HMIs when they come on line.
- Keeps track of the network address (or name) of each drop on the network.
- Maintains the system configuration files for each drop.
- Provides a mechanism for backing up and restoring important files.
- Provides a way to reload workstations.
- Keeps a master copy of the graphics for the OWs.
- Keeps copies of the control logic for each of the controllers.
- Allows users to log in at various security levels to perform tasks.
- Has controls and structure for remote access by other computers.
- Prevents users from accessing software without a valid license key.
- Continuously logs activity on the system and writes errors to files in appropriate locations.
- Allows users to install a software upgrade and security patch in an orderly fashion.

Technicians generally do not perform administrative tasks on the server, but it is important for them to know how the software server functions if they work on the system at all. At times they may be required to perform some duties through the server, so they should have an overall understanding of its operation and purpose.

2.2.3 Database Server

A database server is typically part of a DCS system. Not only must the database store information about the system itself, but it stores information about the ongoing process that can be used for executive level decision-making. The database server will probably be running one of the industry standard database platforms such as Microsoft SQL Server®, Oracle®, or Sybase®.

When a project is developed for a DCS, a database is usually laid out with a program such as Microsoft Access®. The database will have most of the parameters for each point (a channel on an I/O card) as determined by the PIDs or loop sheets. It is the project engineer's job to accurately fill in the database with the appropriate information. All DCS points will need some of the following basic information:

- A point ID number may include alpha characters, and it will be size-limited. This is generally not the tag name found on the loop sheet. A point ID often includes mnemonics containing drop numbers, plant location codes, and other information.
- A plain-language description of the point, such as "steam sealing pressure"
- The drop in which the point is located
- The point's address (which card and channel)
- A termination point (which field wires go where)
- The type of card that is processing the raw data
- High and low alarm limits (analog)
- Alarm state (if digital, does a 0 or 1 equal an alarm)
- A scaling factor that could convert the voltage measured on the card to a value in some engineering unit.

While the DCS will have a mechanism for building individual points, the database is generally imported into the DCS. Once the database has been completed and loaded into each controller, it is also updated on the software server, which maintains the global database for the DCS. Subsequent point parameter changes or additions of database points made in the controller will also be automatically updated on the server.

2.2.4 Other Servers

Other servers that perform DCS functions may be present as well. In many cases, these perform non-critical functions and don't require the same level of reliability as other servers. An example of such a server is a historian.

The job of the historian is to track data values of process points and write these to a hard drive so that they can be retrieved later for analyzing trends. Trends are graphical representations of process data that can be displayed on an OW or used for management-level decisions. Trends can always be displayed in real time, but this is a resource-intensive process.

Historical data is useful in a number of ways. Control technicians use it to help troubleshoot plant problems. Engineers can use it as a baseline of equipment performance and to track degradation over time (to determine when to shut down the equipment or the production line for maintenance). Operators can use it to look back and see how equipment behaved on a previous shift or to help determine what caused a plant trip or failure.

A historian does not have a keyboard and monitor, so personnel don't interact with it directly. It may be its own computer or it may be an application running on another server. A similar application, called a *logger*, can perform historian-like tasks, storing data in tabular format after specific events.

Another kind of server found on larger DCSs provides enterprise-level services. These are high-level functions that company management uses to make large-scale decisions. Enterprise servers collect and format data in forms that can be used for analysis and reports. They also can provide functions that allow multiple sites in geographically distant locations to interact and be managed.

2.2.5 The Function of a Workstation

A workstation is usually a client computer that is used in a day-to-day environment to perform one or more specific tasks. Workstations are usually less powerful than servers but are more numerous. In a conventional office environment, ordinary computers are clients on the corporate network. They access file servers and web servers as a part of their normal activities, and they perform a wide variety of workstation tasks. In a DCS environment, workstations are a bit more specialized. Broadly, they fall into two categories: engineering workstations and operator workstations.

2.2.6 Engineering Workstations (EWs)

Engineers use certain types of software to manage the DCS. The applications most frequently involved are listed here:

- A program to build and modify operator workstation graphics
- A program to build and modify control logic
- Software to configure and maintain drops in the system
- An I/O builder
- A program to load controllers with control logic
- Various diagnostic tools

Engineering workstations allow engineers and technicians to perform these tasks. In some systems, the software server can also function as an engineering workstation. But in many cases, this isn't convenient or desirable, so the EWs will be separate computers.

Because a DCS will vary in how the different programs are used, details about how to use them will not be explained here. Technicians should use vendor documentation and attend classes to gain proficiency in a particular system.

2.2.7 Operator Workstations (OWs)

The primary job of an operator workstation (*Figure 10*) is to enable an operator to control plant processes. Because the output of many plants is worth millions of dollars per day, safe and economical operation of the plant from the control room is critical. The operator workstation is loaded with all the graphics and alarm screens that an operator needs to see everything that is happening in the plant. A well-designed control room will allow an operator to monitor all the plant processes on the DCS, quickly handle emergencies, and communicate with attendants in the field. Operator workstations will be discussed in greater detail later in this module.

2.2.8 Server and Workstation Operating Systems

Servers and workstations run conventional operating systems just like the ones used in general-purpose IT environments. Older DCSs frequently used Unix®-based operating systems for servers and some workstations, but these have largely been replaced with Microsoft® operating systems. Servers typically run Microsoft Server 2008 or 2012, while workstations usually employ ordinary desktop versions of Microsoft Windows. While DCSs tend to be Microsoft-centric, other servers on the network that may interact with the DCS could use operating systems such as Linux®. Mobile OWs based on smartphone or tablet technologies run on the operating systems normally associated with those devices.

2.3.0 DCS Communications

A distributed control system is all about communications. Without it, control functions couldn't be distributed across multiple devices spread over physically distant areas. For this reason, DCSs make extensive use of networking technologies. A typical DCS hierarchy has at least two levels of networks. One is fairly low level (the fieldbus), while the other is usually much more powerful and sophisticated (the corporate building network). Before exploring the different types of DCS networks, it's essential to understand some basic networking concepts.

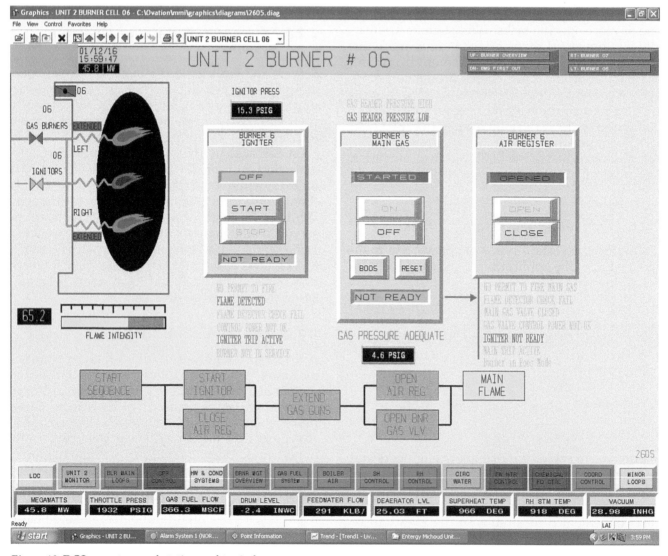

Figure 10 DCS operator workstation and typical screen.

2.3.1 Basic Networking

There are many different kinds of computer networks. All do essentially the same thing—they connect devices together so they can share information. Networking is a unique profession, and it cannot be taught in a single module. But since networking is so significant in today's instrumentation workplace, it's necessary to understand some basic networking concepts.

Network engineers commonly describe networks in terms of layers. Each layer performs some part of the task of sending information from point A to point B. There are a number of network models with different numbers of layers. A very common one is the OSI Model, which has seven layers (*Figure 11*). While coverage of all seven of its layers is beyond the scope of this module, a brief summary of layers 1, 2, and 3–7 is presented here.

All networks have what is commonly called a physical layer (Layer 1 in the OSI Model). This layer includes cables, connectors, electrical/optical/RF specifications, and the way in which binary data is represented by electrical, light, or radio-frequency pulses. Networking technologies like Ethernet, Wi-Fi, RS-485, and RS-422 all have a physical layer aspect because they define many of the physical specifications for the network hardware.

The layers above Layer 1 are essentially protocol layers. A protocol is a set of rules for organizing binary data, sending and receiving it, and handling problems that may occur as it flows through the network. In a sense, protocols have to do with software since some type of software has to arrange the data according to the protocol's rules in order for the physical layer to send it.

At the lowest protocol level (Layer 2) are the rules for arranging the data into groups of bits called *frames*. These are then given an identifier, called an address, that's unique to the device. Networking technologies like Ethernet and Wi-Fi also have a Layer 2 presence since they perform this function as well.

Higher protocol levels (Layers 3 through 7) handle more complex tasks, like controlling where data goes, performing error detection and correction, and managing higher-level interactions between devices. Well-known protocols like TCP/IP, the protocol on which the global Internet is based, operate on some of these layers. Be aware that not all of the layers are used in every circumstance.

An important concept to grasp is that protocols can travel on top of different physical layers. That's the reason that Internet traffic travels over so many different networking technologies, such

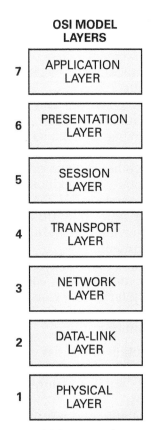

Figure 11 The OSI Model for networks.

as Ethernet, fiber optic, satellite, and wireless networks. In each case, the physical layer is different but the crucial protocols are the same. In many DCS networks, industry standard protocols define how devices communicate, but the physical layers that actually move the bits around can vary from installation to installation.

Finally, protocols can be embedded inside of other protocols. This happens frequently in DCS environments. Consider an example in which a device needs to communicate with a computer that's in another building half a mile away. Both devices speak a common DCS protocol but the physical layer that the devices would normally use has a distance limit of a few hundred feet. But both buildings have a regular corporate network that's also connected to the Internet. By embedding the DCS protocol inside the Internet protocol, it becomes possible to send it back and forth over the Internet.

2.3.2 Fieldbuses

In the early days of DCSs, almost all networking technology was proprietary (controlled by a single vendor). Eventually, as the technology matured, there was pressure to develop industry standards so users would have more choices. The result was a collection of protocols that allow the creation of a low-level network called a fieldbus. It's crucial to understand that the term *fieldbus* is both a generic name for a class of technology and, in certain circumstances, a specific brand (FOUNDATION fieldbus).

Fieldbuses link sensors and actuators at the process level with DCSs, PLCs, and other control devices. Employing a fieldbus reduces system complexity by using a simple cabling system to link devices together. Compared to the networks used for corporate IT, fieldbuses are relatively simple and low-speed. But they are more than fast enough for what they do and they're very well designed for the demanding and hostile industrial environment.

Fieldbuses must meet strict guidelines regarding modularity, maintenance, ease of installation, reliability, and interference resistance. They offer the following advantages over older techniques and analog communications:

- Higher resolution of process values
- Reduction of planning costs and maintenance
- Bidirectional digital communication
- Simple cabling
- Easy expandability and retrofitting
- Higher safety through self-monitoring

Fieldbuses rely on digital communications techniques. But because analog technologies are still common in many facilities, a fieldbus must be able to co-exist, at least to some degree. For a fieldbus to be universal, it would have to cover problems inherent in both factory automation and process control.

There are a variety of solutions to the concerns of process control and automation. One approach co-exists with existing technology. Another approach focuses on a new, more-powerful technology that stresses speed and reliability. Smart transmitters using the HART® and Modbus® protocols belong to the first group. FOUNDATION fieldbus and PROFIBUS® are in the second.

2.3.3 Modbus

Modbus is an older technology, but it is still common in industry. It uses a serial communications physical layer and a protocol developed for the Modicon line of PLCs in 1979. There are three versions of Modbus, all of which share these operating principles:

- There is one communications master.
- All other devices (nodes) are slaves.
- Each node has a numeric address.
- Nodes cannot transmit until given permission by the master.

- There can be up to 247 nodes on the network.
- All nodes hear all network traffic but each node responds only to its own address.
- A command sent to node 0 is a broadcast, so all nodes accepts the command.

Although designed in the 1970s, Modbus is a good tool to use with DCSs that communicate with PLCs and other similar devices. There is also a newer version of Modbus called Modbus TCP/IP that can be embedded inside the TCP/IP protocols used by more powerful corporate networks and the Internet. It has a number of advantages over traditional Modbus, speed in particular. It also has overcome some of the limitations imposed by the original system.

2.3.4 PROFIBUS

PROFIBUS (Process Bus) was first developed in 1989 in Germany. A group of 21 companies and institutes developed this fieldbus technology. There are two versions of the protocol, PROFIBUS DP and PROFIBUS PA. It can run over a number of different physical layers at several different speeds.

One advantage to PROFIBUS is that it has self-diagnostic capabilities. The physical layer specifications for the PA version also include a cabling system that can be used in explosive environments, a definite advantage in some industries. PROFIBUS is prevalent in Europe, and is supported by DCS vendors such as Siemens and Yokogawa. *Figure 12* shows a DCS that's a combination of PROFIBUS and Modbus.

2.3.5 FOUNDATION Fieldbus

FOUNDATION fieldbus, which was developed to replace 4–20 mA analog technology, is very similar to PROFIBUS. Currently, there are two versions designed for different physical layers: the low-speed H1 and the high-speed HSE. The H1 version can supply power using the same cable that handles communications.

FOUNDATION fieldbus is predominant in North America. Emerson Process Management's DeltaV™ and Foxboro® DCS systems have used it extensively. A recent Emerson white paper claims that terminations can be reduced 75 percent by using this technology. The DeltaV FOUNDATION fieldbus incorporates junction boxes that collect wiring from six instruments. A single cable from each of the junction boxes is brought back to the I/O cards. These junction boxes are shown in *Figure 13* as the lower three spur connections on each side of the junction blocks.

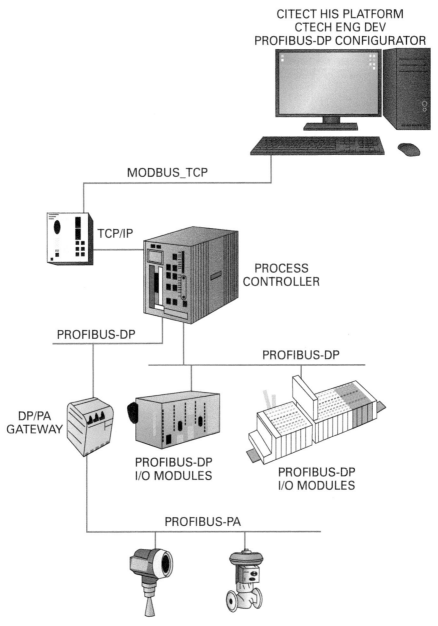

Figure 12 DCS using both Modbus and PROFIBUS fieldbuses.

A single cable from the junction blocks is brought back to the I/O cards. This is the H1 trunk line. An I/O card can handle up to 64 signals.

2.3.6 Higher-Level Networks

Fieldbus networks tend to be fairly simple and not very fast—usually less than 15 megabits per second (Mbps). Often they are limited to a few hundred devices on a single network. However, they are more than sufficient for what they do. Even more importantly, they're highly reliable. But in many cases, a more robust network is needed to handle some of the other pieces of the DCS system, such as the servers, OWs, and EWs.

In almost all environments, a standard IT-type network is already present in the building. It may connect hundreds, or even thousands, of general-purpose computers and servers together. It may also link physically distant branches of the company over the Internet or dedicated leased lines. These kinds of networks are usually very high speed (gigabits per second, or Gbps). So, it makes sense for a DCS to take advantage of what's already there, along with the expertise of the IT personnel and network engineers.

Corporate networks generally rely on Ethernet technologies running over copper or a combination of copper and fiber optic. Fiber has the advantage of high speeds over long distances,

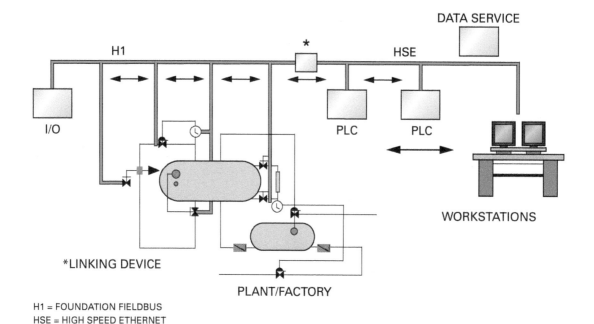

Figure 13 FOUNDATION fieldbus network.

and it is not subject to problems due to electrical interference (*Figure 14*). Many corporate networks also provide wireless hot spots (Wi-Fi) for mobile device access.

2.3.7 Ethernet Networks

Ethernet is a technology that's been around for a relatively long time. It has steadily improved to keep pace with contemporary needs, and newer versions support very high speeds. Ordinary office environment Ethernet running over UTP wiring reaches speeds up to 1 Gbps. Higher-speed versions are available (up to 100 Gbps), but these run mostly over fiber optic cables or over copper for very short distances (less than a few meters).

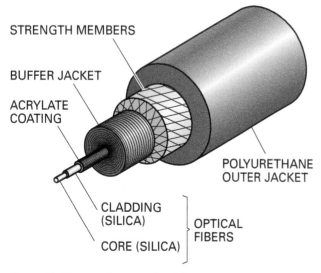

Figure 14 Fiber optic network cables.

Ethernet technology is popular due to its great flexibility and relatively low cost. It usually runs over UTP, but fiber is frequently used as a backbone to join network sections over longer distances. Fiber is desirable in some industrial environments because it is virtually immune to EMI. However, it is much more costly than copper, and requires specialized training to install. Generally, IT personnel or outside contractors perform fiber installations.

A typical Ethernet network includes a number of devices that manage network traffic. At the bottom are network switches (*Figure 15*). Every device on the network has a cable that runs to a switch, which is usually located in a wiring closet or rack room. The network switch manages traffic between different devices so data flows efficiently and doesn't go to places where it isn't needed. Switches can be small, with just a few ports; these are used in homes and small offices. But they can be large as well, with hundreds of ports.

In most corporate networks, a router connects switches together, essentially combining several smaller networks into a larger one. Routers can also be used to join a corporate network to the Internet. A router is a more sophisticated traffic manager than a switch. It examines all network data before forwarding it to the correct smaller network (or to the Internet). Routers keep the network as a whole from becoming congested by traffic. By sending data to the proper location, routers make the best use of the network's capacity. Large corporate networks have many routers since they carry vast amounts of data.

Figure 15 Ethernet switch.

A firewall is a device designed to protect parts of the network from unauthorized access. Firewalls are often built into other pieces of equipment, such as routers. Firewalls block traffic from certain locations and also prevent specific protocols from crossing from one part of the network to the other. A firewall is often part of a DCS that's sharing a corporate network. Since the DCS may be controlling mission-critical processes, it's essential to keep unwanted or potentially malicious traffic away. If a corporate network is connected to the global Internet (most are), firewalls prevent outsiders from gaining inappropriate access to the network, where they could steal sensitive information or interfere with the operation of critical systems like DCSs.

Finally, many Ethernet networks include wireless hot spots (*Figure 16*) that allow mobile electronics to access the network. Wireless networks use a different physical layer and lower-level protocols than Ethernet, but because they can carry most of the industry standard protocols, wireless networks interact smoothly with wired ones. Wireless networks must be properly secured by passwords and encryption to prevent them from being exploited by malicious users.

2.3.8 Industrial Ethernet

Years ago, as Ethernet became popular in the general IT community, industry decided that it was worthwhile to create a ruggedized version. This would allow industrial devices to take advantage of existing network infrastructures and expertise but still have the durability required for hostile plant environments.

Industrial Ethernet has the same basic specifications as ordinary Ethernet and it's compatible with the standard devices and wiring systems.

Figure 16 A wireless hot spot.

The difference is that industrial versions of these components are housed in protected enclosures, have heavy-duty connectors, and better shielding. Copper cabling is usually STP instead of UTP for better EMI rejection. Terminations have extra shielding and sometimes locking or liquidtight shells (*Figure 17*). Fiber often joins physically distant components to keep EMI interference to a minimum.

Many newer industrial control devices are already equipped with Ethernet connections. Older devices can use a gateway that translates their physical layer and its associated protocols to industrial Ethernet. While this approach seems highly desirable, it's not always a good idea. Many older industrial networks and fieldbuses were optimized for DCS traffic, where the amount of data involved was small but real-time response was required. Ethernet was not designed for this kind of work and can be insufficiently responsive in some situations. Consequently, understanding process control requirements is essential before making technology decisions.

2.4.0 Human-Machine Interfaces

If you have ever been in the cockpit of a commercial airliner, you were probably impressed by the incredible array of gauges, switches, and electronic indicators that you saw. Before the advent of the DCS, plants used a similar approach with dedicated switches, indicators, and recorders for monitoring and controlling each part of the process. Controls were located in logical groupings, and operators developed an intimate familiarity with the process. Process upsets could be handled quickly and predictably in most cases.

The downside of this arrangement was the unreliability of the equipment. Recorder pens would run out of ink, indicators would stick, and dirty switch contacts would cause intermittent signals. Much of the equipment, in particular the alarm modules, needed illumination. Bulb replacement was never-ending. Maintenance was time consuming, and there never seemed to be enough time to tune controls or maintain field equipment.

DCSs reduced the maintenance overhead by removing much of the hardware, but they created another problem—operators could not monitor the process on just one screen. This could be alleviated with multiple screens, but graphics and alarming still had to be designed so that operators could process information as quickly as possible in order to safely and effectively control the plant.

The human-machine interface (HMI) of a DCS is usually called the operator workstation (OW). OWs can include the following components (not all stations have each of these features):

- Process graphics, called diagrams, to monitor the process
- Control diagrams with M/A stations
- Informational screens
- A status diagram showing the controllers and network
- A function-specific keyboard for activating controls or a conventional computer keyboard and mouse
- Large, high resolution monitors
- An alarm screen
- Screens with detailed point information
- Diagrams with real-time display of the control logic
- A trending component
- An interface for displaying historical information
- A printer for general use

Operator workstations are updated by controllers and servers across the network. When an operator workstation first comes online, it gets the portion of the database it needs to operate from the software server. If the software server goes offline, it will have little effect on the workstation, since it already has the database. If the workstation boots up without a software server present, it will use a copy of the database from the last time it was running.

Figure 17 Industrial Ethernet connectors.

2.4.1 Operator Workstation Graphics

Graphics tend to fall into three categories: process, control, and informational diagrams. Colorful and highly representational displays can help the operator recognize equipment quickly, but extensive details can overload the operator with too much information. One way to keep the number of graphics to a minimum is to put a great deal of information in one graphic. Depending on the size of the process, a better solution is to have graphics displayed on two or more screens, in resizable windows. This gives the operator greater flexibility in arranging the workstation.

2.4.2 Process Graphics

A process graphic (*Figure 18*) depicts the process equipment and the flow of control. Equipment changes color depending on its current condition. In some plants, red means Running and green means Off. However, this is not an industry standard; the opposite may be true in other plants. Orange indicates a trip or overload.

The Scientific Apparatus Manufacturers Association (SAMA) has guidelines on how symbols should be displayed on P&ID drawings.

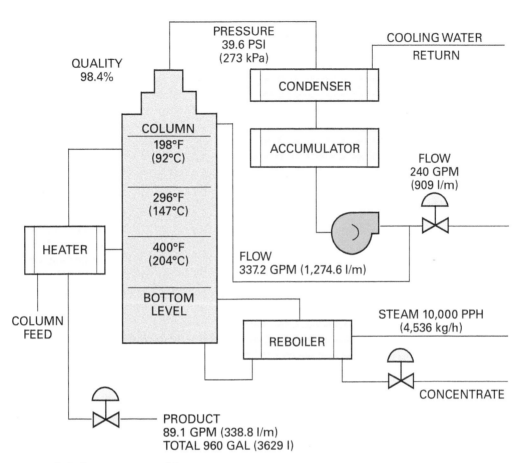

Figure 18 Operator workstation process graphic.

Plant graphics often use these same symbols. Some process graphics include control stations as well as equipment displays and process indications. For clarity, the designer may keep the control station from displaying until the operator selects a piece of equipment.

2.4.3 Analog Control Graphics

Whether control is integrated into a process graphic or on its own diagram, control stations follow conventions going back to the original hand/auto stations on benchboards. A control station (often called a *manual/auto station*) typically has three or four bar graphs arranged with a scale in the middle. Each bar is scaled depending on its function. Because displays of control logic execute from left to right, the station displays are arranged the same way.

The process bar will be scaled 0 to 100 percent of the process. For example, if a process pressure has a full-scale value of 3,000 psig, then the bar will represent 1,500 psig at its midpoint. The setpoint bar determines where the operator wants to control the process. It is also scaled to the process. The operator can drive the bar upscale and downscale using the setpoint arrow in the lower portion of the display.

To the right of the scale, there may be a bias bar. It will be scaled to some small percentage of the process. The bias bar is normally used on a process in which the setpoint is determined by the control logic, not the operator. The operator may be allowed to bias the setpoint within the safety limitations of the process. Note that bias bars are driven either upscale or downscale from the midpoint of the scale. This is because the bias can be either negative or positive in relation to the setpoint. If they are zero-based, process indicators will also drive from the midpoint of the scale.

The output bar is always on the far right and is always 0 to 100 percent. Since it is the command signal being sent to the final control element, it is not scaled to a process value. When in manual, the output may be driven upscale or downscale by the arrow buttons in the lower portion of the display. In addition to this, a control station will also have an auto or manual mode indication. There will also be buttons to change the mode. An operator will have to select a station to gain control of it. Methods for doing so may vary.

Refer to *Figure 19*. In the fourth column from the left, note the control stations for the "A" fan inlet vanes (FD). The process variable bar is on the left and is presently at 86 percent airflow. The controller is in AUTO, as indicated below the percentage readouts. Since the setpoint bar matches the process variable, the two outputs on the right are steady at about 68 percent. In the lower portion of the display, the arrows with the stems are for changing the setpoints. The arrows without the stems are for raising or lowering the outputs when in manual control.

2.4.4 Discrete Control Graphics

Additional stations are commonly used to provide discrete control. These stations have Start/Stop buttons and indications of the state of the equipment (Running or Off). Discrete indications for valve or damper positions may also be shown. With current industry standards, these will likely be red indications for Fully Open and green for Fully Closed.

> **NOTE:** Not all plants follow the red/green standard. Be sure you know the standard being used.

When the valve is mid-travel—neither fully open nor closed—both indicators will be lit. This may also be done by changing the color of parts of a valve on a process diagram, such as the one shown in *Figure 18*. A problem that often arises with discrete valve indications occurs when the limit switch sticks. For example, a valve may indicate a fully closed limit, but the output bar shows an intermediate position. If the operator can manipulate the process with the output in manual and there is no change in limit indication, it means that the limit switch has failed. If properly set, a limit switch provides a critical indication. On important valves, it can show if they are fully open or closed. This is a valuable feature because instrument demand signals can fail due to loss of air or malfunctions in I/P transducers and valve positioners.

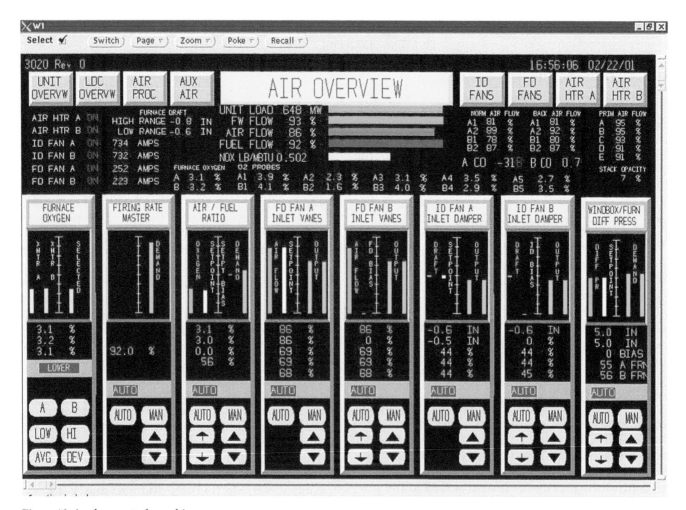

Figure 19 Analog control graphic.

2.4.5 Informational Screens and Navigation

Tabulated lists of process values on a graphic can display failed instruments or hot spots in a process. Startup graphics are also useful. One such graphic is a permissive/trip graphic for showing the operator the permissives that need to be satisfied before starting a unit or piece of equipment. If the equipment fails to start or trips during normal operation, the trip display will show what caused the trip. It does this by means of first-out logic. The first-out logic traps the first trip contact state change to be scanned. It disregards other related conditions or contacts that also changed state after the trip. These displays are usually in a list format, with the pertinent information highlighted.

Graphic displays always have a main menu in which all the graphics are selectable. They can be arranged by plant areas, phases of the process, or process modes, such as startup and shutdown. Menu buttons might also appear on the perimeter of the diagrams to allow the operator to rapidly change to other graphics, which often are related to the one displayed.

2.4.6 The Alarming System

The alarming system provides the operator with an immediate notification of an abnormal event in the plant. *Figure 20* shows an example of an alarm display. The alarming system also provides notification when the process returns to normal. This is called a return. In *Figure 20*, "DEBRIS FILT BACKFLUSH" in the Description column has a green background, the color for an unacknowledged return. Note the word "return" in the third column from the left. Once the operator acknowledges the alarm, it will disappear but can still be viewed on an alarm history screen. The three red lines are currently in alarm.

Traditional hardware alarm systems provided levels of alarm severity using sounds and visual flashing. Colored lenses and bulbs were also used. A DCS can replicate this functionality and add many more alarm levels (called *priorities*) using colors and sounds. The unacknowledged highlighted points on the sample screen would have colored backgrounds indicating their severity levels. Once acknowledged, the unhighlighted

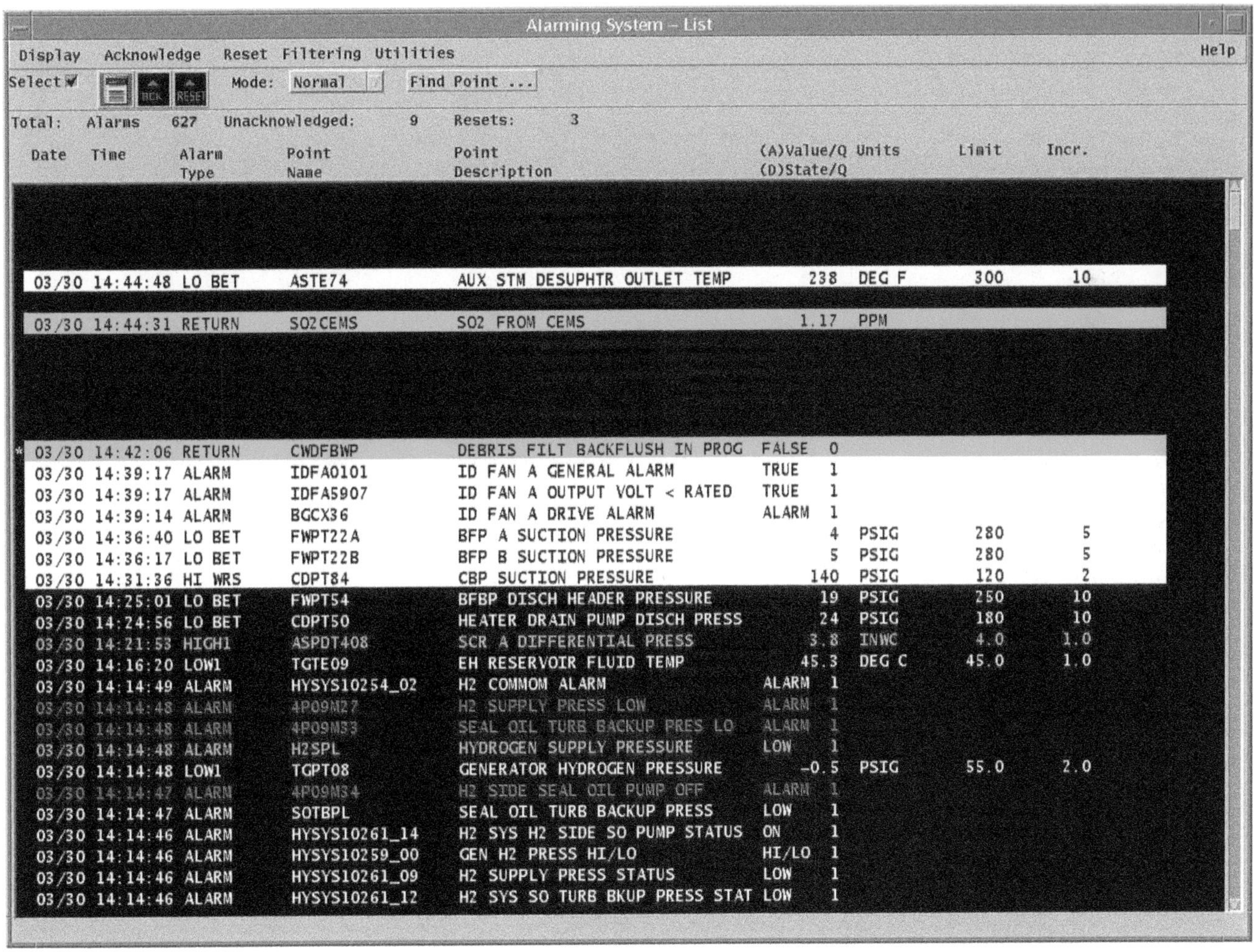

Figure 20 Alarm screen.

text would retain the same color that the background formerly had.

One significant advantage of a DCS alarm system is the ability to reprogram it and time-stamp alarms. Operator acknowledgement of alarms is logged and can assist in troubleshooting. This feature can also be used to assign operator responsibility for actions taken in response to process upsets.

Too many alarms and priorities become a distraction to the operator. An operator can refrain from constantly scanning a busy process graphic, but cannot ignore the alarm screen. As the plant grows, more alarms are added. A point can be reached at which there are too many alarms coming in all the time, and the operator may not be familiar with some of them. This is an all-too-common scenario in plants. For added insurance, some plants continue to rely on the panel-mounted (non-DCS) system for visual and audible alarms on critical conditions.

Besides giving operators the ability to acknowledge alarms, alarm setpoints can also be changed. This allows the operator to continue to monitor an extreme condition on a process that cannot be maintained within its normal operating parameters. Once the abnormal condition has been removed, the alarm setpoint is returned to its normal setting.

Nuisance and intermittent alarms can be turned off in the controller while it continues to display the process value. It is important to note that any changes to alarm monitoring are done in the controller and not in the operator's workstation. This way they are reflected system-wide.

2.4.7 Detailed Point Displays

Detailed point displays, such as the one in *Figure 21*, indicate all the parameters that have been programmed for a point. In this screen, the following real-time information is available:

- If the point is in alarm, and when it last went into alarm
- If alarm checking is On or Off

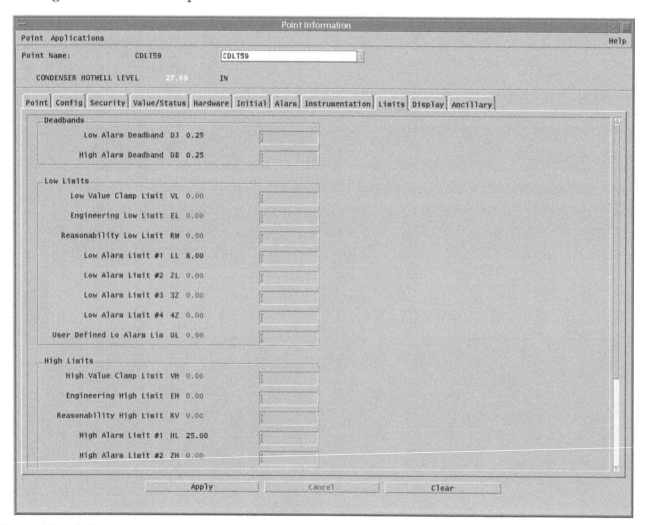

Figure 21 Detailed point screen.

- What the alarm limits are set for on an analog point
- What the alarm deadband is set to on an analog point and its present value
- If scan has been removed
- Its forced value, if taken off scan and made to read a temporary value
- If the channel reading is outside the normal measurable range on a sensor or I/O card

In *Figure 21*, each tab selects another page of information relating to the point. Shown is the Limits tab, which the operator would use to change the alarm setpoint. There are also many other programmable parameters that can be displayed, such as how often the point is to be scanned and the I/O configuration information. While detailed point displays are not frequently accessed by the operator, they are essential to everyday operation and maintenance of the plant.

2.4.8 Trends

Trend displays (*Figure 22*) replace the functionality of plant chart recorders. Trend displays continuously monitor and display values of analog and digital points. DCS trends have great flexibility. A top and bottom scale anywhere within the measurable range of the point can be chosen. The HOURS trend has a top scale of 250. Multiple points can be added to a single trend. Times are variable from minutes to days.

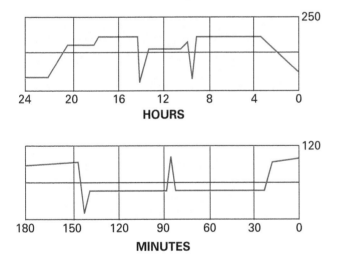

Figure 22 Trend display.

The rate of the data collection is determined by the system processing power, workstation memory, and network speed. If a point is being broadcast every 100 milliseconds across the network, 600 data samples would be taken on a one-minute trend. This performance is close to that of an analog chart recorder. Accuracy is a function of the analog-to-digital converters on the I/O cards, which provide a very high resolution. Too many one-minute trends would overload workstation memory, so the number of simultaneous trends is limited.

Unlike the charts, which were saved when the end of the roll was reached, trends usually run in the temporary memory of the workstation. If the trend is stopped or deleted and rebuilt later, there will be no way to go back and look at values from the old trend. A DCS may have the capability to set up a list of points to continuously trend in workstation memory. If a trend is built with one of those points, trend data will be available as far back as when the workstation was last placed online.

2.4.9 Historical Trends

If a plant has a historian on the network, a historical trend can be built for a point. If the historian has a storage device, such as a hard drive or other similar medium, the trend can be displayed for as long as the device has been collecting data. In some cases, this may be many years. Unfortunately, a historian has limitations. If it were to collect data at high accuracy and speed, its storage would fill up quickly. A deadband value is usually selected for each data point. This means that the point's value would have to change by some percentage before the historian would record the change. As a result, historical data displayed with too narrow a range will often appear like a series of stair steps.

Sometimes a record of startup data is needed with extremely high resolution. This record could be used as a baseline following a turbine overhaul, for example. In this case the historian deadbands could be rolled back to minimums for key points. Later, they could be reset at standard deadbands. Another alternative is to print logs and live trends of the data to refer to later. These logs can generally be exported to other applications.

Additional Resources

The following websites offer resources for products and training:

ABB, a global leader in power and automation technologies, **www.abb.com**

Emerson Process Management, **www.emersonprocess.com**

Fieldbus Foundation, part of the FieldComm Group™, **www.fieldcommgroup.org**

Honeywell International, **www.honeywellprocess.com**

PROFIBUS®, the world leaders in industrial networking, **www.profibus.com**

2.0.0 Section Review

1. DCS controllers generally require an operating system that's _____.
   a. graphical
   b. portable
   c. real time
   d. Unix-like

2. DCS maintenance and programming is usually done on a(n) _____.
   a. OW
   b. EW
   c. historian
   d. fieldbus

3. Which fieldbus includes specifications for working in explosive environments?
   a. HART
   b. Modbus
   c. Industrial Ethernet
   d. PROFIBUS

4. Cables from Ethernet devices like computers and controllers connect to a _____.
   a. network switch
   b. firewall
   c. router
   d. fieldbus

5. To communicate information about process components, most operator workstations use _____.
   a. lights
   b. printers
   c. graphics
   d. chart recorders

Section Three

3.0.0 Maintaining a DCS

Objective

Describe common considerations for the maintenance of DCS technology.

a. Describe various considerations for preventive and/or periodic instrument maintenance.
b. Describe considerations and approaches to the calibration and repair of instrumentation.
c. Explain the importance of expertise in the servicing of instrumentation and how information can be obtained.
d. Identify security issues associated with a DCS and explain how they can be addressed.

Trade Terms

Blacklist: A list of network addresses that may not cross a firewall.

Eavesdropping: Using an unauthorized electronic device to monitor wired or wireless network traffic in order to steal information.

Malware: A class of malicious software that enters a system surreptitiously and that can cause damage, steal information, or spread to other systems. Malware includes viruses, spyware, Trojans, and other similar programs.

Reformed: A maintenance step performed on electrolytic capacitors that have been stored too long and have dried out. A gradually increasing voltage is applied over a period of time to restore the electrolyte to normal condition.

Vulnerability: A flaw in a piece of commercial software that can be exploited as a means of gaining access to a system.

Whitelist: A list of network addresses that may cross a firewall.

A DCS is a complex collection of technologies. An instrument technician's responsibilities could encompass virtually any part of the system. In many cases, they might be focused on the lower hierarchies, such as the sensors, actuators, controllers, and fieldbuses. But it's also possible that they could include networking or server-related tasks. This section examines some of the issues surrounding DCS maintenance, repair, and security.

3.1.0 The Importance of Maintenance

One of the most crucial functions performed by an instrument technician is that of maintenance. Servicing the instrumentation includes preventative tasks, calibration, failure repair, and component replacement. Keeping the system working so production downtime is minimized or eliminated and product quality is maintained is a top priority. Therefore, efforts aimed at failure repair and component replacement take priority over routine maintenance and calibration activities. On the other hand, tasks geared toward prevention can keep problems from happening in the first place.

3.1.1 Preventative Maintenance Scheduling

Production downtime can be minimized by servicing instruments according to a scheduled maintenance program. There are some operations that consider it less costly to perform breakdown maintenance; in other words, as long as the instruments are working properly, they will not be worked on except during a general overhaul when operations are in shutdown. However, it is generally acknowledged that regularly serviced equipment will fail less often and provide more accurate control of the process.

A scheduled maintenance program is typically planned around the following elements:

- *The operation's production schedule* – The production schedule dictates which instruments are critical and when service can be performed with minimal interruption to production. Many maintenance activities are planned around scheduled downtimes or when the process is running at less than optimum levels.
- *Maintenance staff capabilities* – Personnel with suitable technical capabilities are scheduled to service the appropriate instruments in a timely manner.
- *History files* – Maintenance department instrument history files provide valuable information about the types of instruments used, their failure frequency, and the reasons that they failed.

Be aware that some operations have no formal production schedule; their goal is to make as much product as possible for as long as possible. You may encounter situations where maintenance tasks are assigned as if all personnel have identical know-how, or you might find yourself working in a plant where there are no instrumentation histories except the original documentation from when the plant was built. As an instrumentation technician, you must be prepared for anything.

12407-16 Distributed Control Systems

3.1.2 History Files

The format of history files varies from one operation to another. Essential information that is usually available either in written format, electronic format, or both includes the following:

- The instrument name
- Name of the manufacturer
- Model number or serial number
- Location in the plant
- Date when first placed into service
- Instrument tag number
- A diagram showing the process loops
- Associated logic ladder or wiring diagrams
- Associated installation drawings
- File showing the instrument's operating history
- File showing each control loop's history

The operating history file typically includes the following information:

- A description of all troubleshooting and maintenance work
- The date when a failure occurred
- The date on which the instrument was repaired or maintained
- A description of the repair or maintenance work done
- Identification of the technician who performed the work

The loop history file provides all related data for each control loop and a record of historical changes made to the loop.

3.1.3 Maintenance

On-site maintenance in most operations is typically restricted to performance checks, routine adjustments, and cleaning. Ideally, maintenance should be carried out in the instrument shop. Instruments are brought to the shop for routine maintenance based upon the manufacturer's recommended maintenance intervals, manufacturer's service bulletins, or failure mode analysis reports from the maintenance department. The shop should be equipped with the spare parts needed to perform the necessary work. Maintenance work, which is normally scheduled to avoid interference with production, has a lower priority than repair work.

3.2.0 Calibration and Repair

If instruments are not kept calibrated, control of production quality will degrade, and some product may have to be discarded. Therefore, regular instrument calibration is typically scheduled either on site or at the instrument shop.

3.2.1 Calibration

On-site calibration is typically done when the instrument is too large, has complex connections to the process, or when time constraints prohibit the instrument's being taken out of service for any longer than necessary to calibrate it on the spot. On-site calibration requires portable calibration equipment.

If connections to the process are minimal and the instruments are small enough, they are typically inspected and repaired at the shop, where the work can be performed in a controlled environment. However, many engineering guidelines and specifications require a final check with the instrument installed. All calibrations must be traceable to a national standards body, such as the National Institute of Standards and Technology (NIST).

3.2.2 Repair

In order to minimize process downtime, the most common method of repairing or maintaining instruments involves removing the instrument from the line, temporarily replacing it with a spare, and repairing or servicing the damaged instrument in the shop. This method minimizes the impact on production and allows the repair to be done in an environment equipped with appropriate tools. In this way, work can be done without the pressure of the production line, and in a more efficient manner.

On-site repair or maintenance is normally done when it is critical to production or if no spare in-kind instrument is available. Before performing on-site work, you must first consult with the operator in charge of the area in which you will be working. Operators must be aware of and approve all activity within the areas for which they have responsibility. The operator in charge will know if the proposed work will directly affect production downtime; whether or not there is other work going on in the area that could affect the technician's access to the instrument; if any extra safety precautions need to be taken; whether or not the work is likely to affect other production processes; and whether or not the process operation can safely proceed in a bypass mode.

After being given the all-clear from operations, the technician must follow proper lockout/tagout procedures before proceeding with the work.

3.2.3 Troubleshooting a DCS

In most repair situations, the instrument technician begins by troubleshooting the failure in order to find out what has malfunctioned. Troubleshooting a DCS is relatively straightforward because DCS maintenance work orders almost always result from one of three causes:

- A field device failure
- Failure of an I/O module or a point on the module
- Failure of wiring or network communications

The sections that follow address these issues. The quickest way to find a problem is check the field device first, then the I/O modules, and finally the wiring or communication link. Controller programming usually does not cause problems unless the program has recently been changed.

Problems can sometimes result from unique combinations of conditions in a plant that were not anticipated or tested during the factory acceptance test. A technician should never overlook the obvious. If six points all fail at once and there are six-channel I/O modules, it's likely that a module has failed. If a whole rack has stopped communicating, either the communication link is not working, or the controller itself may have a fault.

3.2.4 Field Device Failures

The best place to start when addressing a device failure is with the operator and graphical interface. Always obtain the operator's clearance before you work on any equipment. The operator will indicate which point is not reading properly, and you may already be familiar with the location of the device. If the equipment that the device is mounted on is running, you probably will not be able to work on it. You still should determine if its output is unstable.

The DCS provides a mechanism for removing points from scan. Before you take a device out of service, you must first take its input point out of scan. It will document when this occurred, and by whom. Otherwise, a log book should be used. When a point is removed from scan, the DCS value freezes. Equipment that is dependent on a process value that is off scan should be run in manual mode. Another indication of the same process value can be used instead, if available. Otherwise the operator will have to have an attendant keep track of the local indication. If there is no indication available, the equipment should be shut down.

The next step is to identify the device causing the problem. Take a reading at the device, and see if it matches the local indication or another point on the same process. There are many conditions that could cause it to read differently. Be sure that you're reading the correct device; use your plant documentation and do not hesitate to ask your fellow workers for help. When you are sure that you have the right instrument, you may begin to troubleshoot. Do not simply replace it. Check to make sure that it is valved-in and wired correctly. If it is a smart instrument, it may have been left in the test mode and is therefore not responding to changes in the process. You might even simulate the process value itself in place. If it responds correctly, it may be that the impulse leg is stopped up.

If you must replace the instrument, you will have to remove power from the loop. Fieldbus devices can simply be disconnected. The wiring should be secured and taped and the power source tagged out. When you have secured a calibrated replacement, reverse the procedure.

Returning the point to scan is a critical step. Make sure that all manual/auto stations whose controllers respond to this process point are in manual. There may be high- or low-signal monitors that feed digital logic as well. The online control logic is a tremendous asset to use for determining this. You can also compare the point's actual value to the frozen value and prepare the operator for what it is presently reading.

Once you return the point to scan, monitor the process for a time, and through load changes if time permits. The conditions that caused the original instrument to fail or read incorrectly may still be present (for example, a bad connection or an intermittently open wire).

3.2.5 DCS Component Failures

The most common devices to fail in the DCS itself are the I/O modules. They are connected directly to the instruments and are subject to noise, induced voltage, and wiring errors. A module failure is usually easily spotted. All the points on it will have bad readings. Well-designed DCS installations allow for modules to be removed without disturbing the field wiring. In PLC systems, the connection to the module is called a removable terminal board (RTB). There is not a standard design for a DCS, however. In some cases, field wiring may have to be removed.

Before replacing a module, you must determine the function of each point on the module in the process. You may encounter another point on the same module that plays such an important role in the process that the module replacement must be handled as an off-line job. Another critical concern is how to remove power from the module if it is externally sourced. Its power supply may also source other points that have critical roles in the process. Finally, make sure that you have the right module before you unplug it. There could be many modules of the same type and appearance in the cabinet.

If the module can be safely replaced, make sure that the correct replacement is available. Always be aware that a replacement part could be bad as well. Just because a replacement is new or was certified by the vendor as being repaired is no guarantee that it will function correctly. Modules often become damaged or bad ones get accidentally swapped with good ones during troubleshooting. Technicians should become familiar with the DCS spare parts inventory.

3.2.6 Power Supply Failures

Power supplies are often redundant in a DCS. This makes online replacement less of an issue. The same concerns as with I/O modules are valid. Make sure that you have identified the right power supply and confirm that a spare is available. A breaker will feed the power supply and should be labeled. There is always the chance that it has been incorrectly labeled. Unless you have opened that breaker in the past and are absolutely sure that it controls the power supply you're replacing, it is a good idea to wait until the process is down before opening it. Since the backup supply is providing power already, the chances of its failing in the interim are slim.

There may be a way to test the power supply before using it. You will need some way to load it. Sometimes lamps can provide good loads for this purpose. Some large power supplies may need to have their capacitors gradually reformed before use if they have been on the shelf for a long time. The manufacturer's documentation will have a procedure for this. Do not use a reforming procedure on a power supply unless you've confirmed that it's appropriate. Modern switching power supplies have different maintenance guidelines than older linear designs.

3.2.7 Controller Failures

Generally, the least likely element to fail in a DCS is the controller. Some controller failures stem from memory problems. If a controller's nonvolatile memory relies on a battery backup, the battery may need replacing. Flash-type nonvolatile memory can fail in a number of ways and may or may not be replaceable. A sign of a memory fault is a controller that is still online but with no program running.

After correcting a memory failure, you must be familiar with the proper procedure for restoring a controller's program. Unlike a PLC, the software server will not permit multiple copies of the application program to be stored and loaded from different locations. Hopefully, it will only take minimal effort to reload the controller. Once you load it, there may be a mismatch between the backup controller (now in the lead) and the controller that you've just reloaded. It is possible that someone made changes in that controller and did not back them up properly. Unless you can determine exactly what those changes were, it is best to wait until the process is down before resolving the mismatch.

3.2.8 Network Failures

Since networks contain a lot of different components, they are a common point of failure. Fieldbus failures generally come down to cabling and connector problems. These should be checked first. If an entire group of field devices on a fieldbus fails, it's likely that the controller or network module has malfunctioned.

Problems on a corporate network can be extremely complex (*Figure 23*), depending on the network design. In many cases, you may have to work with an IT department that has overall charge of the network. Some IT personnel do not understand the industrial side of the network as well as they understand the office side. For this reason, it's crucial that you confirm that they understand the issues associated with performing network diagnostics when the process is running. Also confirm that they understand that network

Figure 23 Large corporate networks can be very complex.

devices cannot be connected, disconnected, or restarted without knowing what the consequences to the process might be. Control operators should be informed of diagnostic action and permission obtained before working around a live process.

Network failures frequently come down to cabling problems and malfunctions at the network switch. Depending on the network's size and relationship to the rest of the corporate network, it's also possible that a change made to some other part of the network may have caused an unanticipated problem on the DCS network. Again, it's possible that IT personnel's marginal understanding of industrial networks can lead to problems during upgrades and network reconfigurations. Always consider this possibility when trying to track down mysterious and sudden network failures.

3.3.0 Acquiring Expertise

The cornerstone of repairing and maintaining instruments is having knowledge of the equipment to be serviced, the skills necessary to operate test equipment and repair tools, and the ability to think logically. In support of these primary skills, you must have access to process instrument maintenance and repair records (history files) and supportive product literature.

3.3.1 Equipment Knowledge

In order to repair any instrument, you must have an in-depth knowledge of how it works. In other words, you must know its theory of operation. A superficial knowledge of the equipment will not be sufficient.

Because there are many types of instruments used in a production process, acquiring instrument knowledge is a constant challenge. There are, however, several sources for acquiring such knowledge, and plant management usually recognizes the need and supports their use. The following sources should be used to obtain product-specific knowledge:

- Manufacturer's service training programs
- Self-study service training programs provided by the manufacturer
- Manufacturer's operating and troubleshooting literature
- Manufacturer's service technicians
- Independently published training material (such as the material in this module)

3.3.2 Repair Tool Skills

What has been said about acquiring equipment knowledge also applies to acquiring the skills and knowledge necessary to operate mechanical, power, and electronic tools used to repair instruments. Training in the use of tools must also be pursued, and practice makes perfect. The same resources exist to acquire the necessary skills and knowledge for operating repair tools.

3.4.0 DCS Security

Once something is connected to a network, the potential for malicious remote control can become a reality. In theory, any Internet-connected device can connect to any other device. Similarly, putting a device on a wireless network invites the possibility of remote access, signal jamming, or eavesdropping. For these reasons, security on any network is of paramount importance. With a DCS, security is especially important since malicious access could disrupt vital services, cause safety failures, or result in severe financial loss.

Your company may have its own measures, rules, and guidelines to keep crucial systems secure. Certain industries, such as those working with energy generation and management, have well-defined policies that member companies must observe. While security measures can be annoying and cumbersome, they are usually necessary due to the ingenuity of those trying to thwart them.

3.4.1 Access Control

At the most basic level, DCS devices, particularly servers and workstations, require a means of keeping unauthorized users out. Plants have procedures in place to prevent unauthorized access. For example, you may have to pass through a security checkpoint to enter the plant. The National Institute of Standards and Technology (NIST) currently has standards that define plant security. RFID access cards or keycodes may be required to enter certain areas or rooms (*Figure 24*).

Crucial workstations and servers are protected both by physical means (locked rooms) and by software. In order to access these machines, users must have an account and password. User accounts also define what files a user may access, which applications may be run, and what kinds of remote activity are permitted. Many companies require users to have fairly complex passwords and to change them on a regular basis (every 30, 60, or 90 days). While complex passwords are a bit of a nuisance, they are far harder to guess than simple ones. Passwords should never be obvious, such as the names of family members, birthdates, or anniversary dates.

3.4.2 Attacks from Without

Keeping a network secure is a full-time profession. IT departments and industrial networking personnel specialize in security measures. There are a variety of strategies that may be used to protect a network. Firewalls are a major line of defense. A firewall prevents certain protocols from crossing from one side to the other. It can also permit or deny specific network addresses from crossing by consulting a whitelist and a blacklist. A firewall only works properly if it's configured correctly, so setting one up is a task for an IT professional.

Encryption is another way to prevent security problems. Encrypted data looks like gibberish so it helps to thwart spying and eavesdropping. It also prevents users from controlling DCS devices with unauthorized technology, such as portable electronic devices. Virtually all

Figure 24 Physical access control is the first step toward network security.

wireless communication is encrypted. Even fieldbus technology is starting to make use of network security. Wireless fieldbus protocols such as WirelessHART® use encryption to prevent eavesdropping and unauthorized access. Again, wireless technology must be configured by a specialist to ensure that it won't be set up in such a way that its security is weak or nonexistent.

3.4.3 Attacks from Within

A network is always easier to attack from within than without. One way that malicious users can gain access to or disrupt a system is through malware. These are programs that gain access through an apparently innocent means, such as an email or through a website visit. They can also enter a system by contact with an outside infected computer through USB memory devices and portable hard drives.

For this reason, many companies are very restrictive about allowing users to connect portable memory devices to DCS machines. Similarly, Internet accesses involving social media, websites, and email may be forbidden on some machines. Anti-malware software is usually installed to constantly scan crucial systems on the network, such as servers. Software vendors regularly release security patches to close any known vulnerability. These are quite common, particularly in operating systems.

Additional Resources

The following websites offer resources for products and training:

ABB, a global leader in power and automation technologies, **www.abb.com**

Emerson Process Management, **www.emersonprocess.com**

Fieldbus Foundation, part of the FieldComm Group™, **www.fieldcommgroup.org**

Honeywell International, **www.honeywellprocess.com**

PROFIBUS®, the world leaders in industrial networking, **www.profibus.com**

3.0.0 Section Review

1. When scheduling preventative maintenance, always consult _____.
   a. the history file
   b. the simulator
   c. the fieldbus traffic log
   d. the IT department

2. Ultimately, all calibration must be _____.
   a. done with in-house standards
   b. done every six months
   c. traceable to the NIST
   d. done off-line

3. Before doing anything that might affect a running process, you should check _____.
   a. with the IT department
   b. with the control operator
   c. the logbook
   d. all power supplies

4. A good way to acquire expertise with equipment is to _____.
   a. take it apart and then put it back together
   b. examine broken units
   c. attempt a repair on a non-critical unit
   d. take a manufacturer training program

5. What keeps undesirable traffic out of a DCS network?
   a. A packet sniffer
   b. A protocol analyzer
   c. A firewall
   d. A time-domain reflectometer

SUMMARY

This module has provided an overview of the complex topic of industrial control using a DCS. A DCS is similar to other control technologies such as PLCs and SCADA, but a DCS has its own unique qualities. DCSs contain a large number of components distributed across the process but linked by various networking technologies.

In this ever-changing world of sophisticated control systems, your role as an instrumentation technician will include traditional tasks like maintenance, calibration, and repair. But it will also include keeping up with new technologies, learning to interact with IT and network professionals, and developing new skills. Continual learning is a way of life in the world of digital control.

Review Questions

1. A DCS is hierarchical, meaning that it distributes control in _____.
   a. networks
   b. layers
   c. fieldbuses
   d. field devices

2. A connection to a network is known as a _____.
   a. highway
   b. RS-485
   c. server
   d. drop

3. What is the most common type of operating system for the server and workstation programs of current DCSs?
   a. UNIX®
   b. RTOS
   c. Microsoft®
   d. RT-11

4. Which of the following statements about modern DCS, PLC, and SCADA systems is true?
   a. The distinctions between them aren't as great as they once were.
   b. They serve similar functions but have little in common.
   c. DCS and PLC systems have pretty much replaced SCADA.
   d. SCADA relies on pneumatic control, while DCSs and PLCs are electronic.

5. Spreading critical I/O across several different modules or drops is called _____.
   a. redundancy
   b. RAID
   c. partitioning
   d. feedforward

6. For better performance and reliability, servers often use multiple hard drives called _____.
   a. RAID arrays
   b. SCSI banks
   c. SATA packs
   d. EIDE strings

7. A master database of all key DCS components is stored on the _____.
   a. historian
   b. engineering workstation
   c. software server
   d. security server

8. The lowest layer in all networks, which includes cabling and connectors, is the _____.
   a. presentation layer
   b. physical layer
   c. session layer
   d. data link layer

9. What kind of heavy duty network is very similar to regular corporate networking technology?
   a. Modbus
   b. PROFIBUS
   c. FOUNDATION fieldbus
   d. Industrial Ethernet

10. What part of the DCS can alert the operator of problems through colors, sounds, priorities, and severity levels?
    a. The process graphic
    b. The alarm system
    c. The navigation screens
    d. The trend screen

11. Maintenance works best when it's _____.
    a. scheduled
    b. daily
    c. weekly
    d. reactive

12. In order to keep a process under control and production quality high, instruments must be regularly _____.
    a. disassembled
    b. repaired
    c. calibrated
    d. swapped

13. A controller that's online but has no running program has probably had a _____.
    a. processor failure
    b. memory failure
    c. network failure
    d. server failure

34 NCCER – Instrumentation Level Four 12407-16

14. The most important key to being effective at repairing instruments is _____.
    a. strong supervision
    b. basic intuition
    c. Internet access
    d. proper training ✓

15. Properly setting up network security generally involves _____.
    a. NFPA standards
    b. ITU documents
    c. IT personnel ✓
    d. NIST traceability

Trade Terms Introduced in This Module

Algorithm: In the context of computer programming, this refers to code that performs a specific task. In analog control systems, it refers to symbols for components that perform various analog functions, such as PIDs and summers.

Backbone: A high-speed network channel designed to connect multiple networks together, usually across a facility, campus, or geographically distant area. Backbones often use fiber optic technology.

Blacklist: A list of network addresses that may not cross a firewall.

Cloud-based storage: Remote file storage accessed through the Internet and provided by a service company that maintains one or more servers in various locations.

Drop: A controller, server (see *server*), workstation, or other DCS component connected to a network.

Eavesdropping: Using an unauthorized electronic device to monitor wired or wireless network traffic in order to steal information.

Encryption: A security measure against data interception in which the data is turned into an unreadable form using cipher techniques prior to being transmitted. It is returned to its original form upon receipt.

Enterprise-level: Technology designed to provide high-level information services for company executives in order to facilitate large-scale planning and decision-making.

Fail safe: A system that puts itself into a safe or minimally hazardous condition in the event of a failure.

Fiber optic network: A networking technology in which data is transmitted through fiber optic cables as pulses of light.

Fieldbus: A general term for an industrial process network and protocols used to connect instruments, sensors, and controllers together using simple cabling systems.

Field device: A general term for a sensor or actuator that interacts directly with a process. Field devices are at the bottom of the DCS hierarchy.

Firewall: A network device that divides a network into two sections and prevents unauthorized accesses or protocols from crossing between the sections.

First-out logic: The first element (possibly of several) that caused a process trip.

Functional: Diagram of a digital logic system expressed as connected logic components.

Gateway: A network device that translates from one type of physical layer and/or protocol to another. Gateways allow normally incompatible systems to communicate.

Hierarchical: Arranged in layers, with higher layers being more sophisticated and performing more supervisory functions than lower ones.

Historian: A program that keeps a record of process data for later analysis and to reveal trends. Historians can run on dedicated computers or on a server along with other programs.

Malware: A class of malicious software that enters a system surreptitiously and that can cause damage, steal information, or spread to other systems. Malware includes viruses, spyware, Trojans, and other similar programs.

Network switch: A kind of network hub to which many devices connect. Switches perform intelligent traffic management, channeling data between their ports based on hardware addresses.

Partitioning: Dividing process I/O across multiple drops or multiple I/O cards in order to prevent a failure from causing a process shutdown.

Patch: A fix for defects discovered in software after its release. Many patches are security-oriented.

Permissive: A condition in a control system that must be met before an action can occur. For example, a group of series-wired sensor switches would be permissives in an electrical control system.

Protocol: A set of rules for communications between two devices. Using a common protocol enables devices to communicate, even if they are made by different manufacturers.

RAID array: Stands for Redundant Array of Independent Disks; a group of hard drives acting as a unit in order to improve performance and/or provide error correction and redundancy.

Real time: A computer system that responds with minimal and predictable delays to events as they happen. Critical control systems must be real time.

Reformed: A maintenance step performed on electrolytic capacitors that have been stored too long and have dried out. A gradually increasing voltage is applied over a period of time to restore the electrolyte to normal condition.

Router: A network traffic manager that connects multiple networks together. Routers examine data at the protocol levels to decide where it should be forwarded.

Server: A networked computer that provides information or services to other "client" machines.

Supervisory Control and Data Acquisition (SCADA): An industrial control system that emphasizes data collection and display. Human operators make and execute high-level control decisions over the process through the system.

Third-party solutions: Products developed by an outside company to solve a problem or meet a need associated with a particular manufacturer's technology.

Transmission Control Protocol/Internet Protocol (TCP/IP): A family of communications protocols used to connect many different types of networked computers together. It is the protocol basis of the global Internet.

Vulnerability: A flaw in a piece of commercial software that can be exploited as a means of gaining access to a system.

Whitelist: A list of network addresses that may cross a firewall.

Additional Resources

This module presents thorough resource for task training. The following resource material is recommended for further study.

The following websites offer resources for products and training:
    ABB, a global leader in power and automation technologies, **www.abb.com**
    Emerson Process Management, **www.emersonprocess.com**
    Fieldbus Foundation, part of the FieldComm Group^{TM}, **www.fieldcommgroup.org**
    Honeywell International, **www.honeywellprocess.com**
    PROFIBUS®, the world leaders in industrial networking, **www.profibus.com**
    Wi-Fi® is a registered trademark of Wi-Fi Alliance, **wi-fi.org**

Figure Credits

©Yongnian Gui/Dreamstime.com, Module Opener, Figure 4

Photos Courtesy of Entergy New Orleans, Inc. ©2016 All rights reserved, Figures 2, 10

Automationdirect.com, Figure 3

SIXNET - Solutions for Your Industrial Networking Challenges, www.sixnet.com, Figure 5

©Pongmoji/Dreamstime.com, Figure 9

©Panitan Kanchanwong/Dreamstime.com, Figure 15

©Stanislav Slepuschenko/Dreamstime.com, Figure 16

©Emel82/Dreamstime.com, Figure 17

Emerson Process Management, Figures 19–21

©Plus69/Dreamstime.com, Figure 23

©Boonsom/Dreamstime.com, Figure 24

Section Review Answer Key

Answer	Section Reference	Objective
Section One		
1. b	1.1.0	1a
2. d	1.1.1	1a
3. d	1.2.0	1b
4. c	1.3.2	1c
Section Two		
1. c	2.1.4	2a
2. b	2.2.6	2b
3. d	2.3.4	2c
4. a	2.3.7	2c
5. c	2.4.0	2d
Section Three		
1. a	3.1.1	3a
2. c	3.2.1	3b
3. b	3.2.2	3b
4. d	3.3.1	3c
5. c	3.4.2	3d

NCCER CURRICULA — USER UPDATE

NCCER makes every effort to keep its textbooks up-to-date and free of technical errors. We appreciate your help in this process. If you find an error, a typographical mistake, or an inaccuracy in NCCER's curricula, please fill out this form (or a photocopy), or complete the online form at **www.nccer.org/olf**. Be sure to include the exact module ID number, page number, a detailed description, and your recommended correction. Your input will be brought to the attention of the Authoring Team. Thank you for your assistance.

Instructors – If you have an idea for improving this textbook, or have found that additional materials were necessary to teach this module effectively, please let us know so that we may present your suggestions to the Authoring Team.

NCCER Product Development and Revision
13614 Progress Blvd., Alachua, FL 32615

Email: curriculum@nccer.org
Online: www.nccer.org/olf

❏ Trainee Guide   ❏ Lesson Plans   ❏ Exam   ❏ PowerPoints   Other _____

Craft / Level: _____   Copyright Date: _____

Module ID Number / Title: _____

Section Number(s): _____

Description: _____

_____

_____

_____

_____

Recommended Correction: _____

_____

_____

_____

_____

Your Name: _____

Address: _____

_____

Email: _____   Phone: _____

12409-16
Analyzers and Monitors

Overview

Modern industrial plants are equipped with numerous analyzers and monitors that measure the chemical characteristics and compositions of the different materials used in industrial processes. Instrumentation technicians are required to install, calibrate, and maintain this equipment. This module introduces the instruments used to measure and monitor the characteristics and chemical compositions of process materials. It reviews basic chemistry concepts and methods associated with analytical measurements, and discusses common uses and basic principles of operation for specific types of analyzers and monitors.

Module Seven

Trainees with successful module completions may be eligible for credentialing through the NCCER Registry. To learn more, go to **www.nccer.org** or contact us at 1.888.622.3720. Our website has information on the latest product releases and training, as well as online versions of our *Cornerstone* magazine and Pearson's product catalog.

Your feedback is welcome. You may email your comments to **curriculum@nccer.org**, send general comments and inquiries to **info@nccer.org**, or fill in the User Update form at the back of this module.

This information is general in nature and intended for training purposes only. Actual performance of activities described in this manual requires compliance with all applicable operating, service, maintenance, and safety procedures under the direction of qualified personnel. References in this manual to patented or proprietary devices do not constitute a recommendation of their use.

Copyright © 2016 by NCCER, Alachua, FL 32615, and published by Pearson Education, Inc., New York, NY 10013. All rights reserved. Printed in the United States of America. This publication is protected by Copyright, and permission should be obtained from NCCER prior to any prohibited reproduction, storage in a retrieval system, or transmission in any form or by any means, electronic, mechanical, photocopying, recording, or likewise. To obtain permission(s) to use material from this work, please submit a written request to NCCER Product Development, 13614 Progress Blvd., Alachua, FL 32615.

From *Instrumentation Level Four, Trainee Guide*, Third Edition. NCCER.
Copyright © 2016 by NCCER. Published by Pearson Education. All rights reserved.

12409-16
ANALYZERS AND MONITORS

Objectives

When you have completed this module, you will be able to do the following:

1. Describe basic chemistry concepts and identify key characteristics of compounds and solutions.
   a. Identify and describe basic properties of elements and compounds.
   b. Define and describe chemical bonding and reactivity.
   c. Define and describe solutions and concentration.
   d. Define and describe acids, bases, pH, and salts.
2. Define the physical properties of density, specific gravity, viscosity, and turbidity, and identify methods used to analyze them.
   a. Define the properties of density and specific gravity and identify methods used to analyze them.
   b. Define the property of viscosity and identify methods used to analyze it.
   c. Define the property of turbidity and identify methods used to analyze it.
3. Define the properties of flash point, pH, conductivity, and oxidation-reduction potential, and identify methods used to analyze them.
   a. Define the property of flash point and identify methods used to analyze it.
   b. Define the property of pH and identify methods used to analyze it.
   c. Define the property of conductivity and identify methods used to analyze it.
   d. Define the property of oxidation-reduction potential and identify methods used to analyze it.
4. Identify methods used to analyze air and determine its content of O_2, CO, CO_2, H_2S, and THC.
   a. Identify methods used to analyze air and determine its content of oxygen (O_2).
   b. Identify methods used to analyze air and determine its content of carbon monoxide (CO).
   c. Identify methods used to analyze air and determine its content of carbon dioxide (CO_2).
   d. Identify methods used to analyze air and determine its content of hydrogen sulfide (H_2S).
   e. Identify methods used to analyze air and determine its content of total hydrocarbons (THC).
5. Define the properties of particulate count, chemical composition, infrared radiation, and UV light absorption, and identify methods used to analyze them.
   a. Define the property of particulate count and identify methods used to analyze it.
   b. Define the property of chemical composition and identify methods used to analyze it.
   c. Define the property of infrared radiation and identify methods used to analyze it.
   d. Define the property of UV light absorption and identify methods used to analyze it.

Performance Task

Under the supervision of your instructor, you should be able to do the following:

1. Determine the pH of a given solution and propose the proper adjustment.

Trade Terms

- Acid
- Atom
- Atomic mass
- Atomic number
- Base
- Bond
- Buffer solution
- Chemical reaction
- Chromatography
- Colorimeter
- Compound
- Conductivity
- Covalent bond
- Density
- Dissociate
- Electrolytes
- Electrons
- Electron shells
- Element
- Emissivity
- Flash point
- Formula unit
- Hydrocarbon
- Hydrometer
- Ion
- Ionic bond
- Ionize
- Isotopes
- Metallic bond
- Molarity (M)
- Mole (mol)
- Molecules
- Neutrons
- Nucleus
- Oxidation-reduction potential (ORP)
- Parts per million (ppm)
- Periodic table
- pH
- Poise (P)
- Protons
- Salt
- Solute
- Solvent
- Specific gravity
- Total hydrocarbon content (THC)
- Turbidity
- Unified atomic mass unit (u)
- Valence shell
- Viscosity

Industry Recognized Credentials

If you are training through an NCCER-accredited sponsor, you may be eligible for credentials from NCCER's Registry. The ID number for this module is 12409-16. Note that this module may have been used in other NCCER curricula and may apply to other level completions. Contact NCCER's Registry at 888.622.3720 or go to **www.nccer.org** for more information.

Contents

- 1.0.0 Chemistry Concepts 1
  - 1.1.0 The Science of Matter 2
    - 1.1.1 Elements 2
    - 1.1.2 The Periodic Table 2
    - 1.1.3 Compounds 4
  - 1.2.0 Chemical Reactions 4
    - 1.2.1 Chemical Bonding 4
    - 1.2.2 Atomic Mass and Chemical Reactions 7
  - 1.3.0 Concentration 8
  - 1.4.0 Acids, Bases and pH 8
    - 1.4.1 Ions 8
    - 1.4.2 Acids and Bases 9
    - 1.4.3 pH 9
    - 1.4.4 Neutralization and Salts 10
- 2.0.0 Analyzing Physical Properties 12
  - 2.1.0 Density and Specific Gravity 12
    - 2.1.1 Hydrometer 13
    - 2.1.2 Air Bubble Measurement 13
    - 2.1.3 Displacement Measurement 13
    - 2.1.4 Densitometer 13
    - 2.1.5 Nuclear Detectors 14
  - 2.2.0 Viscosity 14
    - 2.2.1 Viscometers 15
  - 2.3.0 Turbidity 16
    - 2.3.1 Jackson Turbidimeter 17
    - 2.3.2 Transmission Analyzer 17
    - 2.3.3 Reflection Analyzer 18
    - 2.3.4 Ratio Analyzer 19
- 3.0.0 Analyzing Chemical Properties 21
  - 3.1.0 Flash Point Determination 21
    - 3.1.1 Establishing the Flash Point 21
    - 3.1.2 *OSHA 1910.106(a)* 22
  - 3.2.0 Measuring pH 23
    - 3.2.1 pH Instruments 23
    - 3.2.2 pH Probes 23
    - 3.2.3 Calibration and Compensation 24
    - 3.2.4 pH Analyzer/Controller 25

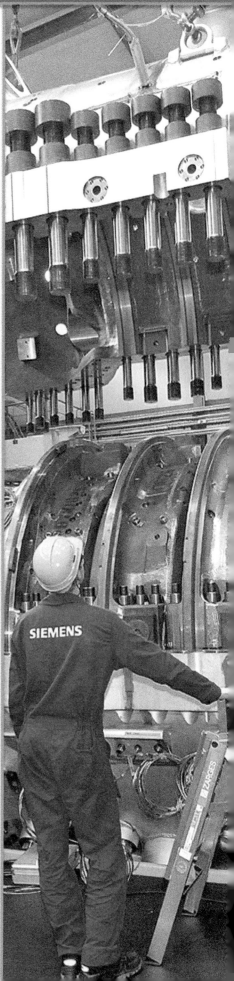

3.3.0	Measuring Conductivity	26
3.3.1	Conductivity Instruments	26
3.3.2	Additional Factors in Measuring Conductivity	26
3.3.3	Inductive Conductivity Instruments	27
3.4.0	Measuring Oxidation-Reduction Potential	27
3.4.1	ORP Instruments	27
3.4.2	Probe Calibration	28
3.4.3	Probe Maintenance	29
4.0.0	SGas Analysis	31
4.1.0	Measuring Oxygen	31
4.1.1	High-Temperature Electrochemical Sensors	31
4.1.2	Paramagnetic Analyzers	32
4.1.3	Galvanic Sensors	33
4.2.0	Measuring Carbon Monoxide	33
4.3.0	Measuring Carbon Dioxide	33
4.3.1	Monitoring Carbon Dioxide	34
4.4.0	Measuring Hydrogen Sulfide	34
4.4.1	Personnel Protection Indicators	34
4.4.2	Semiconductor Sensors	34
4.4.3	Electrochemical Sensors	35
4.5.0	Measuring Total Hydrocarbons	35
4.5.1	Flame Ionization Detector	35
5.0.0	Section Five Other Analysis Techniques	37
5.1.0	Measuring Particulates	37
5.1.1	Optical Microscopy	37
5.1.2	Discrete Particle Counters (DPCs)	38
5.2.0	Determining Chemical Composition	38
5.2.1	Gas Chromatography	38
5.3.0	Measuring Infrared Radiation	40
5.3.1	Basic Theory	40
5.3.2	Factors Affecting Infrared Imaging	40
5.3.3	Sensing Equipment	40
5.3.4	IR Spectrometry	42
5.4.0	Measuring Ultraviolet Radiation	42
5.4.1	UV Analysis	42
5.4.2	Applications	43
5.4.3	Advantages	43
5.4.4	Calibration	44
5.4.5	Flame Detectors	44

Figures and Tables

Figure 1	Carbon atom	2
Figure 2	The periodic table	3
Figure 3	Periodic table entry for nitrogen	4
Figure 4	Periods in the periodic table	5
Figure 5	Groups in the periodic table	6
Figure 6	Ionic bonding	7
Figure 7	Covalent bonding	7

Figures and Tables (continued)

Figure 8 Dissociation and ionization 9
Figure 9 The pH scale 9
Figure 10 A hydrometer 13
Figure 11 Dual bubblers used to measure density 14
Figure 12 Displacement measurement of density 14
Figure 13 Density measurement by a radiation detector 15
Figure 14 Oil and water change in viscosity with temperature variation 15
Figure 15 Basic falling ball viscometer 16
Figure 16 Basic rotating disc viscometer 16
Figure 17 Basic capillary viscometer 17
Figure 18 The Jackson turbidimeter 17
Figure 19 Transmission-type turbidity analyzer 17
Figure 20 Colorimeter calibration curve for chlorine 18
Figure 21 Typical reflection-type, in-line turbidity analyzer 18
Figure 22 Light beam traveling through a ratio turbidity analyzer 19
Figure 23 The pH scale range 23
Figure 24 A pH indicator scale 23
Figure 25 Handheld pH instrument with single probe 24
Figure 26 A pH electrode 24
Figure 27 Reference electrode 24
Figure 28 In-line pH probe and meter 24
Figure 29 Simplified diagram of the essential parts of a
          pH measuring instrument 25
Figure 30 A pH analyzer/controller 25
Figure 31 Simple conductivity meter 26
Figure 32 Inductive conductivity measurement 27
Figure 33 ORP probe assembly 28
Figure 34 Piping design including an ORP in-line system 29
Figure 35 High-temperature electrochemical oxygen detector 31
Figure 36 Magneto-dynamic oxygen analyzer 32
Figure 37 Principles of a galvanic oxygen sensing cell 33
Figure 38 Bacharach handheld CO analyzers 33
Figure 39 Personnel protection indicator 35
Figure 40 Flame ionization detector for hydrocarbon measurement 35
Figure 41 Optical particle counter (OPC) 38
Figure 42 Simplified chromatography analyzer 39
Figure 43 Chromatograph capillary tube 40
Figure 44 A typical chromatogram 40
Figure 45 The electromagnetic spectrum 41
Figure 46 Handheld IR imager 41
Figure 47 Simple infrared spectrometer 43
Figure 48 Simple ultraviolet analyzer 43
Figure 49 Ultraviolet flame detector 44

Table 1 pH/pOH Relationship 10
Table 2 Standardized Systems for Flash Point Determination 22
Table 3 Common Particulates and Their Sizes 37
Table 4 Typical Emissivity Values 41

SECTION ONE

1.0.0 CHEMISTRY CONCEPTS

Objective

Describe basic chemistry concepts and identify key characteristics of compounds and solutions.
a. Identify and describe basic properties of elements and compounds.
b. Define and describe chemical bonding and reactivity.
c. Define and describe solutions and concentration.
d. Define and describe acids, bases, pH, and salts.

Trade Terms

Acid: A substance that produces hydrogen ions (H^+) in solution.

Atom: The smallest particle with a unique chemical identity.

Atomic mass: The average mass of the naturally occurring isotopes of a given element. Expressed in unified atomic mass units.

Atomic number: The number of protons in a particular element. This value is unique to each element.

Base: A substance that produces hydroxide ions (OH^-) in solution.

Bond: An attraction between atoms that enables elements to join together, forming compounds.

Chemical reaction: A process that occurs when substances combine or break apart at the chemical level.

Compound: A chemical combination of two or more elements.

Covalent bond: A chemical bond in which atoms share some of their valence shell electrons.

Dissociate: Occurs when ionic compounds dissolve and separate into ions.

Electrolytes: Substances that dissociate/ionize into ions when dissolved in water. Electrolytes make solutions electrically conductive.

Electrons: The negatively charged parts of an atom that orbit around the nucleus.

Electron shells: The paths in which electrons orbit around the atom's nucleus.

Element: A substance composed of just one type of atom.

Formula unit: The most fundamental unit of an ionic compound.

Ion: An atom, or group of atoms, that has a positive or negative electrical charge.

Ionic bond: A chemical bond formed by the transfer of electrons from one atom to another.

Ionize: To convert into an ion or ions; ionization occurs when covalent compounds break down into ions in solution.

Isotopes: Forms of the same element but with different numbers of neutrons in the nucleus.

Metallic bond: A chemical bond in which the valence electrons of many atoms move freely between their atoms, forming a kind of electron "sea."

Molarity (M): The concentration of a solution equal to the number of moles of solute divided by the number of liters of solution.

Mole (mol): A quantity equal to 6.02×10^{23} particles (atoms, formula units, molecules, or ions).

Molecules: The most fundamental units of a covalent compound.

Neutrons: Uncharged atomic particles that are similar in size to protons and are present in the nucleus of an atom.

Nucleus: The central part of an atom. The nucleus contains the protons and neutrons.

Parts per million (ppm): The concentration of solute in a solution, expressed as the parts of solute per million parts of solution.

Periodic table: A grid arrangement in which the elements are ordered in a way that shows their similarities and repeating properties.

pH: A measurement scale that expresses how acidic or basic a solution is based on its concentration of hydrogen (H^+) ions; values typically range from 0 to 14.

Protons: Positively charged atomic particles that are similar in size to neutrons and are present in the nucleus of an atom.

Salt: An ionic compound composed of at least one metallic and one nonmetallic element. Salts are formed from a neutralization reaction between an acid and a base.

Solute: A substance that is dissolved in another substance, forming a solution.

Solvent: The substance in which another substance is dissolved to form a solution.

Unified atomic mass unit (u): A measurement unit of atomic mass based on $\frac{1}{12}$ the mass of a carbon-12 atom. One u equals about 1.660×10^{-27} kg.

Valence shell: The outermost electron shell of an atom. Chemical reactions involve the valence shell.

Instrumentation technicians routinely install, calibrate, and maintain analyzers and monitors that provide analytical measurements of the various materials used in process systems. Technicians are better equipped to perform these tasks if they have a sound understanding of some basic chemistry concepts and the terminology associated with the instruments used to make chemical measurements. This section provides a review of some basic chemistry concepts.

> **NOTE**
> Virtually all science, including chemistry, is done using metric units of measure only. For this reason, only metric units (not US measurements) are shown throughout the first section of this module.

1.1.0 The Science of Matter

Chemistry is the study of matter and the ways in which it changes. Matter is anything that occupies space and has mass. The smallest particle of matter with a distinct chemical identity is the atom (*Figure 1*). Atoms contain up to three different particle types: protons, neutrons, and electrons. Protons carry a positive electrical charge, electrons carry a negative charge, and neutrons have no charge. Protons and neutrons form the central part of the atom, called the nucleus. Neutrons contribute to the atom's mass. Electrons orbit the nucleus in paths called electron shells. Normally, the number of electrons equals the number of protons. As a result, the atom is electrically neutral because the positive and negative charges balance each other out.

1.1.1 Elements

The number of protons in an atom determines its chemical properties. A substance composed of just one type of atom (meaning that all of its atoms have the same number of protons) is called an element. Chemists divide elements into three broad categories: metals, metalloids, and nonmetals. Electrical wiring, railroad tracks, and water pipes are made from metals. The oxygen and nitrogen that you're breathing are nonmetals. The silicon making up the integrated circuits in your smartphone is a metalloid.

While the number of protons in a given element is always the same, the number of neutrons can vary. These variations of the same element are called isotopes. They behave in the same way chemically, but their masses are slightly different since they have different numbers of neutrons. For example, oxygen comes in three main isotopes (oxygen-16, oxygen-17, and oxygen-18). Each has 8 protons, but the different isotopes have 8, 9, or 10 neutrons, respectively.

Chemists use alphabetic symbols to identify elements. For example, the symbol for carbon is C, aluminum is Al, and iron is Fe. Some symbols are obvious, while others come from the historic name for the element. For example, iron was called *ferrum* in Latin.

At the present time, there are 118 known elements, 98 of which occur naturally. The others are manmade and generally exist for only very brief periods of time. Since each type of atom has a unique number of protons, elements are also identified by a numeric value, called the atomic number. The atomic number corresponds to the number of protons in the nucleus of the atom. For example, Uranium (U) has 92 protons, so its atomic number is 92.

An element's atomic mass (also called the atomic weight) is an average based on the masses of the different isotopes of the element. This is why it is always a decimal number. The measurement used for atomic mass is the unified atomic mass unit (u), which equal $\frac{1}{12}$ the mass of a carbon-12 atom (1 u equals about 1.660×10^{-27} kg). Scientific standards bodies publish updated atomic mass numbers at regular intervals as it becomes possible to measure atomic masses with greater precision.

1.1.2 The Periodic Table

The periodic table (*Figure 2*) is a tool that organizes the elements in a way that reflects their properties. All versions of the table are basically the same—a framework of squares arranged in rows and columns with the atomic numbers moving upwards from 1 to 118. Each square contains the element's symbol, atomic number, and atomic mass. As shown in *Figure 3*, nitrogen has the symbol N, an atomic number of 7 and an atomic mass of 14.0067.

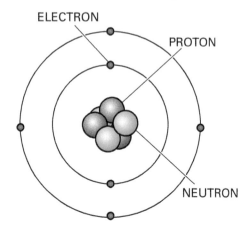

Figure 1 Carbon atom.

Figure 2 The periodic table.

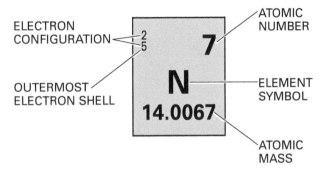

Figure 3 Periodic table entry for nitrogen.

The horizontal rows of the periodic table are referred to as *periods*. As shown in *Figure 4*, the first period consists of the two elements hydrogen (H) and helium (He). The second period starts with lithium (Li) and ends with neon (Ne), and so on. Each period in the table ends with a very unreactive gaseous element called a noble gas.

The vertical columns are referred to as *groups*. As shown in *Figure 5*, Group 1 includes the elements Li (lithium) through Fr (francium), all of which are very reactive metals. Group 2 includes the elements Be (beryllium) through Ra (radium), all of which are another class of metals. Group 17 includes the elements fluorine (F) through astatine (At), which are all very reactive nonmetal elements called halogens.

The elements within a given group behave in similar ways chemically. For example, sodium (Na) and potassium (K) are very much alike. That's why potassium chloride can be used by people with high blood pressure as a replacement for sodium chloride (table salt). It tastes and behaves similarly, but it contains no sodium.

1.1.3 Compounds

Elements combine together to form substances known as chemical compounds. A compound contains at least two different kinds of atoms, but may contain many more. While there are just 118 elements, there are hundreds of thousands of known compounds.

Elements combine to form compounds in specific proportions. Much of what chemists do revolves around creating and analyzing different compounds. The atoms that make up compounds combine and break apart in a process known as a chemical reaction. They are held together by interactions known as bonds. (A bond is an attraction between atoms that enables elements to join together, forming compounds.) In the next section, you will learn about some of the rules by which elements combine.

1.2.0 Chemical Reactions

A chemical reaction centers upon the atom's outermost electron shell, known as the valence shell. Atoms are most stable when their valence shells contain the maximum number of electrons permitted (usually 8). When atoms react, they tend to gain, lose, or share electrons with other atoms in order to complete their valence shells.

An atom with a mostly full valence shell tends to gain electrons from other atoms, thereby completing its own valence shell. An atom with an almost empty valence shell tends to lose electrons to another atom. This action empties the original valence shell, causing the next-lowest electron shell to become the new valence shell. Since it's already full, the result is a stable valence shell. Atoms with valence shells that are neither almost empty nor almost full often share multiple electrons with other atoms. The periodic table entry for each element includes an electron configuration list of numbers. The bottom number is the valence shell and indicates how full it is.

1.2.1 Chemical Bonding

As an example, sodium (Na) has one electron in its valence shell. When it reacts with other elements, it tends to lose this electron. Chlorine (Cl) has seven electrons in its valence shell. When it interacts with other elements, it tends to gain an electron to complete its valence shell. If sodium and chlorine combine, this is exactly what happens. The gain/loss process forms a bond between the sodium and chlorine, creating a compound called sodium chloride (NaCl).

When a metallic atom reacts with a nonmetallic atom, an electron gets transferred from one atom to the other. This process links the atoms together in what is called an ionic bond and forms an ionic compound. The most fundamental unit of an ionic compound is called a formula unit. Ordinary table salt (NaCl, or sodium chloride) is an ionic compound. Sodium (a metal) gives up an electron to chlorine (a nonmetallic gas) forming an ionic bond (*Figure 6*).

When two nonmetallic elements react, they share electrons between themselves, essentially passing them back and forth. This process forms a covalent bond. Compounds formed by covalent

PERIODIC TABLE OF THE ELEMENTS

Period	1	2	3	4	5	6	7	8	9	10	11	12	13	14	15	16	17	18
1	1 H 1.0080																	2 He 4.00260
2	3 Li 6.941	4 Be 9.01218											5 B 10.81	6 C 12.0111	7 N 14.0067	8 O 15.9994	9 F 18.9984	10 Ne 20.179
3	11 Na 22.9898	12 Mg 24.305											13 Al 26.9815	14 Si 28.086	15 P 30.9738	16 S 32.06	17 Cl 35.453	18 Ar 39.948
4	19 K 39.102	20 Ca 40.08	21 Sc 44.956	22 Ti 47.90	23 V 50.941	24 Cr 51.996	25 Mn 54.9380	26 Fe 55.847	27 Co 58.9332	28 Ni 58.71	29 Cu 63.54	30 Zn 65.37	31 Ga 69.72	32 Ge 72.59	33 As 74.9216	34 Se 78.96	35 Br 79.904	36 Kr 83.80
5	37 Rb 85.4678	38 Sr 87.62	39 Y 88.906	40 Zr 91.22	41 Nb 92.906	42 Mo 95.94	43 Tc 99	44 Ru 101.07	45 Rh 102.906	46 Pd 106.4	47 Ag 107.870	48 Cd 112.40	49 In 114.82	50 Sn 118.69	51 Sb 121.75	52 Te 127.60	53 I 126.9045	54 Xe 131.30
6	55 Cs 132.91	56 Ba 137.34	57-71 Rare Earths (See Below)	72 Hf 178.49	73 Ta 180.95	74 W 183.8	75 Re 186.2	76 Os 190.2	77 Ir 192.2	78 Pt 195.09	79 Au 196.967	80 Hg 200.59	81 Tl 204.37	82 Pb 207.2	83 Bi 208.9806	84 Po 209	85 At 210	86 Rn 222
7	87 Fr 223	88 Ra 226.03	89-103 Radioactive Rare Earths (See Below)	104 Rf 261	105 Db 262	106 Sg 263	107 Bh 264	108 Hs 265	109 Mt 266	110 Ds 271.15	111 Rg 272.15	112 Cn 277	113 Uut ?	114 Fl 289.2	115 Uup ?	116 Lv 293	117 Uus ?	118 Uuo ?

Lanthanides	57 La 138.91	58 Ce 140.12	59 Pr 140.91	60 Nd 144.24	61 Pm 147	62 Sm 150.4	63 Eu 151.96	64 Gd 157.25	65 Tb 158.93	66 Dy 162.50	67 Ho 164.93	68 Er 167.26	69 Tm 168.93	70 Yb 173.04	71 Lu 174.97
Actinides	89 Ac 227	90 Th 232	91 Pa 231	92 U 238	93 Np 237	94 Pu 242	95 Am 243	96 Cm 248	97 Bk 249	98 Cf 249	99 Es 254	100 Fm 257	101 Md 258	102 No 259	103 Lr 262

Note: Atomic weights based on 12/6C = 12.

Figure 4 Periods in the periodic table.

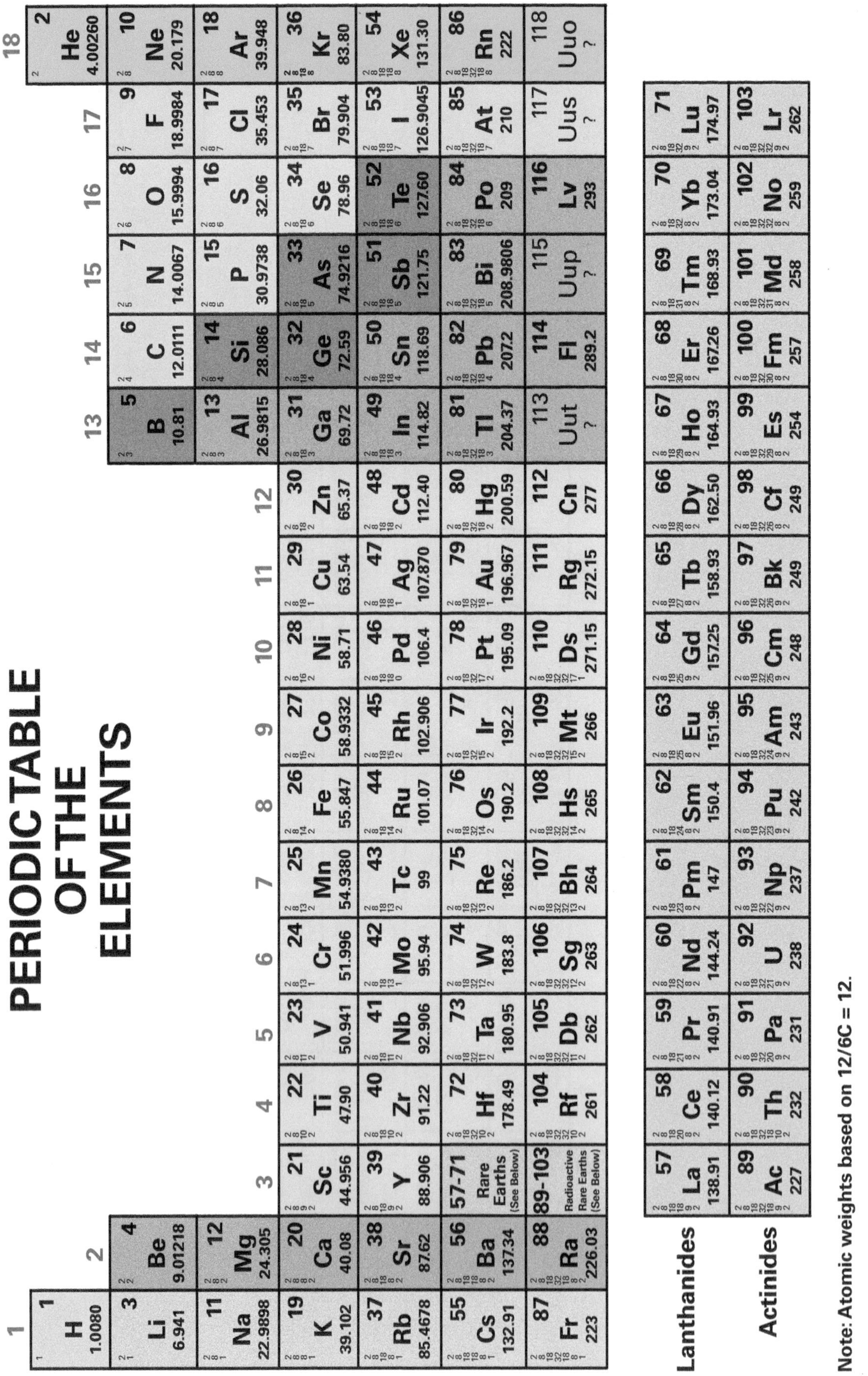

Figure 5 Groups in the periodic table.

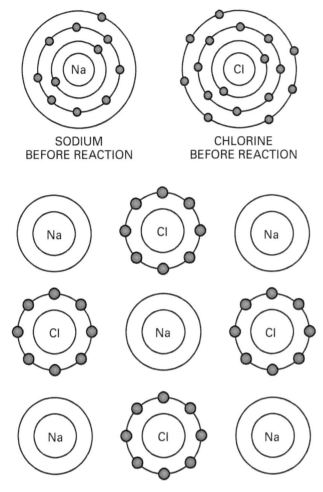

Figure 6 Ionic bonding.

A mole consists of 6.02×10^{23} particles. These may be atoms, formula units, or molecules. One mole of any element has a mass (in grams) equal to that element's atomic mass value. So, 1 mol of sodium (atomic mass of 22.9898) atoms has a mass of 22.9898 g. Similarly, 1 mol of chlorine (atomic mass of 35.453) has a mass of 35.453 g. If you combine 1 mol of sodium and 1 mol of chlorine, you will get 1 mol of sodium chloride (NaCl). Its mass will equal 58.4428 g, the sum of the masses of the original elements.

> NOTE
> Notice that combining 1 mol of sodium and 1 mol of chlorine doesn't result in 2 mol of sodium chloride. That's because the mol is a fixed number of particles. Combining 6.02×10^{23} atoms of sodium with 6.02×10^{23} atoms of chlorine yields 6.02×10^{23} formula units of NaCl—in other words, 1 mol.

Example 1:
How many grams of carbon and oxygen are required to form 1 mol of carbon dioxide (CO_2)? What will the carbon dioxide's mass be when the reaction is complete?

Since each atom of carbon must react with two atoms of oxygen to form one molecule of carbon dioxide, 1 mol of carbon and 2 mol of oxygen will be required to produce 1 mol of CO_2 molecules. Referring to the periodic table in *Figure 2*, notice

bonding exist in units called molecules. Water (*Figure 7*) is one example in which two hydrogen atoms form covalent bonds with a single oxygen atom (H_2O). It's also possible for atoms of the same element to form covalent bonds. For example, oxygen will covalently bond with another oxygen atom (O_2).

Finally, metal atoms can form bonds with each other in which their valence electrons get shared among a huge number of atoms and form a kind of electron "sea." This type of bond is called a metallic bond. The metal usually forms crystals as a result of a metallic bond.

1.2.2 Atomic Mass and Chemical Reactions

Because elements always combine in specific proportions, chemists need to be able to deal with quantities in a practical way. Since atoms are very tiny and the quantities involved in most reactions are huge, it's easier to work with masses rather than individual atoms. Chemists do this by way of a unit called the mole (mol).

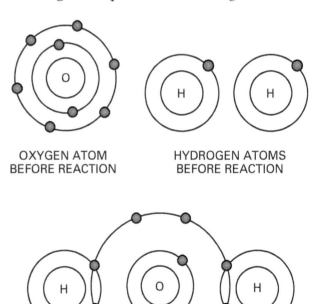

Figure 7 Covalent bonding.

that 1 mol of carbon (atomic mass of 12.0111) has a mass of 12.0111 g. Similarly, 1 mol of oxygen (atomic mass of 15.9994) has a mass of 15.9994 g. Since you need 2 mol of oxygen, however, the total mass is twice that amount—31.9988 g. Once combined, the result is 1 mol of CO_2, which has a mass of 44.0099 g.

1.3.0 Concentration

The fundamental reason for most analytical measurements is to answer the basic question, *"How much?"* Frequently, this question is answered in terms that express the concentration of solutions.

When crystals of salt are stirred into a sufficient quantity of water, the salt disappears and a clear mixture of salt and water is formed. The salt is said to have dissolved or gone into solution in the water. The solution consists of two components, the solute (the dissolved salt) and the solvent (the water). In a solution, the solute is uniformly distributed among the molecules of the solvent. A solution containing a small amount of solute compared to the overall amount of solution is said to be dilute. As more solute is added to the solution, the solution becomes more concentrated. Two terms commonly used to describe the concentration of a solution are molarity (M) and parts per million (ppm).

Molarity is the ratio of the number of moles of solute present per liter of the solution. (Thus, 1 mol/L is equal to 1 M.) Many different chemical solution concentrations are expressed in molarity, particularly industrial chemicals such as acids and bases. Incidentally, the unit for molarity, M, is pronounced *molar*. So, a bottle of 14 M sulfuric acid would be described as *"14 molar sulfuric acid"*.

Example 2:
What is the molarity of a saltwater solution composed of 2 L of water and 50 g of salt?

Step 1 Begin by calculating how many grams are in a mole of salt. As previously discussed, 1 mole of any element in grams is equal to that element's atomic mass value (which can be found on the Periodic Table of Elements). Since 1 mol of sodium (Na) is equal to 22.9898 g, and 1 mol of chlorine is equal to 35.453 g, these values can be combined to determine that 1 mol of salt (NaCl) is equal to 58.4428 g. This is illustrated as follows:

22.9898 g/mol of Na + 35.453 g/mol of Cl =
58.4428 g/mol of salt

Step 2 Calculate the number of moles of salt present in the solution. This is done by dividing the grams of salt in the solution by the number of grams in 1 mol of salt, as follows:

$$\text{moles of salt} = \frac{\text{grams of salt present}}{\text{g/mol of salt}}$$

$$\text{moles of salt} = \frac{50\text{g}}{58.4428 \text{ g/mol}}$$

moles of salt = 0.8555 mol

Step 3 Since molarity is a ratio of the number of moles of solute per liter of the solution, and there are 0.8555 moles of salt in 2 liters of water in this example, the molarity can be calculated as follows:

$$\text{molarity} = \frac{\text{moles of solute}}{\text{liters of solution}}$$

$$\text{molarity} = \frac{0.8555 \text{ mol}}{2\text{L}}$$

molarity = 0.4278M

The term *parts per million* is frequently used to report small concentrations of impurities in solutions. Parts per million (ppm) is a ratio of solute to solution. As with most ratios, both the solute and solution should be expressed in the same units. The size of the solution term in the ratio should be a factor of one million larger than the solute term. A concentration of 10 ppm chlorine, for instance, means that for every 10 g of chlorine there are 1×10^6 g of solution.

1.4.0 Acids, Bases and pH

Acids and bases figure prominently in industrial processes, so it's important that you understand what they are and how their presence is measured and classified.

1.4.1 Ions

When certain substances dissolve in water, they break up into components, each of which has an electric charge. A charged particles is called an ion. When an ionic compound dissolves, it dissociates into a metal ion and a nonmetal ion. For example, when sodium hydroxide (NaOH) dissolves in water, it dissociates into a sodium ion (Na^+) and a hydroxide ion (OH^-).

Certain covalent compounds also break down in water, but the process is called ionization. For example, if you pour hydrochloric acid (HCl) into water, it will ionize into a hydrogen ion (H⁺) and a chloride ion (Cl⁻). *Figure 8* illustrates both dissociation and ionization.

Because ions have an electric charge (identified by the + or − sign), they enable an electric current to flow through the solution. Substances that dissociate or ionize in solution are called electrolytes.

1.4.2 Acids and Bases

An acid is a compound that produces hydrogen ions (H⁺) in a water solution. In water solutions, all ions are surrounded by the water molecules (H_2O). Hydrogen ions in water attach to water molecules and form hydronium ions (H_3O^+). The more hydronium ions present, the more acidic the solution becomes.

A base is a compound that produces hydroxide ions (OH⁻) in a water solution. An example of a common base is sodium hydroxide (NaOH). In solution, it dissociates to form sodium ions (Na⁺) and hydroxide ions (OH⁻). The more hydroxide ions present, the more basic (alkaline) the solution becomes.

An acid is classified as strong if it ionizes into a lot of hydrogen ions that form hydronium ions when added to water. For example, hydrochloric acid ionizes essentially 100 percent. Consequently, it's a strong acid. Acetic acid (found in vinegar) does not ionize completely. Only a small percentage of its hydrogen turns into free hydrogen ions. Consequently, it's a weak acid. Similarly, strong bases ionize or dissociate into many hydroxide ions, while weak bases produce relatively few.

> **NOTE**
>
> Be careful not to confuse the terms *strong* and *weak* with *concentrated* and *dilute*. These terms have distinct meanings. A solution that is concentrated has a fairly large amount of solute present with respect to the total amount of solution. It does not automatically follow that the solution is strong. Remember, strong and weak refer to the degree of ionization/dissociation. It is possible to produce a concentrated solution of a weak acid or a dilute solution of a strong acid. On the other hand, it is often true that concentrated solutions pose greater risks than dilute ones. The key is knowing what you're working with and following safety guidelines at all times.

SODIUM HYDROXIDE EXAMPLE OF DISSOCIATION

HYDROCHLORIC ACID EXAMPLE OF IONIZATION

Figure 8 Dissociation and ionization.

1.4.3 pH

A convenient way to state the acidity or basicity (alkalinity) of a solution is the pH scale. A solution's pH value is a measurement of the quantity of hydrogen ions (H⁺) present. As the quantity of hydrogen ions increases, the solution becomes more acidic. Conversely, if the number of hydrogen ions falls and the number of hydroxide ions (OH⁻) increases, the solution becomes more basic.

The pH scale (*Figure 9*) typically ranges from 0 to 14. Values less than 7 are acidic, with smaller numbers being more acidic. Values greater than 7 are basic, with larger numbers being more basic. A value of 7 is considered neutral—neither acidic nor basic. Pure water has a pH of 7 at room temperature because it has an equal number of hydronium and hydroxide ions.

The pH scale is unusual in that it's logarithmic (based on powers of ten). In other words, each change of unit jumps by a factor of 10 rather than 1. So, a pH value of 9 is ten times as alkaline as 8.

A related measurement similar to pH is pOH, which is a measurement of the quantity of hydroxide ions in a solution. Essentially, the pOH scale is backwards in relation to the pH scale. A pOH of 0 is very basic, while a pOH of 14 is very acidic. *Table 1* shows the relationship between pH and pOH.

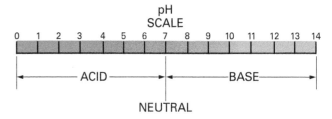

Figure 9 The pH scale.

1.4.4 Neutralization and Salts

When an acid and a base combine, the solution's pH moves closer to 7 (neutral). This process is called *neutralization*. In industrial settings, acids and bases are often neutralized prior to disposal since neutral solutions are far less harmful. A neutralization reaction normally produces an ionic compound called a salt. You're probably used to thinking of salt as something that you eat, but there are actually many different salts. When salts dissolve in water, they dissociate into ions. Since ions are electrically charged, they enable the solution to carry an electric current. Salts are electrolytes, as are acids and bases.

For example, if you take a dose of sodium bicarbonate (baking soda) dissolved in water to cure a case of heartburn, you're performing a neutralization reaction. The excess hydrochloric acid (HCl) in your stomach combines with the sodium bicarbonate ($NaHCO_3$) producing carbon dioxide (CO_2), water (H_2O), and a salt—in this case, sodium chloride (NaCl).

Table 1 pH/pOH Relationship

Hydrogen Ion Concentration (H^+) (gm moles/⁻)	Hydrogen Ion Concentration (OH^-) (gm moles/⁻)	pH	pOH
1	10=–14	0	14
10–1	10–13	1	13
10–3	10–11	3	11
10–4	10–9	5	9
10–7	10–7	7	7
10–9	10–5	9	5
10–11	10–3	11	3
10–13	10–1	13	1
10–14	1	14	0

Additional Resources

Basic Chemistry, Karen C. Timberlake and William Timberlake. Fourth Edition. 2013. Upper Saddle River, NJ: Prentice-Hall.

Chemistry: Concepts and Problems: A Self-Teaching Guide, Clifford C. Houk and Richard Post. Second Edition. 1996. New York, NY: John Wiley & Sons, Inc.

1.0.0 Section Review

1. The smallest particle with a distinct chemical identity is a(n) _____.
   a. ion
   b. molecule
   c. atom
   d. formula unit

2. The atomic number of an element indicates _____.
   a. the element's mass
   b. the number of protons in the nucleus
   c. the number of neutrons in the nucleus
   d. the number of protons and neutrons in the nucleus

3. In the periodic table, elements with similar chemical properties are found _____.
   a. in the same group
   b. in the same period
   c. near elements with similar atomic numbers
   d. near elements with similar atomic masses

4. Chemical reactions always involve the electrons in the _____.
   a. ionic solution
   b. nucleus
   c. innermost shell
   d. valence shell

5. A chemical bond in which atoms share electrons is called a(n) _____.
   a. ionic bond
   b. covalent bond
   c. metallic bond
   d. balanced bond

6. A technician mixes 5.0 mol of sodium hydroxide with water to make a 3.0 L solution. What is the molarity of the resulting solution?
   a. 0.60 M
   b. 1.7 M
   c. 3.0 M
   d. 5.0 M

7. An acid that ionizes into only 5 percent of its total hydrogen atoms when mixed with water is classified as _____.
   a. weak
   b. strong
   c. dilute
   d. concentrated

 12409-16 Analyzers and Monitors

Section Two

2.0.0 Analyzing Physical Properties

Objective

Define the physical properties of density, specific gravity, viscosity, and turbidity, and identify methods used to analyze them.

a. Define the properties of density and specific gravity and identify methods used to analyze them.
b. Define the property of viscosity and identify methods used to analyze it.
c. Define the property of turbidity and identify methods used to analyze it.

Trade Terms

Chromatography: A family of analytical laboratory techniques in which substances are separated for individual identification.

Colorimeter: A device that measures the quantity of specific suspended particles in a solution by measuring how particular wavelengths of light are absorbed when they pass through the solution.

Density: A physical property of matter defined as mass per unit volume.

Hydrocarbon: A chemical compound composed of hydrogen and carbon atoms.

Hydrometer: A device used to measure the specific gravity of a solution.

Poise (P): A unit of dynamic viscosity. It is often used with the prefix *centi-*; one centipoise is one one-hundredth of a poise. It is derived from French and pronounced *pwaz*.

Specific gravity: A physical property of matter defined as the ratio of a substance's mass to the mass of an equal volume of a reference substance, usually water or air.

Turbidity: The cloudiness of a solution containing suspended particles.

Viscosity: The internal friction of a fluid that opposes flow.

A process analyzer is an unattended instrument that monitors a physical or chemical property of a process. Most process analyzers operate on laboratory principles, but have added mechanisms and circuitry to work in an industrial environment, providing data in the necessary form.

Process analyzers may be classified in various ways. Some classification schemes are as follows:

- Operating principle (such as infrared, ultraviolet, or chromatography)
- Type of analysis (such as oxygen or carbon dioxide)
- Selectiveness (monitoring just one or several components)

Process analyzers are almost always calibrated by applying standard samples prepared and analyzed by a laboratory. For this reason, process calibration will be no better than the laboratory analysis because any errors will be compounded. Repeatability is essential in process analyzers to eliminate any human error due to variances and ambient conditions.

2.1.0 Density and Specific Gravity

The term density is defined as mass per unit volume and is most commonly expressed in either grams per cubic centimeter (g/cm^3) or pounds per cubic foot (lbs/ft^3). A good illustration of density is a one-inch cube of steel and a one-inch cube of balsa wood. Both have the same volume, but their masses are very different. Density is calculated using the following equation:

$$\text{Density} = \frac{\text{Mass}}{\text{Volume}}$$

Consequently, the steel cube has a greater density than the balsa cube since it has much more mass in the same volume. Liquids also have densities, with pounds per gallon (lbs/gal) being one possible unit. As an example, one gallon of gasoline is lighter than one gallon of water, so water is the denser liquid.

Specific gravity is a term often used interchangeably with density. For liquids, it is expressed as the ratio of the density of the liquid to the density of water. For gases, it is expressed as the ratio of the density of the gas to the density of air. Unlike density measurements, specific gravity values do not have units because they are ratios. Specific gravity is calculated with the following equation:

$$\text{Specific gravity} = \frac{\text{mass of specific volume of substance}}{\text{mass of the same volume of water (or air)}}$$

Keep in mind that, when using the specific gravity equation, the same units of measurement must be used for both quantities in order to obtain an accurate result. Because volume can change with temperature, density and specific gravity measurements must be made at specific temperatures or correction factors must be applied for other temperatures. Since gases are compressible, pressure must also be standardized or corrected for when determining their density or specific gravity.

2.1.1 Hydrometer

One way of measuring a liquid's specific gravity is with a device called a hydrometer. The common hydrometer consists of a glass float that is weighted at the bottom (see *Figure 10*). The float has a hollow stem, inside of which is a graduated scale. To find the specific gravity of a liquid, float the hydrometer in the liquid. The position of the liquid's surface on the hydrometer scale indicates the liquid's specific gravity.

2.1.2 Air Bubble Measurement

The simplest and perhaps most widely used method of measuring density and specific gravity in processes is using bubble tubes and a reference chamber in conjunction with a differential pressure transmitter. The two bubble tubes are installed in a vessel containing the process liquid to be measured but at different levels in the vessel, as shown in *Figure 11*.

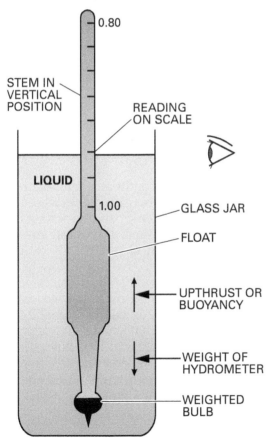

Figure 10 A hydrometer.

The pressure required to cause bubbles to escape into the liquid is a measure of the pressure at the level of that tube's outlet. The difference between the two pressures required to cause bubbles to escape out of each tube is measured by the differential pressure transmitter and is equal to the weight of a constant-height column of the liquid. The change in differential pressure caused by changing density is proportional to the density of the liquid. The adjustable rotameters are used to keep a calibrated continuous purge flowing through the two bubbler lines and can also be used to calibrate the transmitter.

2.1.3 Displacement Measurement

A displacer is sometimes mistakenly called a float. A true displacer submerges in the liquid based on its calculated buoyancy, whereas a true float rides on top of the liquid. *Figure 12* illustrates the application of displacement for measuring density. The force acting on the balance arm is directly related to the density of the fluid displaced by the displacer. Because the displacer is in direct contact with the process liquid, a continuous flow in the process is necessary to purge the unit with fresh liquid. Low flows are necessary through the process lines, and the temperature of the process liquid must remain constant. The process liquid must be clean, to prevent settling of solids in the chamber and buildup of material on the displacer.

2.1.4 Densitometer

Measuring the density of gases is important in the gas transmission and petrochemical industries. This measurement can be accomplished using a densitometer with a probe. The probe includes a vane that is installed in the pipeline that contains the flowing process fluid. The vane oscillates in the fluid, causing an acceleration of the flowing process fluid. The frequency of the oscillation varies with the density of the fluid. As the density increases, the oscillation frequency decreases. The relationship is expressed by the following equation:

$$\text{Density} = \frac{f^2 \times (A - B)}{(f + C)}$$

In this equation, f is the frequency of the oscillation, and A, B, and C are constants related to the size of the probe and the properties of the fluid. The signal is amplified by the transmitter. The transmitter converts the frequency to a 4–20 mA current. The densitometer can also be used with liquids.

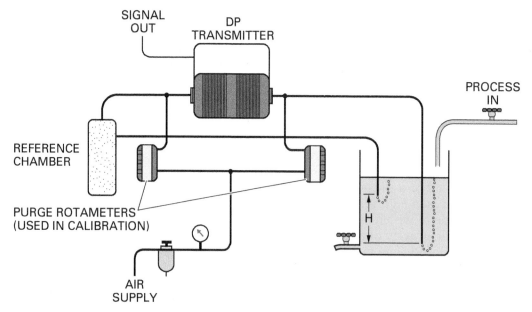

Figure 11 Dual bubblers used to measure density.

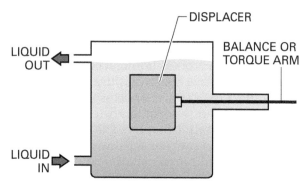

Figure 12 Displacement measurement of density.

2.1.5 Nuclear Detectors

The density of a liquid can be measured using the nuclear radiation from a suitable radioactive isotope, such as cobalt-60. The radioactive source is placed at one side of a vessel. Its radiated particles and/or gamma rays are directed across the vessel. A radiation detector is placed on the opposite side of the vessel so that it senses the amount of energy remaining after the radiation passes through the vessel walls and the liquid, as shown in *Figure 13*. The amount of radiation absorbed by any material varies directly with its density. Such a method is best suited for liquids that do not permit a sensing element to be immersed because of potential corrosion, abrasion, or other similar limitation.

2.2.0 Viscosity

The internal friction, or resistance to flow within a fluid is called viscosity. For example, a thick liquid like honey offers more resistance to flow than water does. Viscosity is responsible for most of the energy loss in liquid and gas pipelines. For this reason, viscosity is a very important factor in fluid flow.

Of all the physical conditions, temperature has the greatest effect on viscosity. An increase in temperature causes a greater separation among the liquid molecules, a reduction in the shear stress, and therefore a decrease in viscosity. Lubricating oils may fail to form a protective film at low temperatures. The graphs in *Figure 14* illustrate the changes in viscosity that occur in oil and water with increasing temperature. Notice the dramatic change in oil versus the very small change in water.

As another example, tar at 15°C has a viscosity of 1.65×10^6 poise (P), whereas at 45°C its viscosity is 1.3×10^3 poise. For a temperature change of just 30°C, the viscosity changes by a factor of a thousand!

Oil Viscosity

The effectiveness of lubricating oil depends on its viscosity, among other factors. Generally, lubricating oils should be sufficiently viscous not to be squeezed out of bearings and yet not so viscous as to increase the resistance to the motion of the moving parts that are being lubricated. Likewise, heavier oils have higher viscosities than lighter oils. The density of an oil, however, does not bear any relation to the oil's viscosity.

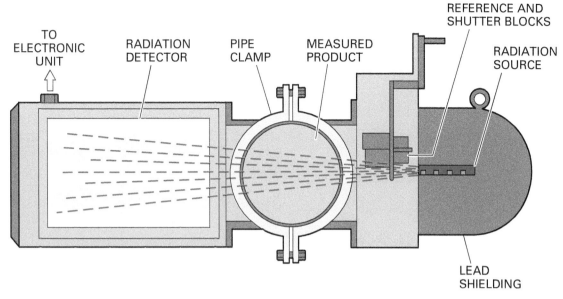

Figure 13 Density measurement by a radiation detector.

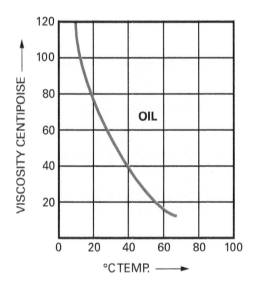

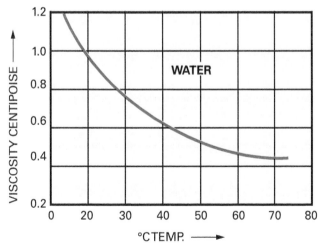

Figure 14 Oil and water change in viscosity with temperature variation.

2.2.1 Viscometers

Viscometers are instruments used for measuring the viscosity and flow properties of fluids at ambient or other specified temperatures. There are many types and models of viscometers available for determining liquid viscosities, most of which are designed for use with specific liquids and viscosity ranges. Commonly used viscometer designs typically fall into one of three categories:

- Falling ball or piston
- Rotating disc
- Capillary

A falling ball viscometer (*Figure 15*) measures viscosity by measuring the time required for a ball to fall through a specific distance of the liquid under test. The ball used for the test has a specific density value and is typically made of glass or stainless steel. Balls of different densities are used for testing different liquids. The precision ball falls through a precision-diameter, temperature-controlled glass tube containing the liquid under test. The fall time through the calibrated distance is typically measured using a timer.

A rotating disc viscometer (*Figure 16*) requires a certain amount of torque to rotate a disc at a constant speed when immersed in a sample of the fluid being tested. The torque is a measure of the test fluid's dynamic viscosity and is proportional to the viscous drag on the immersed disc. A synchronous induction motor drives a container coupled through a calibrated spring (torsion wire) to the disc. During measurement, the spring tends

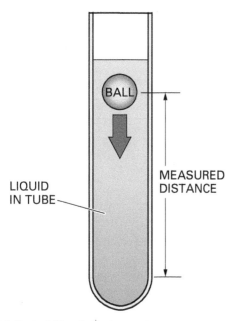

Figure 15 Basic falling ball viscometer.

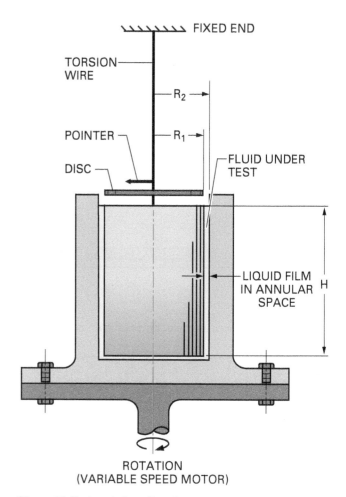

Figure 16 Basic rotating disc viscometer.

to wind up until its force equals the viscous drag on the disc that is immersed in the fluid under test. When this occurs, the container and disc both rotate at the same speed but with an angular difference between them. This angular difference is proportional to the torque on the spring and is converted into a viscosity reading. Depending on the device, the viscosity indication can be accomplished electronically or by a pointer. Other versions of a rotating viscometer that operate in a similar manner rotate a cone, sphere, or spindle in the fluid under test.

Capillary viscometers (*Figure 17*) measure the flow rate of a fixed volume of fluid at a controlled temperature through a small orifice. The time that it takes for a specific volume of fluid to pass through the capillary orifice is proportional to the dynamic viscosity of the fluid. It should be pointed out that it also depends on the density of the fluid, because the denser the fluid, the faster it will flow through the capillary orifice.

Viscometer manufacturers often describe their instruments as being used to measure Newtonian or non-Newtonian fluids. A Newtonian fluid is defined as a fluid that has a viscosity that is independent of shear. Oil, liquid hydrocarbon, beer, and milk are some examples of a Newtonian fluid.

A non-Newtonian fluid is defined as a fluid that has a viscosity that is dependent on shear. This means as the shear rate changes, the fluid velocity changes. Some examples of non-Newtonian fluids are grease, paint, clay slurries, and candy compounds. Note that with an increase in shear rate, the viscosity of some non-Newtonian fluids will increase, while for others, it will decrease. Obviously, the type of viscometer selected for a specific use must be compatible with the type of fluid being tested.

Viscometers can be portable instruments, fixed instruments, or in-line devices. In-line instruments can be inserted into and retracted out of a process line without disturbing the flow of the fluid in the line. Viscometers are available with analog, digital, or graphical displays. Some have no local display and send the measured viscosity data to a remotely located display or to an intelligent device. Operating controls can also be analog or digital. Some viscometers have the capability of being operated under the control of a computer. Other models are pre-programmed and have no user controls. Features available for viscometers include temperature compensation, temperature sensing, and data storage.

2.3.0 Turbidity

The cloudiness of a fluid caused by suspended particles is known as turbidity. The particles scatter light passing through the liquid. Turbidity is influenced by the concentration, size, shape, and optical properties of the particles in addition to the optical properties of the fluid.

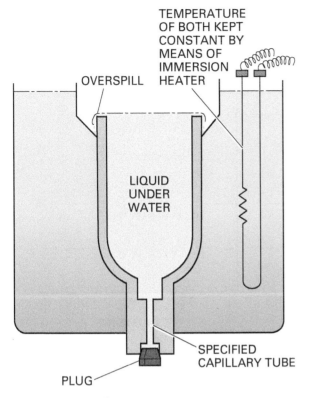

Figure 17 Basic capillary viscometer.

2.3.1 Jackson Turbidimeter

The Jackson turbidimeter, illustrated in *Figure 18*, is a historical device that set the standard for measuring turbidity and also established its unit. It consists of a special candle and a flat-bottomed glass tube graduated in Jackson turbidity units. The technician pours the sample into the tube, watching the candle flame's image from above. When it becomes a diffuse glow rather than a distinct image, the current liquid level is read off the side of the tube.

The Jackson turbidimeter has a number of limitations under certain conditions. Modern instruments that measure light passing through a sample use similar principles, but without many of the weaknesses.

2.3.2 Transmission Analyzer

A transmission turbidity analyzer contains a light source, lens, sample cell, photocell, and a control unit with a readout (*Figure 19*). The lens focuses light from the source through the sample cell, where it is measured by the photocell. Liquids with low turbidity absorb and reflect very little of the light, so the photocell's output is high. Liquids with high turbidity absorb and reflect a significant portion of the light, so the photocell's output is low. The control device amplifies and transmits the photocell's output signal to the readout device.

Turbidity, measured in percent of light transmitted, is calibrated by checking zero and then 100-percent transmission of light with a clear sample. The adjustable shutter shown in the diagram can be used to regulate the amount of light from the source, depending on the degree of turbidity in the sample. As long as the shutter is set before calibrating the analyzer with a clear sample, the specific amount of light will not affect the reading because the reading is based on a percentage of light available.

The transmission-type turbidity analyzer may also be used as a colorimeter to measure changes to the light as a result of the presence of color. Occasionally, it is desirable to monitor one specific component in the sample. The light wavelength of the transmission-type turbidity analyzer may be calibrated and limited based on a colorimeter calibration curve, like the one shown in *Figure 20*, so that the instrument responds only to changes in a particular component of the sample.

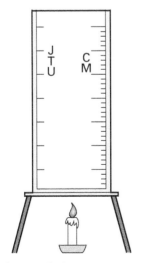

Figure 18 The Jackson turbidimeter.

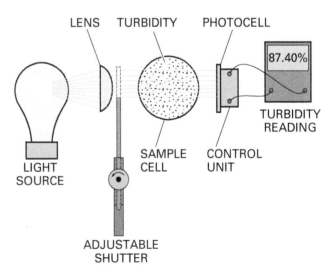

Figure 19 Transmission-type turbidity analyzer.

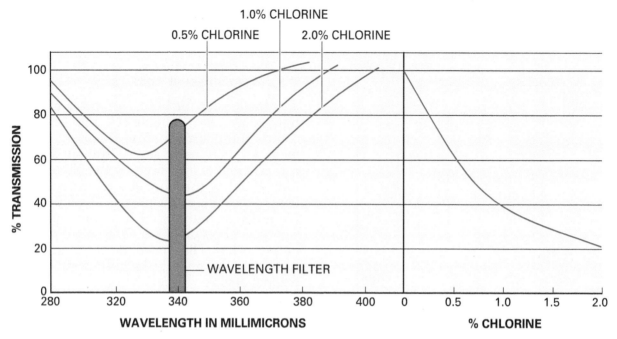

Figure 20 Colorimeter calibration curve for chlorine.

For example, if the light wavelength is limited to a narrow band near 340 nanometers (nm), the instrument will be sensitive only to changes in chlorine content. This is because chlorine absorbs strongly at that wavelength. The transmission-type turbidity analyzer can be sensitized in this manner by selecting the proper light source, filters, and photocell.

Gauging color in a lab is usually accomplished by holding a container filled with liquid in front of a light and comparing its color to a group of color standards. The transmission-type process analyzer uses the same method for gauging color, but it substitutes more reliable and calibrated photocells for the human eye.

2.3.3 Reflection Analyzer

Another analyzer that is similar to the transmission type is the reflection-type turbidity analyzer, which measures the amount of light reflected by the suspended particles in a sample stream. A simplified reflection analyzer (*Figure 21*) has a shield between the light source and photocell to prevent the transmission of light directly to the cell. Particles in the stream reflect light around the shield to the photocell. The quantity of light reaching the photocell is directly proportional to

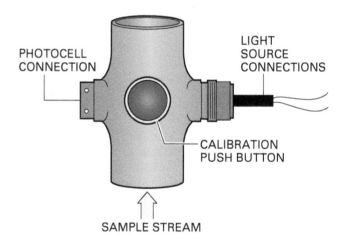

Figure 21 Typical reflection-type, in-line turbidity analyzer.

the concentration of the particles present in the stream. The electrical signal from the photocell is typically connected externally to either an indicating or recording instrument that is calibrated in parts per million of turbidity.

The turbidity sensor is designed with a self-contained calibration feature consisting of a push button that actuates a slide plate, exposing a precisely machined opening in the light shield. This allows an amount of light exactly equivalent to a turbidity of 5 ppm to be transmitted directly to the photocell. When this button is pressed, it changes the output of the sensing cell by an amount that is equal to 5 ppm turbidity. This signal change permits the system to be quickly and accurately calibrated.

2.3.4 Ratio Analyzer

A ratio turbidity analyzer uses a more complex system to measure turbidity, sharing technology from both the transmission-type and reflection-type of analyzers. See *Figure 22*, in which light from a common source passes through a rotating chopper disc that alternately blanks out one light beam at a time. The transmitted beam passes through the sample cell to photocell A, followed by the scattered beam at right angles. Photocell A sees the incoming beam during one half of the chopper disc's cycle and the reflected beam during the last half of the cycle.

Photocells B and C generate pulses that are synchronized with the transmitted and scattered beams and are only used as control signals for the electronic circuit associated with this analyzer. The two light beam signals are amplified, switched to the output circuit, and transmitted to the readout device. The readout device is usually calibrated in parts per million.

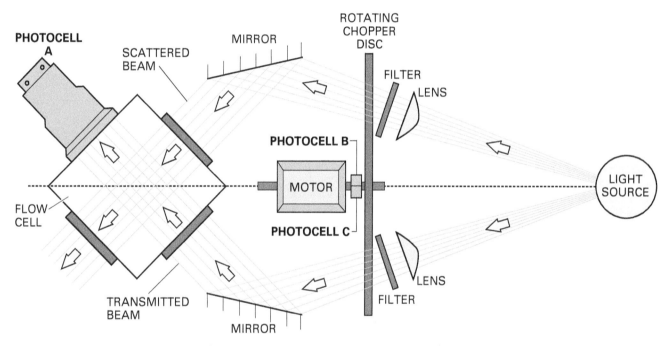

Figure 22 Light beam traveling through a ratio turbidity analyzer.

Additional Resources

Industrial Pressure, Level, and Density Measurement, Donald R. Gillum. Second Edition. 2008. Research Triangle Park, NC: The International Society of Automation.

Instrument Engineers Handbook, Volume 1: Process Measurement and Analysis, Béla G. Lipták (editor). Fourth Edition. 2003. Boca Raton, FL: CRC Press.

Instrumentation Reference Book, Walt Boyes (editor). Fourth Edition. 2009. Oxford, UK: Butterworth-Heinemann.

Measurement and Instrumentation Principles, Alan S. Morris. Third Edition. 2001. Oxford, UK: Butterworth-Heinemann.

The Condensed Handbook of Measurement and Control, N.E. Battikha. Third edition, 2006. Research Triangle Park, NC: The International Society of Automation.

The International Society of Automation, **www.isa.org**

2.0.0 Section Review

1. A block of wood 6.0 cm on each side has a mass of 160 g. What is its density?
   a. 0.74 g/cm^3
   b. 1.4 g/cm^3
   c. 4.4 g/cm^3
   d. 27 g/cm^3

2. Crude oil has a specific gravity of 0.92. If one gallon of water has a mass of 8.3 lbs at room temperature, what is the mass of one gallon of crude oil at the same temperature?
   a. 1.1 lbs
   b. 7.6 lbs
   c. 9.0 lbs
   d. 9.2 lbs

3. The instrument that works together with a differential pressure transmitter to measure density is the _____.
   a. hydrometer
   b. displacer
   c. bubbler
   d. densitometer

4. What class of instrument measures a fluid's resistance to flow?
   a. A viscometer
   b. A densitometer
   c. A turbidimeter
   d. A ratio analyzer

5. The measure of the cloudiness of a liquid is called _____.
   a. density
   b. viscosity
   c. turbidity
   d. opacity

Section Three

3.0.0 Analyzing Chemical Properties

Objective

Define the properties of flash point, pH, conductivity, and oxidation-reduction potential, and identify methods used to analyze them.

a. Define the property of flash point and identify methods used to analyze it.
b. Define the property of pH and identify methods used to analyze it.
c. Define the property of conductivity and identify methods used to analyze it.
d. Define the property of oxidation-reduction potential and identify methods used to analyze it.

Performance Task

1. Determine the pH of a given solution and propose the proper adjustment.

Trade Terms

Buffer solution: A solution with a specific pH value that strongly resists changes in pH and so maintains its value reliably.

Conductivity: The ability of a solution to carry an electric current due to the presence of ions. Measured in siemens per centimeter (S/cm).

Flash point: The lowest temperature at which a liquid gives off vapors that will mix with air and be able to ignite.

Oxidation-reduction potential (ORP): The ability of a solution to act as an oxidizer or reducing agent. Determined by measuring voltage polarity across a pair of electrodes.

In the previous section, you learned about several physical properties commonly measured in industrial process settings. Physical properties determine how a substance behaves mechanically as it moves through the process. In this section, materials are examined from the perspective of how they behave based on their chemical properties. These properties determine such things as the conditions required for a substance to catch fire, or they can indicate how corrosive a substance is, or whether an electric current will pass through it. In each case, instrumentation measures the property of interest so those in control can make informed decisions and keep the process running smoothly.

3.1.0 Flash Point Determination

OSHA 1910.106(a) defines flash point as the lowest temperature at which a liquid gives off vapor within a test vessel in sufficient concentrations to form an ignitable mixture with air near the surface of the liquid. A closely related and less common term associated with flash point is the term *fire point*. This is defined as the temperature at which the flame becomes self-sustained and continues burning the liquid. At the flash point, the flame does not need to be sustained. The fire point is usually a few degrees above the flash point.

Materials with a flash point below 100°F (38°C), such as solvents and solvent-borne coatings, are considered dangerous. Manufacturers provide flash point data about their products in the related safety data sheet (SDS). For flash point testing purposes, a test sample flashes when a large flame appears and instantly spreads across the surface of the sample. It should be pointed out that a blue halo does not constitute a flash point. The US Department of Transportation requires that all transported substances have their flash points determined and that any materials with flash points lower than 140°F (60°C) be handled with extra caution.

3.1.1 Establishing the Flash Point

The flash points of various materials can be determined by performing flash point tests using the appropriate instruments and procedures. Always follow correct sampling methods and standardized procedures precisely. Flash point testing methods have become standardized by a number of national and international organizations over the years. In general, the two methods for determining flash points are the closed-cup and open-cup methods. The closed-cup method prevents vapors from escaping and, therefore, usually results in a flash point that is a few degrees lower than in an open cup. Because the two systems give different results, always identify the method that you use for testing.

There are several types of commonly used flash point analysis systems, including the following:

- Abel
- TAG
- Rapid
- Pensky-Martens
- Cleveland
- Auto-ignition

The point of ignition with the Abel, TAG, Pensky-Martens, and Cleveland systems of flash point testing is determined by increasing the temperature of the sample and applying an igniter. You can use the rapid flash test system to quickly determine if a flash point occurs below or above a specific temperature; however, it does not find the actual flash point temperature. The auto-ignition system works at extremely high temperatures. The hot surfaces and air inside the container cause the sample to ignite.

When performing flash point testing, always operate the flash point test instrument in accordance with the manufacturer's instructions, and perform the test in accordance with the applicable standard.

Flash point testers typically operate by one of the four following methods:

- Standard instruments with gas heating and gas ignition
- Electrically heated instruments with gas or electric ignition
- Partially automated instruments with electric heating and automated gradient control
- Fully automatic instruments with automatic flash point detection

Table 2 shows the applicable standards associated with each of the most common standardized systems used to determine flash point.

> **WARNING!** Be sure to follow all safety procedures when using a flash tester. Fire extinguishers, safety visors, and breathing apparatus must be close at hand and should be used as needed. Proper ventilation is extremely important because heating can produce toxins, such as PCBs.

> **NOTE** The Deutsches Institut für Normung (DIN) standard is an international standard developed in Germany.

3.1.2 OSHA 1910.106(a)

OSHA 1910.106(a) defines flash point and the methods of testing flash point in some detail. Regarding the methods of testing, the standard states that if a liquid has a viscosity of less than 45 SUS at 100°F (38°C), does not contain suspended solids, and does not have a tendency to form a surface film while under test, then the test procedure specified in *ASTM D56-70* must be used. However, if a liquid has a viscosity of 45 SUS or more at 100°F (38°C), or contains suspended solids, or has a tendency to form a surface film while under test, the standard method of test for flash point by Pensky-Martens closed tester specified in *ASTM D93-71* must be used. There are some exceptions to this rule, which may be found in *ASTM D93 Notes*.

Table 2 Standardized Systems for Flash Point Determination

Low Temperatures (–20°C to +80°C)	
System	Standard
Abel	Institute of Petroleum (IP) 33, IP 170
Tag	American Society for Testing and Materials (ASTM) D 56, IP 304
Rapid	ASTM D 3228, ASTM D 3243, ASTM D 3278, IP 303, International Standards Organization (ISO) 3679, ISO 3680
Pensky-Martens	ASTM D 93
Medium Temperatures (+60°C to +360°C)	
Pensky-Martens	ASTM D 93, IP 34, Deutsches Institut für Normung (DIN) 51 758, ISO 2719, ASTM D 3228, ASTM D 3243
Rapid	ASTM D 3228, ASTM D 3243, ASTM D 3278, IP 303, ISO 3679, ISO 3680
High Temperatures	
Cleveland	ASTM D 92, ISO 2592
Highest Temperatures	
Auto-ignition	DIN 51 794, ASTM E 659

> **WARNING!**
> Know the flash point of any material you work with by referring to the SDS. Always avoid heat, open flame, sparks, or other sources of ignition when a material is near, at, or above its flash point. A common error is failure to pay attention to flash points when using a heating bath in a lab or on a calibration bench.

3.2.0 Measuring pH

Determining pH is one of the most widely used analytical measurements in industrial applications. In the refining and petrochemical industry, pH measurement is applied extensively for the following purposes:

- Quality control, where pH control determines product uniformity
- Corrosion inhibition, where pH monitoring is used in conjunction with oxygen control to minimize corrosion in high-pressure boilers and other water systems
- Effluent pH control, in the neutralization of liquids discharged into public streams

3.2.1 pH Instruments

As shown in *Figure 23*, the numbering system for a pH scale ranges from 0 to 14. A value of 7 is at the center of this span and indicates a neutral solution. The values below 7 represent acid levels, with the smaller numbers indicating the highest acid levels. The numbers above 7 represent the base scale, with the larger numbers indicating the highest base levels.

The pH level of a solution can be determined by direct measurement of the DC voltage developed between two electrodes immersed in the solution. The circuits in the instrument convert the voltage developed by the electrodes into a pH indication. Hand-deflecting instruments, chart recorders, and digital readout displays are commonly available. Recorders provide a permanent record of pH levels over a variety of different time spans.

Figure 24 shows a typical pH indicator scale for a hand-deflection instrument. There are ten small graduations or divisions between every two numbers on the scale. Each division, therefore, represents 0.1 pH unit. If the scale's indicating hand (not shown in the figure) is deflected to the third small graduation to the right of the number 5, the pH level would be 5.3. A pH meter can also be used to indicate positive and negative voltage values. For the indicator shown in *Figure 24*, full-scale deflection is +700 to –700 mV. In practice, the deflecting hand should come to rest at zero when measurements are not being taken.

Most meter scales include a curved mirror surface that runs parallel to the scale. This is designed to produce a reflection of the indicating hand. In practice, an operator looks at the scale in such a way that the indicating hand and its reflection are exactly in line. When this occurs, the indication should be quite accurate because the operator is not looking at the scale from an angle. Consequently, there should be no distortion or parallax effect.

Today, most pH instruments have direct digital readouts rather than meter scales. These are quicker and simpler to read. However, instruments with meter scales may be found in older installations.

3.2.2 pH Probes

Instruments that measure pH do so by immersing two electrodes into the solution. Many pH instruments use a single probe, which contains both electrodes in a single assembly. Handheld instruments usually use this arrangement (*Figure 25*). Other systems use two physically separate probes, called the pH electrode (*Figure 26*) and the reference electrode (*Figure 27*), respectively. In either case, the actual technology involved is much the same.

The pH electrode is often viewed as a battery whose voltage varies with the pH level of the solution. This component is sensitive to hydrogen

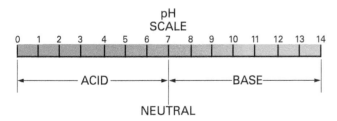

Figure 23 The pH scale range.

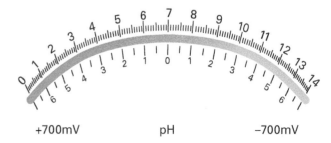

Figure 24 A pH indicator scale.

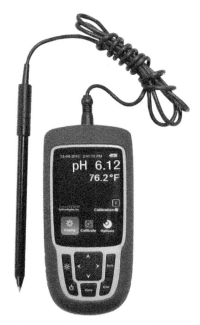

Figure 25 Handheld pH instrument with single probe.

Figure 26 A pH electrode.

Figure 27 Reference electrode.

(H^+) and is often referred to as the *measuring half-cell*. A special glass bulb or membrane material allows H^+ ions to pass into the tube. When the pH electrode is placed into the solution, a voltage proportional to the hydrogen ion concentration develops between an inner electrode and the outer electrode or glass bulb material.

The reference electrode, or *reference half-cell*, is responsible for producing a stable voltage that is independent of the solution's properties. The reference half-cell develops a fixed voltage value when placed into the solution. There are four types of reference electrode junctions. Each is best suited for a particular type of solution or a particular set of operating conditions. The four types of junctions are as follows:

- *Quartz fiber* – Maintains a rapid flow rate for fast response and minimum tendency to clog. It is the best general-purpose junction.
- *Ceramic frit* – Recommended for minimal potassium and chloride contamination of the sample.
- *Inverted sleeve* – Provides a very high flow rate for troublesome samples. May be disassembled for cleaning. Recommended for pastes, creams, colloids, sludge processes, and highly viscous samples.
- *Double* – Has the same characteristics as the ceramic frit junction. It is recommended where the flowing electrolyte must be matched to the sample and/or the sample must be completely isolated from the filling solution. The inner chamber of the double-junction electrode is not refillable. The outer chamber can be refilled with a custom filling solution to match the sample.

The two electrodes acting together produce the complete probe circuit. *Figure 28* shows a typical pH probe and meter.

Figure 29 is a diagram of the essential parts of a pH measuring instrument. In this case, pH is observed on a numeric display. The instrument produces quantitative measurements of pH. A majority of the pH instruments used in industry today are of this type. They range from small portable units, housed in convenient carrying cases, to larger stationary units.

3.2.3 Calibration and Compensation

Due to the way they work, pH electrodes require regular calibration. This is accomplished using a buffer solution. There are a number of different buffer solutions, each with a precise known pH value. Solutions with a pH of 4.01 and 10.0 are typical since they offer good midpoints for acidic and basic pH values. Buffer solutions with different pH values may be required for electrodes designed to measure narrower ranges.

Depending on the process that they are measuring, pH probes and electrodes also may require cleaning to restore their responsiveness.

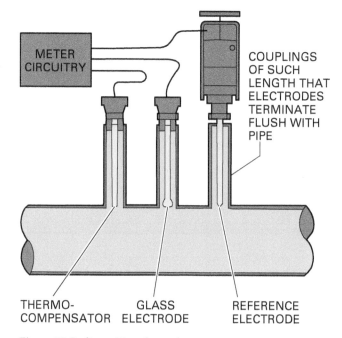

Figure 28 In-line pH probe and meter.

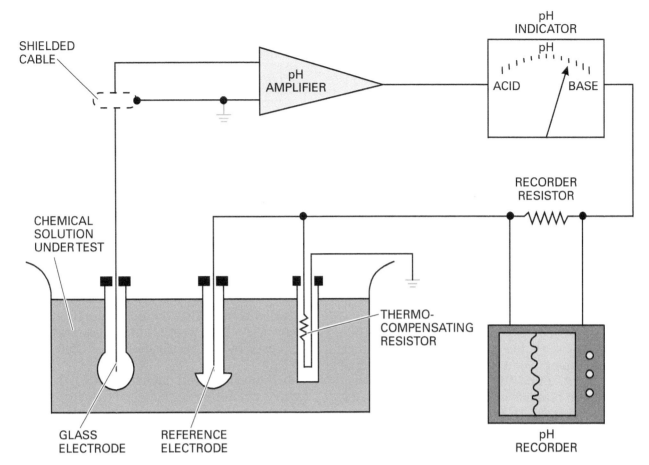

Figure 29 Simplified diagram of the essential parts of a pH measuring instrument.

If pH measurements start to become sluggish, it's likely that the electrode needs cleaning. This is accomplished by using a special cleaning solution, often a strong acid. Consult the manufacturer's guidelines before attempting to clean probes or electrodes. Similarly, pH probes must be stored properly to keep them from degrading. Generally, the tip must be kept immersed in a storage solution. Again, consult the manufacturer's guidelines.

Since pH can vary with temperature, many pH measuring instruments include a temperature sensor or probe that allows them to compensate for temperature variations. Compensation may be automatic or it may require some instrument configuration, depending on the setup.

3.2.4 pH Analyzer/Controller

Many pH analyzers are microprocessor-based instruments that also function as controllers. An example is shown in *Figure 30*. This analyzer/controller still requires an input from an electrode or probe system, but it can easily be calibrated by using a dedicated calibration key. The key permits technicians to make rapid calibrations by reading standardized buffer solutions.

Figure 30 A pH analyzer/controller.

The display shows the pH value of the buffer solution. Instrument technicians can then use the keypad to make changes to obtain the correct value for the buffer solution. A Hold key is provided to maintain analyzer output level and alarm status during calibration procedures. The display shows the calculated value of pH to within 0.01 pH unit, at an update rate of 0.5 second per reading. The useful limits for this display are –2 pH and 14 pH, although higher displays are available.

This type of pH analyzer also features extensive online diagnostics, as well as test functions. A keypad-entered security code prevents access to instrument settings, if desired. The analyzer is designed for panel, pipe, or wall mounting and is typically enclosed in a watertight and corrosion-resistant industrial case.

3.3.0 Measuring Conductivity

A pH measurement system is designed to be sensitive to just one ion, the hydrogen ion. Similar systems can be designed to measure other specific ions. There are applications, however, where it is necessary to determine the total concentration of all ions. This can be accomplished by measuring the solution's conductivity.

With pure water, the only ions in solution are those produced by the self-ionization of water into hydroxide (OH^-) and hydronium (H_3O^+) ions. Because current flow through a solution relies on ions as the charge-carrying conductors, very little current will flow in a solution of pure water. As impurities are added to the water, however, the number of available ions increases, reducing the electrical resistance to current flow.

Conductivity is a measure of how easily a material or solution conducts electricity. The flow of current in liquid solutions is different from that in metals. It is accomplished by the drift of positive and negative ions, rather than the movement of free electrons in metallic conductors. So, the conductivity of a solution is directly proportional to the total number of ions in the solution.

Conductivity is sometimes referred to as *specific conductance* and is defined as the reciprocal of the resistance (in ohms) of a one-centimeter cube of the solution at a specific temperature. The symbol for conductance is the letter G. It is represented by the following formula:

$$G = 1 \div R$$

Conductivity is measured in siemens/cm (S/cm), although older instruments may use mhos/cm, an equivalent unit. For poorly conductive solutions, the microsiemen/cm (μS/cm) version of the unit may be used instead.

3.3.1 Conductivity Instruments

A conductivity cell is a basic device used to measure the conductivity of a solution. The device has two parallel metallic electrode plates immersed in the solution (*Figure 31*). The external circuitry must provide a source of voltage to the electrodes and a method for determining the current flow

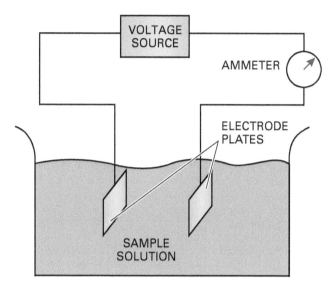

Figure 31 Simple conductivity meter.

between them. Most conductivity cells are driven by an AC voltage in order to avoid errors that can be caused by the polarization associated with DC voltage.

If the voltage and current flow are accurately known, the resistance between the electrodes can be determined using Ohm's law. This law states that the current through a conductor between two points is directly proportional to the potential difference across the two points. Once the resistance of the sample is known, its conductivity can be calculated. Since the distance between the plates (cell length) and the plate size factor in to the calculation, manufacturers normally provide a numeric cell constant that gets integrated into the calculation. Cells with high constants usually have small, widely spaced electrodes and are used to measure solutions with high concentrations of ions. Conversely, low cell-constant electrodes are used to measure low concentrations of ions and have large, closely spaced electrodes.

Technicians may use a buffer solution to check the cell constant of a conductivity cell or determine the cell constant of an unknown cell. An acceptable method of determining a cell constant is given in the American Society for Testing and Materials (ASTM) *Standard D 1125, Standard Test Methods for Electrical Conductivity and Resistivity of Water*.

3.3.2 Additional Factors in Measuring Conductivity

Temperature strongly influences a solution's conductivity. For this reason, manual or automatic temperature compensation must be provided. The actual change in conductivity with respect to temperature is determined by the type of ions in the solution. Therefore, automatic compensation has to make assumptions about the composition of the electrolyte.

The actual compensation measurement is often done with a thermistor. The nonlinear characteristics of the thermistor are matched to the nonlinear changes in conductivity that occur in the sample solution. Technicians generally use standard reference tables for different electrolytes to manually compensate conductivity-measuring instruments that may be used on many different solutions and that provide direct conductivity readouts.

3.3.3 Inductive Conductivity Instruments

Inductive conductivity probes operate on the principle of magnetically inducing an electric current. They usually consist of two enclosed electrical coils. The instrument transmits an AC current of known value through the primary coil, as shown in *Figure 32*. As this current passes through the primary coil, it creates a magnetic field and induces an electric current in the process liquid. This current then induces a current in the pickup coil. The value of the induced current is directly proportional to the conductivity of the liquid.

No direct contact is required between the coils and the solution, so maintenance problems are minimized. The inductive conductivity analyzer typically transmits a 4–20 mA signal that is proportional to the measured conductivity.

Calibration of the inductive conductivity cell is performed in much the same way as any probe calibration. A liquid containing a known standard measurement of conductivity is used. When using standard liquids for conductivity calibration, first immerse the probe in the standard, and then adjust the instrument to display that conductivity. It is recommended that you calibrate the instrument using two or more known liquid standards that are within the instrument's range. If only two standards are used, it's best to use one that is at the low end and one that is at the high end of the instrument's range. The low end is often at or near 0.0 µS/cm and is referred to as the *zero standard measurement*. The high end of the range is referred to as the *span standard*.

3.4.0 Measuring Oxidation-Reduction Potential

Oxidation is the loss of electrons by an atom, molecule, or ion, while reduction is the acquisition of electrons. Every liquid has both oxidizing and reducing ions, but their balance varies from liquid to liquid. Those with extra electrons are called *reducing agents*, while those that have a shortage are called *oxidizing agents*.

The oxidation-reduction potential (ORP) of a process liquid indicates by its voltage polarity whether the process liquid has an oxidizing potential or a reducing potential. A liquid with an oxidizing potential has a positive polarity relative to an applied reference voltage. A liquid with a reducing potential has a negative polarity relative to the reference voltage. The unit of ORP measurement, which is the voltage potential measured between two electrodes submerged in the sample, is the millivolt (mV).

3.4.1 ORP Instruments

An ORP probe, shown in *Figure 33*, is made up of two separate electrodes—a reference electrode and a measurement electrode—which are housed in an outer shell. The reference electrode provides a constant voltage potential. The lead wire of a reference electrode is positioned in an inner tube inside the shell. The inner tube typically contains silver metal and silver chloride paste. The paste is in contact with a saturated solution of potassium chloride, which acts as an electrical bridge to the solution being measured.

The potential of the reference electrode depends on temperature and on the concentration of potassium chloride. The potassium chloride slowly migrates from the reference electrode to the solution being measured by means of a liquid junction consisting of a porous ceramic material near the bottom of the electrode. Crystals of solid potassium chloride in the bottom of the electrode

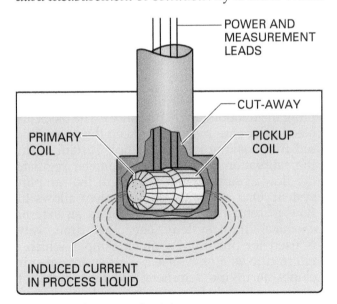

Figure 32 Inductive conductivity measurement.

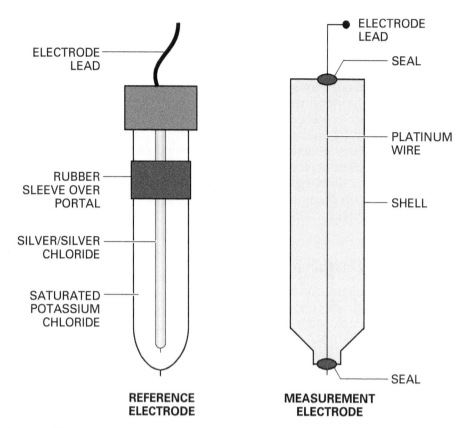

Figure 33 ORP probe assembly.

ensure that the solution stays saturated. A port is provided for replenishing the potassium chloride, and a rubber sleeve protects the port from contamination.

The ORP measurement electrode is a platinum wire that is exposed at its bottom end to the process liquid. This electrode reacts to variations in voltage. Voltage variations caused by changes in the activity of the oxidizing and reducing ions in the process liquid are amplified and displayed as the liquid's ORP.

3.4.2 Probe Calibration

An ORP probe can be calibrated by immersing both the measurement and reference electrodes in a liquid that has a known oxidation-reduction potential. These liquids are used specifically for calibration and are known as *standards*. Calibration happens in one of two ways—by removing the electrodes from the process and immersing them in a suitable container holding the standard or by putting the standard directly into the sample tubing leading to the installed electrodes. The first method is the easiest and the most accurate. A piping design that allows you to remove the electrodes easily without shutting off the process flow is recommended. If you use the second method, you must be extremely careful of the following:

- Do not allow the process liquid to come into contact with the standard, which would change its ORP value.
- Use a sufficient quantity of the standard to rinse the tubing to the electrode so no process liquid remains and affects the reading.
- Flush the standard to a chemical sewer or waste dump after the test. Standards and process liquids are frequently different chemical solutions, so the system must be flushed thoroughly using either the process liquid itself or demineralized water.

Figure 34 is a one-line diagram of a typical sampling installation for ORP analysis. The block and bleed valves are installed to keep the process liquid from leaking into the sample system during the calibration process. The rotameter regulates the flow of sample liquid through the sampling system. One of the two drain valves allows the flow to return to the process or to an external chemical drain system. The other drain valve is used for obtaining a quick sample, which is referred to as a *grab sample*. This system also includes an in-line demineralized flush system to purge the lines before and after calibration.

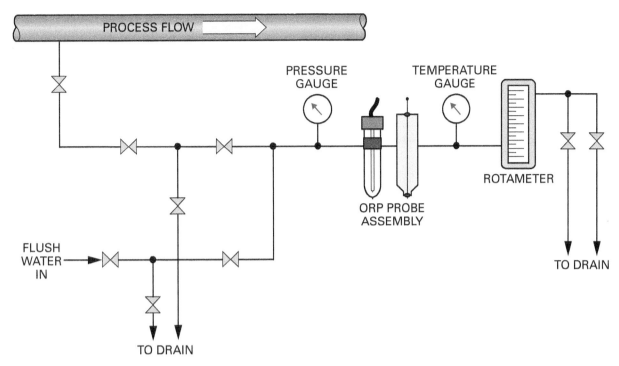

Figure 34 Piping design including an ORP in-line system.

3.4.3 Probe Maintenance

ORP probes usually require little maintenance as long as the process liquid does not contain contaminants that could coat or erode the electrodes or liquid junctions. One important maintenance procedure is replacing the potassium chloride solution in the reference electrode. Although most reference electrodes are refillable, some cannot be refilled and must be replaced when they lose their potassium chloride solution.

You must periodically clean or replace the porous tip of the reference electrode that forms the bridge between the silver/silver chloride electrode and the process liquid. A very small amount of potassium chloride must flow from the electrode into the process liquid to keep the process liquid from contaminating the reference electrode solution.

Additional Resources

American Society for Testing and Materials (ASTM) *Standard D 1125, Standard Test Methods for Electrical Conductivity and Resistivity of Water*. West Conshohocken, PA: ASTM International.

Instrument Engineers Handbook, Volume 1: Process Measurement and Analysis, Béla G. Lipták (editor). Fourth Edition. 2003. Boca Raton, FL: CRC Press.

Instrumentation Reference Book, Walt Boyes (editor). Fourth Edition. 2009. Oxford, UK: Butterworth-Heinemann.

Measurement and Instrumentation Principles, Alan S. Morris. Third Edition. 2001. Oxford, UK: Butterworth-Heinemann.

The Condensed Handbook of Measurement and Control, N.E. Battikha. Third Edition, 2006. Research Triangle Park, NC: The International Society of Automation.

The International Society of Automation, **www.isa.org**

3.0.0 Section Review

1. The temperature at which a liquid will start to form vapors that can ignite is the _____.
   a. fire level
   b. ignition point
   c. flash point
   d. combustion temperature

2. If a liquid has a pH of 4.5, it is _____.
   a. an acid
   b. a base
   c. neutral
   d. a salt

3. Which of the following is used to calibrate a pH electrode/probe?
   a. A calibrator
   b. A buffer solution
   c. A peroxide solution
   d. A hydrocarbon solvent

4. The environmental factor that can influence a conductivity measurement is _____.
   a. pressure
   b. vibration
   c. temperature
   d. humidity

5. If a substance has a shortage of electrons, it can act as a(n) _____.
   a. ORP
   b. buffer
   c. reducing agent
   d. oxidizing agent

Section Four

4.0.0 Gas Analysis

Objective

Identify methods used to analyze air and determine its content of O_2, CO, CO_2, H_2S, and THC.

a. Identify methods used to analyze air and determine its content of oxygen (O_2).
b. Identify methods used to analyze air and determine its content of carbon monoxide (CO).
c. Identify methods used to analyze air and determine its content of carbon dioxide (CO_2).
d. Identify methods used to analyze air and determine its content of hydrogen sulfide (H_2S).
e. Identify methods used to analyze air and determine its content of total hydrocarbons (THC).

Trade Terms

Total hydrocarbon content (THC): A measurement of all hydrocarbon molecules present in a sample.

Many industrial processes use and/or produce gases. Monitoring specific gases is crucial for the sake of the process and also for the safety of plant personnel. Certain gases must not be released into the environment, or may be released only in limited quantities. Gas analyzers determine the presence and level of a specific gas. The information that they provide can then be used to make decisions, or to adjust equipment so levels stay within the desired range.

4.1.0 Measuring Oxygen

The demand for oxygen (O_2) analysis is due, in part, to the essential role that oxygen plays in combustion, oxidation, and other industrial process applications. In industry, the major reason for oxygen analysis is combustion. Different sensor cells are available for measuring the concentration of oxygen, including the following:

- High-temperature electrochemical
- Paramagnetic
- Galvanic

4.1.1 High-Temperature Electrochemical Sensors

High-temperature electrochemical sensors are widely used in combustion control applications to measure exhaust gas oxygen concentrations. One such application is the measurement of the oxygen concentration in the exhaust gases of steel-producing blast furnaces. This type of sensor is also widely used in exhaust stack analyzer applications by manufacturers of chemicals and petrochemicals, ceramics, and glass.

One type of high-temperature electrochemical sensor commonly used in combustion control applications to determine combustion efficiency is the zirconium oxide oxygen sensor. This sensor consists of a cell made of ceramic zirconium oxide combined with an oxide of either yttrium or calcium (*Figure 35*). A coating of porous platinum applied to the inside and outside walls of the cell acts as a pair of electrodes. The cell is then heated to maintain it at a constant temperature.

At high temperatures (typically above 1,200°F/ 650°C), minuscule openings in the cell walls permit the passage of oxygen ions from a sampled gas through the cell. When the oxygen content is equal on both sides of the cell, only a random movement of oxygen ions occurs within the cell. This results in no voltage being generated across the platinum electrodes. However, when there is a difference between the sample and the reference (usually air), oxygen ions will pass through the cell from one electrode to the other.

The rate at which oxygen ions pass through the cell is determined by the temperature of the cell and by the difference in the oxygen content of the sample gas versus the oxygen content of

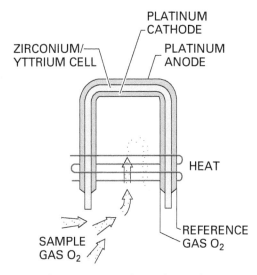

Figure 35 High-temperature electrochemical oxygen detector.

the reference gas. The passage of oxygen ions causes a voltage to be produced across the platinum electrodes. The magnitude of this voltage is a function of the ratio of the sample gas and reference gas oxygen contents. Because the oxygen content of the reference gas is known, the voltage produced by the cell indicates the oxygen content of the sample gas.

The major advantage to using a zirconium oxide oxygen analyzer for high-temperature combustion gas analysis is that the measuring probe can be placed directly into a flue with the probe mounted in virtually any position. One disadvantage is that the zirconium oxide cell has a relatively short life, typically less than 18 months. Frequent thermal cycling can cause stress cracks that can greatly shorten cell life and overall reliability. Another disadvantage is that, as the cell ages, it becomes increasingly more difficult to calibrate.

Zirconium oxide oxygen analyzers are not recommended for use in trace oxygen analysis applications. This is because the presence of gases such as hydrogen, hydrocarbons, or carbon monoxide, will chemically consume oxygen in the sample at the high temperatures necessary for operation and will result in lower-than-actual readings of oxygen content.

4.1.2 Paramagnetic Analyzers

The paramagnetic oxygen analyzer works on the principle that oxygen can be influenced by a magnetic field. Oxygen has an exceptionally high magnetic response compared to other gases. Paramagnetic types of oxygen analyzers are most commonly used in the following applications:

- Analysis of combustion efficiency
- Testing the purity of breathing air and protective atmospheres
- Laboratory instruments
- Medical applications
- Selected industrial process monitoring and control applications

The magneto-dynamic oxygen analyzer described here is the most widely used type of paramagnetic oxygen analyzer. It consists of a small dumbbell-shaped body made of glass and charged with a gas that is not very responsive to a magnetic field (such as nitrogen), a light source, a photocell, a mirror, and a calibrated indicating unit (see *Figure 36*). The dumbbell-shaped body is suspended within the magnetic field of a permanent magnet and is free to rotate in the space between the poles of the magnet. Because of the way that nitrogen responds to the magnetic field, the balled ends naturally deflect away from the point of maximum magnetic field strength.

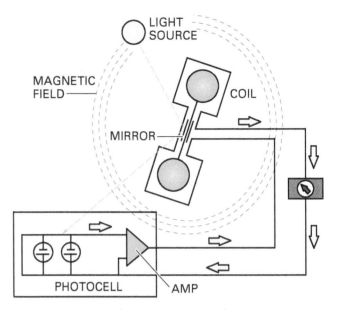

Figure 36 Magneto-dynamic oxygen analyzer.

When a sample containing oxygen is introduced into the test cell, the oxygen in the sample is attracted to the point of maximum field strength, which causes the dumbbell to move. Its displacement is proportional to the amount of oxygen in the sample. The movement of the dumbbell is detected by a light beam from the light source outside of the test cell. The light beam is reflected to an exterior photocell by a mirror attached to the dumbbell body. The amount of light reflected to the photocell depends on the amount of movement. The output of the photocell is then amplified and transmitted to an indicating unit that is calibrated to read out the percentage of oxygen content in the test sample.

An advantage to using a magneto-dynamic oxygen analyzer is that it is not influenced by the thermal properties of the background gas. It also has excellent response speed. Its main disadvantage is that it is a delicate instrument with moving parts. This makes it sensitive to vibration and therefore it requires precise positioning. Also, the magnetic behavior of the sample is influenced by temperature. This characteristic can result in significant errors if the sample temperature deviates from the calibrated temperature. In addition, the accuracy of a dumbbell-type analyzer can be affected by the magnetic behavior of the prevailing background gases.

4.1.3 Galvanic Sensors

The galvanic oxygen sensor, also called a *coulometric oxygen sensor,* is a chemical cell whose oxidation-reduction potential (ORP) is changed by the introduction of the sample gas. The primary element of the galvanic oxygen sensor is an electrochemical cell consisting of two electrodes in contact with a basic electrolyte, usually potassium hydroxide (see *Figure 37*). The cell electrodes are made of dissimilar metals, typically silver and lead.

When a gas sample enters the cell, it diffuses through a membrane usually made out of PTFE (Teflon®). The oxygen in the sample contacts the silver cathode where a reduction reaction occurs, producing hydroxide ions. These ions then flow toward the lead anode, where an oxidation reaction occurs. This reaction results in a flow of electrons that is proportional to the oxygen concentration of the sample. An external metering circuit connected to the cell electrodes measures the electron flow (current) between the electrodes. The system reports either the percentage or the parts per million concentration of oxygen in the sample.

4.2.0 Measuring Carbon Monoxide

The gases used as fuel for furnaces and the various heated vessels in industry, manufacturing, and power-generating facilities are very dangerous. They can cause potential health problems or even death in confined areas where their byproducts can accumulate. One such byproduct of combustible gas is carbon monoxide (CO), which is a colorless, tasteless, and odorless gas that combines with the hemoglobin of human blood and inhibits its oxygen-carrying capacity. It is the same deadly gas that is expelled from the exhaust system in your car and from gas space heaters in your home.

CO sensors are used in industrial applications to warn personnel of dangerous levels of this lethal byproduct of combustion. Handheld carbon monoxide analyzers are available for personnel who must enter confined spaces to work. Examples of these units are shown in *Figure 38*. Analyzers like these can sample CO found in flue gases of residential furnaces, combustion appliances, and commercial and industrial boilers.

CO analyzers are primarily used in the following industrial and manufacturing applications:

- CO safety checks on combustion stack gas and ambient air levels
- Analyzing indoor air quality and safety
- Furnace and boiler servicing or replacement

4.3.0 Measuring Carbon Dioxide

Carbon dioxide (CO_2) is a nonflammable, colorless, odorless gas somewhat heavier than air. Carbon dioxide is one of the gases that make up the earth's atmosphere, along with nitrogen, oxygen, and argon. It is a greenhouse gas, along with methane, water vapor, and many other gases. Carbon dioxide is evenly distributed at a concentration of about 0.033 percent over the earth's surface.

Because of its low concentration in the atmosphere, it is not practical to obtain carbon dioxide gas commercially by extracting it from air. Commercial quantities of carbon dioxide gas are normally produced as byproducts of manufacturing processes, such as the combustion of coal and natural gas, the production of ethanol, and the manufacture of ammonia. Bulk quantities of carbon dioxide are usually stored and shipped as a refrigerated liquid under pressure.

Carbon dioxide is nonreactive (inert) with many materials and is commonly used for applications such as blanketing and purging of

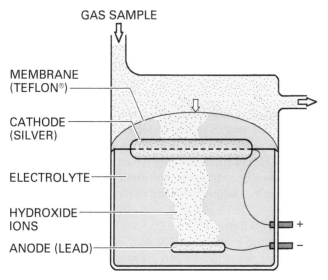

Figure 37 Principles of a galvanic oxygen sensing cell.

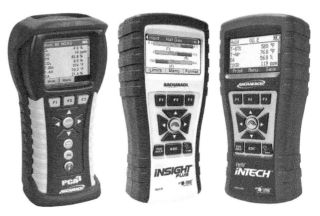

Figure 38 Bacharach handheld CO analyzers.

tanks and reactors, and as a shielding gas in arc welding applications. It is also used in carbonated beverages and to fill certain types of fire extinguishers. At extremely low temperatures (–100°F/–73°C) carbon dioxide solidifies into dry ice, which is often used to freeze and preserve foods.

4.3.1 Monitoring Carbon Dioxide

Combustion processes that use boilers, furnaces, and kilns require proper mixtures of oxygen and fuel in order to burn properly. Because carbon dioxide deprives any environment of oxygen, carbon dioxide levels must be monitored in these types of combustion processes to achieve ideal combustion conditions.

Carbon dioxide analyzers and monitoring are available in many forms, including stack-mounted, portable, and in-situ, depending on the environment and application. Carbon dioxide can be measured by infrared absorption and by electrochemical methods similar to those used for oxygen analysis. The process of electrochemical analysis was discussed earlier, and infrared analysis is described later in this module.

4.4.0 Measuring Hydrogen Sulfide

Hydrogen sulfide (H_2S) is a colorless, very flammable gas. In low concentrations, it smells like rotten eggs. It is heavier than air and very toxic. After just a few minutes of exposure at 100 ppm, people lose their sense of smell and can no longer detect the presence or absence of the gas. When hydrogen sulfide burns, it produces another very toxic gas—sulfur dioxide (SO_2).

Some natural gas contains significant levels of hydrogen sulfide. Because of this, natural gas wells producing this mixture are often called sour wells. Volcanoes also discharge hydrogen sulfide. Anaerobic decay aided by bacteria, such as occurs in sewers, produces hydrogen sulfide. Hydrogen sulfide has a few commercial uses, although it is not broadly used in industry. Most industrial hydrogen sulfide comes from natural gas.

This gas can be detected at levels as low as 2 parts per billion (ppb). To put this number into perspective, 1 ml of the gas distributed evenly in a 100-seat lecture hall is about 20 ppb. The OSHA ceiling for hydrogen sulfide is 20 ppm, and the peak is 50 ppm for 10 minutes. The evacuation point is 65 ppm.

4.4.1 Personnel Protection Indicators

Because hydrogen sulfide is very toxic and can cause you to lose your sense of smell after a brief exposure, it's very important to have an alternative method of detecting the presence of the gas.

One very simple device is a wearable personnel protection indicator (*Figure 39*) that changes color when exposed to toxic gases, including hydrogen sulfide. Badges must be ordered according to the gas being monitored. In the case of hydrogen sulfide, the badge's sensitivity is 10 ppm/10 minutes, and the color strip will change from white to brown. Extra color strips are shipped with the badge and may also be separately ordered for a variety of gases, such as ammonia, carbon monoxide, chlorine, hydrazine, nitrogen dioxide, and ozone. Coincidentally, the ozone strip will also change from white to brown when exposed to ozone, but it has a different sensitivity—0.1 ppm/15 minutes.

4.4.2 Semiconductor Sensors

Semiconductor sensors work well for hydrogen sulfide gas measurements. Although the quality of sensors varies widely from manufacturer to

---

The Greenhouse Effect

There is much scientific discussion and political controversy about global warming and the greenhouse effect. When the sun's energy reaches the earth, some of this energy is reflected back to space, and the rest is absorbed. The absorbed energy warms the earth's surface, which then emits heat energy back toward space as radiation. This outgoing radiation is partially trapped by greenhouse gases, including carbon dioxide, which then radiate the energy in all directions, warming the earth's surface and atmosphere. Scientists are concerned that higher greenhouse gas concentrations will lead to an enhanced greenhouse effect, which may lead to global climate change.

According to the EPA's *Inventory of US Greenhouse Gas Emissions and Sinks: 1990–2013* report, industrial processes account for about 21 percent of the total carbon dioxide emitted in the United States. This amount is significant, although it is less so than transportation (27 percent) and electrical generation (31 percent) outputs.

Figure 39 Personnel protection indicator.

4.5.0 Measuring Total Hydrocarbons

Hydrocarbons—compounds composed of hydrogen and carbon—are present in many forms in atmospheric air. Among these substances, methane (CH_4) has a concentration of approximately 1.7 ppm, which accounts for the majority of the concentration of hydrocarbons. Methane is significant as one of the greenhouse gases. Many other types of organic gases and vapors can also be present in the air at any one time. Volatile organic contaminants can be manmade—such as gasoline vapor, exhaust fumes, cleaning solvents, and lube oil vapor—or can be of biological origin, such as marsh gas, mold, and mildew.

Some of the hydrocarbons are hazardous and may have irritating odors. Because it is impractical to measure each type of organic contaminant present, they are measured as a group and described as total hydrocarbon content (THC). Any instrument used to measure THC responds to the total quantity of hydrogen/carbon molecules present.

4.5.1 Flame Ionization Detector

The instrument most often used to measure vapors for total hydrocarbon content uses the principle of flame ionization detection (FID). For this type of detector, the air sample and a fuel (hydrogen) are sent to the burner nozzle at a controlled flow rate. The concentration of hydrocarbons is determined by ionizing the hydrocarbons in the flame, and then measuring the generated ion current, as shown in *Figure 40*. In principle, the FID method shows responses in proportion to the

manufacturer, semiconductor sensors are among the best for hydrogen sulfide monitoring when sensitivity to low concentrations is required.

For this method of gas monitoring, a semiconductor material is applied to a nonconducting substrate between two electrodes. The substrate is heated to a temperature such that the gas being monitored can cause a reversible change in the conductivity of the semiconductor material. Under zero-gas conditions, it is believed that oxygen molecules tie up free electrons in the semiconductor material, inhibiting electrical flow. As hydrogen sulfide gas is introduced, its molecules replace the oxygen, releasing the free electrons and decreasing the resistance between the electrodes. This change in resistance is measured electrically and is proportional to the concentration of the gas being measured.

4.4.3 Electrochemical Sensors

As described earlier, the electrochemical sensor is widely used as a gas detector. It works well for hydrogen sulfide and is simple, reliable, and inexpensive.

In this application, the sensor is a self-powered fuel cell. The cell consists of a casing containing a gel or electrolyte and two active electrodes. The top of the casing has a membrane that can be permeated by the gas sample. Oxidization takes place at one electrode and reduction at the other. A current is created as the positive ions flow to one electrode and negative ions to the other. This current indicates the presence of hydrogen sulfide.

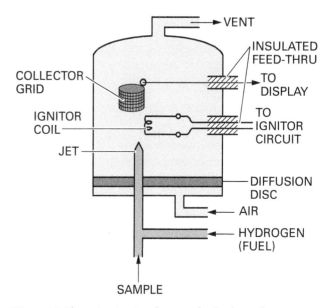

Figure 40 Flame ionization detector for hydrocarbon measurement.

number of carbon atoms in the hydrocarbons. Therefore, concentrations are shown as equivalent to those of methane in ppm by using units called ppmC.

Because the sensitivity of an FID is dependent on the flow rates of the sample gas and the hydrogen fuel, care must be taken to control these rates. The readings and relative sensitivity of the FID method are also adversely affected by oxygen interference. If oxygen is introduced, an error will occur if there is a difference between the oxygen concentrations of the calibration standard gas and those of the sample air. The degree to which these problems affect the results varies depending on the type of organic material and the combustion conditions. This means that the THC value varies depending on the composition of the hydrocarbons and the monitoring system. The FID method should be used bearing these things in mind.

Additional Resources

Instrument Engineers Handbook, Volume 1: Process Measurement and Analysis, Béla G. Lipták (editor). Fourth Edition. 2003. Boca Raton, FL: CRC Press.

Instrumentation Reference Book, Walt Boyes (editor). Fourth Edition. 2009. Oxford, UK: Butterworth-Heinemann.

Measurement and Instrumentation Principles, Alan S. Morris. Third Edition. 2001. Oxford, UK: Butterworth-Heinemann.

The Condensed Handbook of Measurement and Control, N.E. Battikha. Third Edition, 2006. Research Triangle Park, NC: The International Society of Automation.

The International Society of Automation, **www.isa.org**

4.0.0 Section Review

1. The gas that is essential to combustion is _____.
   a. nitrogen
   b. oxygen
   c. carbon dioxide
   d. carbon monoxide

2. The best analyzer for assessing oxygen in medical applications is a _____.
   a. high-temperature electrochemical cell
   b. personnel protection indicator
   c. paramagnetic oxygen analyzer
   d. gas chromatograph

3. Carbon monoxide is a byproduct of _____.
   a. sulfuric acid production
   b. acid/base neutralizations
   c. anaerobic decomposition
   d. combustible gas

4. Carbon dioxide is sometimes used in welding because it is _____.
   a. nonreactive
   b. combustible
   c. easily seen by color
   d. easily detected by odor

5. Which gas ceases to be noticed by people in just a few minutes?
   a. Carbon monoxide
   b. Hydrogen sulfide
   c. Carbon dioxide
   d. Hydrocarbons

6. The most common hydrocarbon is _____.
   a. methane
   b. bromine
   c. hydrogen sulfide
   d. hydrazine

Section Five

5.0.0 Other Analysis Techniques

Objective

Define the properties of particulate count, chemical composition, infrared radiation, and UV light absorption, and identify methods used to analyze them.
  a. Define the property of particulate count and identify methods used to analyze it.
  b. Define the property of chemical composition and identify methods used to analyze it.
  c. Define the property of infrared radiation and identify methods used to analyze it.
  d. Define the property of UV light absorption and identify methods used to analyze it.

Trade Terms

Emissivity: The ratio of the quantity of electromagnetic radiation that an object emits to the amount that a theoretical perfect emitter would emit under the same conditions.

Due to the complexity of industry, there are several miscellaneous analytical techniques used to solve specific problems. Some of these techniques measure physical properties; others involve chemical properties. In some situations, it's useful to measure emitted and absorbed electromagnetic radiation, such as infrared (IR) or ultraviolet (UV) light. This section provides an overview of some miscellaneous analysis techniques used by instrumentation technicians.

5.1.0 Measuring Particulates

Clean rooms are specially built enclosed areas that are specifically designed to control airborne particulates, temperature, humidity, airflow patterns, air motion, and lighting. These sealed facilities use specialized air handling and filtration systems designed to minimize static electricity as well as concentrations of particles and other contaminants that may interfere with manufacturing. Typically, clean rooms produce a smooth vertical flow of air throughout a large area of the space. The air is filtered and contaminants are purged through the large airflow. Because air velocity and environmental factors must be controlled within tightly prescribed limits, proper instrumentation is a must.

Federal Standard 209E establishes standards of cleanliness for airborne particulate levels in clean rooms. This standard also describes methods for monitoring the air and procedures for verifying the classification level of clean rooms. These classifications are established by the number of particulates that are one micron or larger in a cubic foot of space per minute. One micron (1 µm) is equal to one millionth of a meter, which is also $\frac{1}{1,000}$ millimeter (0.001 mm). *Table 3* shows some examples of common object sizes measured in microns.

5.1.1 Optical Microscopy

The method used to monitor and count particulates in a clean room depends on the standards and policies of the facility. If the clean room is involved in any government work, chances are that many of the standards will have originated within a governmental agency.

A common (yet tedious) method often used to determine particulate content is to physically count the particulates in a specified sample. Usually, the method for counting is determined by the size or size range of the particulates of interest. For example, one method specified in *Federal Standard 209E* for counting and sizing the concentration of particles 5 µm and larger in clean rooms is to collect the particles on a membrane filter and then to look through an optical microscope to count them.

For this method, a vacuum draws a sample of air through a membrane filter. The flow rate of the sample is controlled by a limiting orifice or by a flowmeter. The total volume of air sampled is therefore determined by the sampling time.

Table 3 Common Particulates and Their Sizes

Particulate	Size (in microns)
Cigarette smoke particles	0.01 – 1.0
Bacteria	0.3 – 40
Household dust	0.5 – 20

The membrane filter is examined microscopically to determine the number of particles 5 μm and larger collected from the sample of air. Automatic image analysis may replace direct optical microscopy for the sizing and counting of particles.

5.1.2 Discrete Particle Counters (DPCs)

Discrete particle counters (DPCs) provide data on the concentration and size distribution of airborne particles within an approximate range of 0.01 to 20 μm on a near-real-time basis. A DPC will correctly size only those particles within the limits of its range.

Air in the clean room or clean zone to be monitored is sampled at a known flow rate from the sample points of concern. Particles pass through the sensing zone of the DPC, and each particle produces a signal that is related to its size. An electronic system sorts and counts the pulses, registering the number of particles of various sizes that have been recorded within the known volume of air sampled. The concentration and particle size data can be displayed, printed, or processed.

DPC design variations can result in count differences. Potential causes of count differences should be recognized and minimized by using a standard calibration method and by minimizing the variability of sample acquisition procedures for instruments of the same type. Two types of DPC instruments are the optical particle counter (OPC) and the condensation particle counter (CPC).

In the OPC (*Figure 41*), the output from a laser light source is collimated in a beam used to illuminate a sample volume stream flowing through the OPC. The sample enters the OPC through an input nozzle, and it is drawn out by a vacuum source. Light scattered by the various particles contained within the sample is detected by a photodetector. Both the size and the number of particles are measured simultaneously, with the size of each particle being determined by the intensity of the scattered light.

The condensation particle counter (CPC) operates by saturating the heated sample particles with butanol, which is evaporated into the air stream, and then cooling the sample in a condenser. The vapor condenses on the particles, causing them to grow to sizes that are easily detected. The resulting droplets are passed through an optical detector immediately after leaving the condenser.

5.2.0 Determining Chemical Composition

Chemical composition analysis is essential in nearly all of the continuous-process industries. It is particularly important in the petrochemical industry. Incoming raw materials may be sampled to make sure that they meet required chemical specifications or to optimize the yield and minimize production costs. Process analysis is necessary at various stages to ensure that there will not be excessive waste at the end of the production process because of a failure to meet specifications. A process control system can respond quickly to the continuous on-line measurements from analyzing instruments. The control mechanism adjusts the appropriate inputs or production stages to bring the process back into control. This type of continuous analysis permits much greater product throughput with less waste than is possible with a batch-analysis technique.

Chemical composition analysis is also used to ensure public health and safety, as in the case of municipal water treatment, sewage disposal plants, and industrial plants that emit toxic liquids or gases. The objective in all of these applications is to determine which chemical elements are present (qualitative analysis) or to measure how much of one or more of several known elements is present (quantitative analysis). Knowing which chemicals are present in a sample can be crucial in many situations. Some processes require a chemical breakdown for quality-control purposes or to meet EPA requirements.

5.2.1 Gas Chromatography

Gas chromatography is one popular method of chemical analysis. A typical application of gas chromatography is to separate and analyze atmospheric mixtures containing volatile organic com-

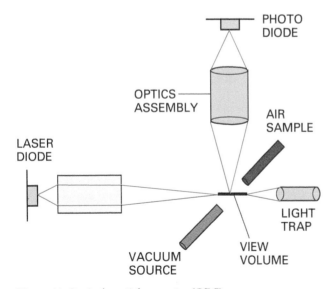

Figure 41 Optical particle counter (OPC).

pounds. Chromatography works on the principle that when a test mixture is processed through the analyzer, the different components that form the sampled mixture separate and exit the chromatography process at different rates. This separation allows the individual components in the sampled mixture, as well as their concentrations, to be identified.

Figure 42 shows a simplified diagram of a gas chromatography analyzer. It consists of a sample mixture injection point, gas cylinder, a chromatography column and oven, and a detector. The oven can produce a fixed temperature, or it can have temperature controls that allow the programming of a specific temperature into the instrument.

The chromatography process involves two phases—mobile and stationary. In the mobile phase, the sample mixture to be analyzed is first vaporized and combined with a moving stream of carrier gas, such as helium, for subsequent passage through the chromatography column. The stationary phase relates to the material used in the column. It is typically an adsorbent—a solid substance that traps other substances on its surface—or a liquid distributed over the surface of the support column. This material separates the components of the sample based on the varying attraction that each material has for the column material. As the sample mixture flows through the column, the different components are adsorbed by the column in varying degrees.

The different components of the sampled mixture adsorbed by the column exit the column at different rates. Those with a weak attraction exit first and are followed by those with a stronger attraction. This action separates the mixture's components. Following this, the detector gathers and analyzes the different components as they exit from the column and produces an electrical output signal.

In some gas chromatographs, packed columns have been replaced by capillary columns. A capillary column is basically a very long tube, which generally has a very small inside diameter. The tube is coiled into a compact helix (see *Figure 43*). The capillary tube has a thin liquid film bonded to the inside surface that performs the adsorption task. Capillary column chromatographs respond quickly—as fast as a few seconds—whereas response times for packed column chromatographs often take many minutes.

The numerous types of detectors used in gas chromatograph analyzers are specifically designed for particular types of applications. Two widely used types of detectors are the thermal conductivity and flame ionization detectors. The output from a gas chromatograph detector can be

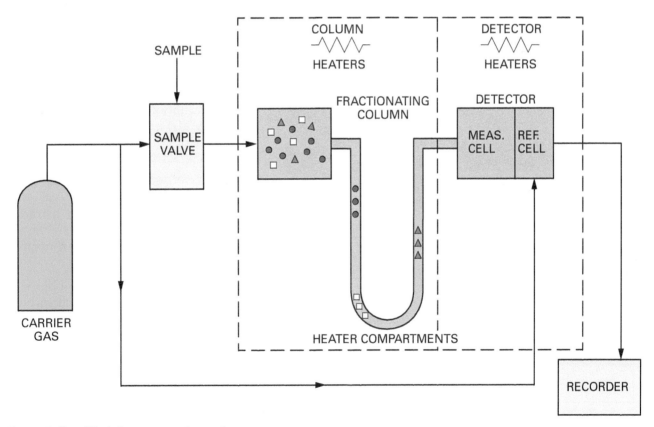

Figure 42 Simplified chromatography analyzer.

Figure 43 Chromatograph capillary tube.

in the form of computer-readable data or an analog voltage, current, or frequency. Often the detector output is used to drive a printer to produce a chromatogram (*Figure 44*). Chromatograms are graphical representations that show a series of peaked waveform responses, each of which corresponds to a different component in the sampled mixture. Another option is to have the output trip an alarm or cause a change in state of a switching device. Some models of gas chromatographs are programmable and have data storage options.

5.3.0 Measuring Infrared Radiation

Infrared sensing of thermal radiation was initially developed by the military to allow soldiers to see in the dark. Later, the technology moved to the public sector. Since that time, infrared inspection has developed into a state-of-the-art technique used to monitor equipment performance. It is a nondestructive, noncontact testing method used to monitor energy losses and identify potential failures. Data collected over a period of time can indicate impending failure of a piece of electrical equipment. The equipment can then be repaired before more costly corrective maintenance is required.

5.3.1 Basic Theory

Thermography is the technique of using an optical and electronic system to convert invisible infrared (IR) radiation into visible light. Thermography is based on the principle that any object having a temperature above absolute zero will emit some infrared radiation. The magnitude of infrared radiation emitted by a source is proportional to its temperature. Infrared radiation is part of the electromagnetic spectrum just below visible red light. Its wavelengths range from 1 mm to 700 nm (*Figure 45*).

Because infrared radiation is invisible to the human eye, infrared sensing equipment is required to perceive it. This equipment collects and focuses infrared energy on sensitive detectors that convert the infrared energy into electrical impulses. The electrical impulses are then amplified, scanned, and used to drive a color display in which different colors represent the relative degrees of thermal energy. The image is called a *thermogram*. Infrared imaging can also be used to measure temperatures at any point on an object.

5.3.2 Factors Affecting Infrared Imaging

Infrared sensing works by measuring the amount of infrared radiation emitted from an object. The emitted infrared radiation must pass through a medium (generally air) and be absorbed by the infrared detectors. Objects can appear to be cooler or warmer than they actually are because of the following factors:

- Emissivity of the object
- Direct sunlight or other background infrared sources
- Atmospheric gases in higher-than-normal concentrations

When infrared radiation strikes an object, it can pass through, bounce back, or be absorbed. Absorbed energy heats the object. Objects, in turn, radiate energy (heat) back into their environment. Not all objects absorb, reflect, transmit, or emit infrared energy equally. Highly polished materials such as metals tend to reflect rather than absorb. On the other hand, a painted surface both absorbs and reflects. An object's tendency to emit IR is called its *emissivity*. Emissivity is

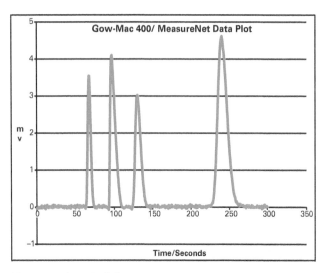

Figure 44 A typical chromatogram.

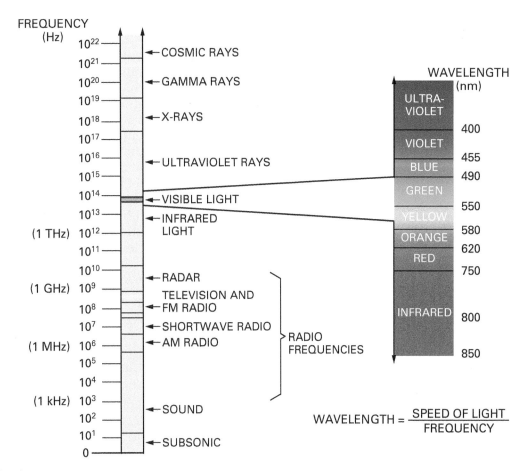

Figure 45 The electromagnetic spectrum.

a ratio between the quantity of IR energy that the object emits and what a theoretical perfect emitter would emit under the same conditions. Small emissivity values indicate objects that emit poorly. *Table 4* shows some typical values.

5.3.3 Sensing Equipment

Infrared thermal inspection can be used to detect potential failures in electrical equipment by sensing differences in infrared emission. A typical handheld infrared imager is shown in *Figure 46*. While limited use of infrared detection for temperature monitoring is not new, using IR thermography as a diagnostic predictive maintenance technique has gained universal acceptance more recently. Infrared thermal inspection is used in industrial and commercial facilities to locate sources of energy loss and potential failure. The following are examples of infrared thermal inspection applications:

- *Electrical equipment* – Poor connections, corrosion of connections, faulty materials and equipment

Table 4 Typical Emissivity Values

Material	Emissivity
Silver (polished)	0.01
Brass (polished)	0.03
Molten steel	0.28
Asbestos cloth	0.90
White paint (ThO^2)	0.90
Red brick	0.93
Carbon	0.95
Black paint (CuO)	0.96

Figure 46 Handheld IR imager.

- *Electronic circuits* – Faulty components and connections
- *Motors* – Hot spots and worn bearings
- *Power transmission equipment* – Bad connections
- *Power factor capacitors* – Overheating
- *Switches and breakers* – Overheating
- *Three-phase circuits* – Unbalanced loads
- *Transformers* – Hot spots
- *Bearings* – Hot spots and wear

Infrared inspection can be used to determine if a bearing is not receiving adequate lubrication and is overheating. Bearings in inaccessible areas are sometimes overlooked. If the bearing can be seen, infrared inspection can be used to determine its temperature and locate possible hot spots.

Infrared imaging is particularly useful for electrical inspections. Overheating in electrical equipment is a major cause of failures. Infrared inspection of electrical equipment can locate problems that may go undetected using conventional methods. In a plant electrical distribution system, electrical connections that are loose, corroded, or deteriorated overheat. Typical examples are substations, bus bars, motor controllers, motor switching gear, power transformers, power factor capacitors, and lighting circuitry. It is critical to minimize failures in these areas to ensure smooth plant operation and safety.

Infrared inspection is a fast, reliable method to spot these problem areas. Infrared inspection can view through cabinets and panels, minimizing the inspection time. In addition to locating overheated connections, infrared inspection can visualize overheated conduit, clogged transformer cooling tubes, and defective electrical components.

5.3.4 IR Spectrometry

Another use of infrared detectors is in the science of spectrometry. Spectrometry is used to determine the chemical composition of an object. If an unknown sample is analyzed using a spectroscope, the chemical makeup of the sample can be determined.

In absorption spectrometry, radiations of a particular wavelength in the appropriate region of the electromagnetic spectrum are employed. An infrared analyzer responds to the absorption of infrared radiation. When light is passed through a light-transmitting solid or fluid, absorption takes place, depending on the chemical identity of the absorbing medium. The number of molecules per unit volume (concentration) and the length of the light path within the medium influence the amount of absorption in terms of the ratio between incoming and outgoing light beams. Thus, each substance has a characteristic absorption spectrum so that a graph of wavelength versus percentage absorption/transmission enables the material to be identified.

In operation, the proper wavelength must first be selected for each component. Then, by setting the instrument successively at each of these wavelengths and comparing the absorption of the unknown with that of a set of standards of known concentrations, the amounts of each component can be determined.

The basic components of an infrared absorption spectroscope are as follows:

- A source of infrared radiation
- A comparison cell having infrared-transparent windows and located between the source and the detector through which the sample passes
- A detector sensitive to infrared radiation
- A signal amplifier and recorder

A simple single-beam infrared spectrometer is shown in *Figure 47*.

5.4.0 Measuring Ultraviolet Radiation

Ultraviolet (UV) radiation is also commonly called *ultraviolet* or *black light*. It has much higher frequencies and much shorter wavelengths than infrared radiation. In fact, it lies just above the violet part of the visible light spectrum (*Figure 45*). Because of this higher frequency and the resulting shorter wavelengths, ultraviolet radiation interacts with materials differently than infrared light waves.

5.4.1 UV Analysis

Process stream analyzers based on the measurement of ultraviolet radiation absorption are used throughout process industries to monitor and control the concentrations of components in both gas and liquid streams. Like the infrared detectors, the UV spectrophotometer compares absorption to wavelength as a means to identify the components in the mixture as well as the concentration.

The UV absorption pattern of a compound is not so distinctive a fingerprint as its infrared counterpart, and fewer compounds absorb in the ultraviolet region than in the infrared. However, several important classes of compounds absorb

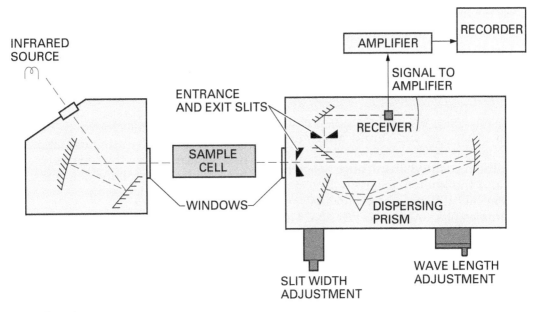

Figure 47 Simple infrared spectrometer.

strongly in the ultraviolet region, whereas water and the usual components of air do not absorb in this region. As a result, ultraviolet absorption analyzers may be more selective and sensitive than infrared and other types of analyzers.

The basic UV absorption analyzer consists of a UV source, optical filters, a sample cell, a detector, and an output circuit. A simple block diagram of a typical UV detector is shown in *Figure 48*. A transmittance measurement is made by calculating the ratio of the output reading with a sample in the cell to the reading obtained with the cell empty. The concentration is then calculated using the Beer-Lambert law, which describes how different materials absorb specific wavelengths.

The source must provide the ultraviolet wavelengths required. Optical filters or spectral dispersing systems are used to screen out radiations of unwanted wavelengths emitted by the source. The sample cell has windows that are transparent at the chosen wavelengths. The path length between windows must be fixed.

5.4.2 Applications

Because of their high sensitivity, accuracy, reliability, and precision, UV absorption analyzers are suited to thousands of process stream analysis applications. UV absorption analyzers are particularly suited for the detection of halogens such as chlorine, benzene compounds, sulfur compounds, oxidizing agents, and many pollutants.

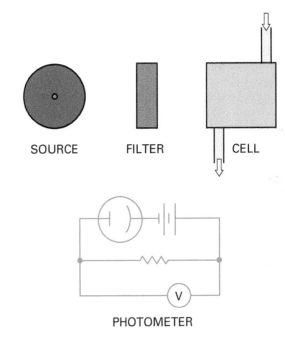

Figure 48 Simple ultraviolet analyzer.

5.4.3 Advantages

Compared to infrared analyzers, UV analyzers are generally more sensitive, yet they provide a comparable level of selectivity. UV analyzers are less sensitive to temperature variations, are more applicable to analyzing liquid streams, and can monitor several important chemicals to which infrared analyzers are completely insensitive. For applications in which both infrared and ultraviolet absorption analyzers are of equal sensitivity, such as for sulfur dioxide monitoring, the insensitivity of UV analyzers to commonly expected components in stack gas—such as moisture and carbon dioxide—is an important advantage.

5.4.4 Calibration

UV absorption analyzers are calibrated in a manner similar to other gas and liquid analyzers. Two methods can be used—absorbance calibration based on the measured absorbance range of the analyzer and the known absorptivity of the sample, or chemical calibration based on analyzer readings of samples of known concentration. The absorbance method is easier than the chemical method for UV absorption analyzers; however, this method usually acts as a secondary standard with the chemical method as the primary standard.

5.4.5 Flame Detectors

In addition to chemical analyzers, UV detectors are also used in furnaces as flame detectors. UV flame detectors use a different type of cell than the device previously mentioned. The cell is a sealed tube containing an ultraviolet-sensitive gas and two electrodes that are connected in an AC circuit. When the gas is exposed to ultraviolet radiation from the flame, it becomes an electrical conductor and can carry electrons from one electrode to the other. When the gas conducts the electrons, the AC circuit is completed and electricity flows through the circuit.

This process occurs with abrupt starting and stopping of the current, which is known as the *avalanche effect*. Each avalanche, or period of electricity flow, is counted to provide a flame signal. When the number of counts exceeds a preset minimum, the presence of a flame is confirmed. Ultraviolet detectors are considered to be superior to infrared detectors in this area because they are not affected by the radiation emitted from hot surfaces inside the furnace. *Figure 49* shows a typical ultraviolet flame detector.

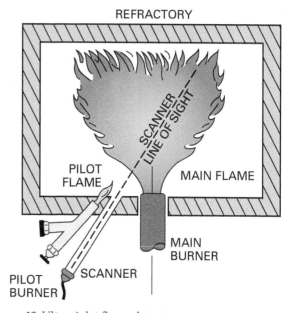

Figure 49 Ultraviolet flame detector.

Additional Resources

Instrument Engineers Handbook, Volume 1: Process Measurement and Analysis, Béla G. Lipták (editor). Fourth Edition. 2003. Boca Raton, FL: CRC Press.

Instrumentation Reference Book, Walt Boyes (editor). Fourth Edition. 2009. Oxford, UK: Butterworth-Heinemann.

Measurement and Instrumentation Principles, Alan S. Morris. Third Edition. 2001. Oxford, UK: Butterworth-Heinemann.

The Condensed Handbook of Measurement and Control, N.E. Battikha. Third Edition, 2006. Research Triangle Park, NC: The International Society of Automation.

The International Society of Automation, **www.isa.org**

5.0.0 Section Review

1. Assessing particulates is often a manual operation when using a(n) _____.
   a. DPC device
   b. OPC device
   c. optical microscope
   d. condensation particle counter

2. A gas chromatograph begins its analysis by _____.
   a. adsorbing different substances
   b. vaporizing the sample
   c. testing the sample with a detector
   d. using a carrier gas

3. Visualizing an object by its infrared emission is called _____.
   a. emissivity
   b. thermalizing
   c. IRDA
   d. thermography

4. In order to identify substances, a spectrograph measures wavelength _____.
   a. absorption
   b. adsorption
   c. scattering
   d. emissivity

5. How does UV absorption analysis compare to IR absorption analysis?
   a. It works for more compounds.
   b. It works for fewer compounds.
   c. It has difficulty if water is involved.
   d. It can't detect chlorine.

Summary

Analytical instrumentation is sometimes considered a specialty trade within the instrumentation industry because of its complexity. Vast amounts of knowledge and equipment are required in order to monitor, detect, and analyze the thousands of variables present in various industries.

Chemistry is the study of the composition and properties of matter and the changes that matter undergoes. It underlies the performance of process systems and other industrial operations, such as power generation and water treatment. Because chemistry plays an essential role in modern industry, this module began with a summary of basic chemistry.

The physical and chemical property measurements performed by various types of analyzers and monitors that are widely used in industrial processes was described, and the basic features and principles of operation of each of these instruments was explained.

Many instrumentation technicians do not have a specialized crew to install, calibrate, and maintain the analyzer instruments for which they are responsible. The information in this module has been presented with these technicians in mind.

Review Questions

1. An element's atomic mass is an average based on the masses of its naturally occurring _____.
   a. neutrons
   b. isotopes
   c. ions
   d. valence shells

2. Noble gases _____.
   a. have incomplete valence shells
   b. are in Group 7 of the periodic table
   c. do not react chemically with other elements
   d. are also known as transition elements

3. When elements interact chemically they tend to gain, lose, or share _____.
   a. atoms
   b. protons
   c. neutrons
   d. electrons

4. What kind of bonds do metals form with nonmetals?
   a. Covalent bonds
   b. Metallic bonds
   c. Ionic bonds
   d. Trivalent bonds

5. Molarity refers to the _____.
   a. viscosity of an oil
   b. concentration of a solution
   c. atomic mass of a gas
   d. gram atomic weight of a liquid

6. Ions are atoms or groups of atoms that _____.
   a. have an electric charge
   b. have more protons than neutrons
   c. are generated by electromagnetic interference
   d. have more neutrons than protons

7. The measure of the acidity or basicity of a solution is _____.
   a. opacity
   b. conductivity
   c. concentration
   d. pH

8. The numbering system for the pH scale typically ranges between _____.
   a. 0 and 7
   b. 0 and 10
   c. 0 and 14
   d. 0 and 100

9. A steel casting has a volume of 500 cm^3 and a mass of 3,875 g. What is its density?
   a. 0.129 g/cm^3
   b. 1.13 g/cm^3
   c. 7.75 g/cm^3
   d. 12.3 g/cm^3

10. A gallon of glycerol has a mass of 10.5 lbs. A gallon of water at the same temperature has a mass of 8.30 lbs. What is glycerol's specific gravity?
    a. 0.795
    b. 1.27
    c. 1.54
    d. 2.21

11. The specific gravity of a liquid can be measured using a(n) _____.
    a. viscometer
    b. pH meter
    c. hydrometer
    d. opacity meter

12. One method for determining viscosity is by measuring the _____.
    a. ratio of solution to solute in a representative sample
    b. temperature difference between two hot wires suspended in the sample
    c. time required to fill a graduated cylinder to a specific height
    d. time required for a ball to fall a specific distance through the sample

13. What do most turbidity analyzers measure to assess the suspended particles in a solution?
    a. Thermal motion
    b. Light
    c. Conductivity
    d. Viscosity

14. What is the relationship between the flash point temperature and the fire point temperature?
    a. They are identical.
    b. The flash point temperature is lower than the fire point temperature.
    c. The flash point temperature is higher than the fire point temperature.
    d. There is no relationship—flash point and fire point are unrelated concepts.

15. Which flash point determination method does *not* find the specific flash point temperature?
    a. Abel
    b. TAG
    c. Rapid
    d. Pensky-Martens

16. A pH meter gives a reading of 13.5 for a process liquid. The liquid is _____.
    a. neutral
    b. basic
    c. acidic
    d. indeterminate

17. In a two-electrode pH measuring system, the pH electrode is paired with a(n) _____.
    a. probe electrode
    b. conductivity electrode
    c. ion-exchange electrode
    d. reference electrode

18. Conductivity is inversely related to _____.
    a. resistance
    b. voltage
    c. current
    d. siemens

19. ORP probes indicate oxidation or reduction potential by _____.
    a. current flow magnitude
    b. resistance values
    c. conductance values
    d. voltage polarity

20. The primary element of a high-temperature electrochemical oxygen sensor is a _____.
    a. mercury oxide cell
    b. uranium lattice chamber
    c. magnetic field
    d. ceramic zirconium oxide cell

21. Which kind of oxygen detector measures a change in ORP caused by the amount of oxygen in the sample?
    a. A galvanic sensor
    b. A high-temperature electrochemical sensor
    c. A paramagnetic analyzer
    d. A semiconductor sensor

22. Which gas is a toxic byproduct of combustion?
    a. Nitrogen sulfide
    b. Carbon monoxide
    c. Hydrogen cyanide
    d. Total hydrocarbon

23. Dry ice is frozen _____.
    a. nitrogen
    b. carbon monoxide
    c. carbon dioxide
    d. hydrogen sulfide

24. Decomposing sewage produces _____.
    a. hydrogen sulfide
    b. carbon monoxide
    c. carbon dioxide
    d. oxygen

25. Total hydrocarbon content (THC) is often measured with a(n) _____.
    a. paramagnetic analyzer
    b. semiconductor sensor
    c. FID
    d. ORP probe

26. What kind of particulate measuring device uses butanol to enhance the particles for counting?
    a. An optical microscope
    b. A membrane filter
    c. A condensation particle counter
    d. An optical particle counter

27. Gas chromatography relies on the principle of _____.
    a. separation
    b. ORP
    c. magnetic response
    d. absorption

28. A gas chromatograph passes vaporized materials through a column or _____.
   a. condenser
   b. magnetic field
   c. capillary tube
   d. double helix

29. An infrared thermometer may indicate an incorrect temperature due to _____.
   a. spectrum adsorption
   b. darkness
   c. improper laser collimation
   d. material emissivity

30. A UV sensor is a better flame detector than an IR sensor because it isn't fooled by _____.
   a. flash point
   b. hot components
   c. emissivity
   d. water vapor

Trade Terms Introduced in This Module

Acid: A substance that produces hydrogen ions (H^+) in solution.

Atom: The smallest particle with a unique chemical identity.

Atomic mass: The average mass of the naturally occurring isotopes of a given element. Expressed in unified atomic mass units.

Atomic number: The number of protons in a particular element. This value is unique to each element.

Base: A substance that produces hydroxide ions (OH^-) in solution.

Bond: An attraction between atoms that enables elements to join together, forming compounds.

Buffer solution: A solution with a specific pH value that strongly resists changes in pH and so maintains its value reliably.

Chemical reaction: A process that occurs when substances combine or break apart at the chemical level.

Chromatography: A family of analytical laboratory techniques in which substances are separated for individual identification.

Colorimeter: A device that measures the quantity of specific suspended particles in a solution by measuring how particular wavelengths of light are absorbed when they pass through the solution.

Compound: A chemical combination of two or more elements.

Conductivity: The ability of a solution to carry an electric current due to the presence of ions. Measured in siemens per centimeter (S/cm).

Covalent bond: A chemical bond in which atoms share some of their valence shell electrons.

Density: A physical property of matter defined as mass per unit volume.

Dissociate: Occurs when ionic compounds dissolve and separate into ions.

Electrolytes: Substances that dissociate/ionize into ions when dissolved in water. Electrolytes make solutions electrically conductive.

Electrons: The negatively charged parts of an atom that orbit around the nucleus.

Electron shells: The paths in which electrons orbit around the atom's nucleus.

Element: A substance composed of just one type of atom.

Emissivity: The ratio of the quantity of electromagnetic radiation that an object emits to the amount that a theoretical perfect emitter would emit under the same conditions.

Flash point: The lowest temperature at which a liquid gives off vapors that will mix with air and be able to ignite.

Formula unit: The most fundamental unit of an ionic compound.

Hydrocarbon: A chemical compound composed of hydrogen and carbon atoms.

Hydrometer: A device used to measure the specific gravity of a solution.

Ion: An atom, or group of atoms, that has a positive or negative electrical charge.

Ionic bond: A chemical bond formed by the transfer of electrons from one atom to another.

Ionize: To convert into an ion or ions; ionization occurs when covalent compounds break down into ions in solution.

Isotopes: Forms of the same element but with different numbers of neutrons in the nucleus.

Metallic bond: A chemical bond in which the valence electrons of many atoms move freely between their atoms, forming a kind of electron "sea".

Molarity (M): The concentration of a solution equal to the number of moles of solute divided by the number of liters of solution.

Mole (mol): A quantity equal to 6.02×10^{23} particles (atoms, formula units, molecules, or ions).

Molecules: The most fundamental units of a covalent compound.

Neutrons: Uncharged atomic particles that are similar in size to protons and are present in the nucleus of an atom.

Nucleus: The central part of an atom. The nucleus contains the protons and neutrons.

Oxidation-reduction potential (ORP): The ability of a solution to act as an oxidizer or reducing agent. Determined by measuring voltage polarity across a pair of electrodes.

Parts per million (ppm): The concentration of solute in a solution, expressed as the parts of solute per million parts of solution.

Periodic table: A grid arrangement in which the elements are ordered in a way that shows their similarities and repeating properties.

pH: A measurement scale that expresses how acidic or basic a solution is based on its concentration of hydrogen (H^+) ions; values typically range from 0 to 14.

Poise (P): A unit of dynamic viscosity. It is often used with the prefix *centi-*; one centipoise is one one-hundredth of a poise. It is derived from French and pronounced *pwaz*.

Protons: Positively charged atomic particles that are similar in size to neutrons and are present in the nucleus of an atom.

Salt: An ionic compound composed of at least one metallic and one nonmetallic element. Salts are formed from a neutralization reaction between an acid and a base.

Solute: A substance that is dissolved in another substance, forming a solution.

Solvent: The substance in which another substance is dissolved to form a solution.

Specific gravity: A physical property of matter defined as the ratio of a substance's mass to the mass of an equal volume of a reference substance, usually water or air.

Total hydrocarbon content (THC): A measurement of all hydrocarbon molecules present in a sample.

Turbidity: The cloudiness of a solution containing suspended particles.

Unified atomic mass unit (u): A measurement unit of atomic mass based on $\frac{1}{12}$ the mass of a carbon-12 atom. One u equals about 1.660×10^{-27} kg.

Valence shell: The outermost electron shell of an atom. Chemical reactions involve the valence shell.

Viscosity: The internal friction of a fluid that opposes flow.

Additional Resources

This module presents thorough resources for task training. The following reference material is recommended for further study.

Basic Chemistry, Karen C. Timberlake and William Timberlake. Fourth Edition. 2013. Upper Saddle River, NJ: Prentice-Hall.

Chemistry: Concepts and Problems: A Self-Teaching Guide, Clifford C. Houk and Richard Post. Second Edition. 1996. New York, NY: John Wiley & Sons, Inc.

Industrial Pressure, Level, and Density Measurement, Donald R. Gillum. Second Edition. 2008. Research Triangle Park, NC: The International Society of Automation.

Instrument Engineers Handbook, Volume 1: Process Measurement and Analysis, Béla G. Lipták (editor). Fourth Edition. 2003. Boca Raton, FL: CRC Press.

Instrumentation Reference Book, Walt Boyes (editor). Fourth Edition. 2009. Oxford, UK: Butterworth-Heinemann.

Measurement and Instrumentation Principles, Alan S. Morris. Third Edition. 2001. Oxford, UK: Butterworth-Heinemann.

The Condensed Handbook of Measurement and Control, N.E. Battikha. Third Edition, 2006. Research Triangle Park, NC: The International Society of Automation.

The International Society of Automation, **www.isa.org**

Figure Credits

Modcon Analyzer Houses, Module Opener
Spectrum Technologies, Inc., Figure 25
Mettler-Toledo, LLC, Figures 26, 27
Courtesy of Honeywell International Inc., Figure 30
© 2015. Photos courtesy of Bacharach, Inc. www.MyBacharach.com, Figure 38

Morphix Technologies®, Figure 39
© Agilent Technologies, Inc. 2015 Reproduced with Permission, Courtesy of Agilent Technologies, Inc., Figure 43
MeasureNet Technology, Ltd., Figure 44
Fluke Corporation, reproduced with permission, Figure 46

Section Review Answer Key

Answer	Section Reference	Objective
Section One		
1. c	1.1.0	1a
2. b	1.1.1	1a
3. a	1.1.2	1a
4. d	1.2.0	1b
5. b	1.2.1	1b
6. b*	1.3.0	1c
7. a	1.4.2	1d
Section Two		
1. a*	2.1.0	2a
2. b*	2.1.0	2a
3. c	2.1.2	2a
4. a	2.2.1	2b
5. c	2.3.0	2c
Section Three		
1. c	3.1.0	3a
2. a	3.2.1	3b
3. b	3.2.3	3b
4. c	3.3.2	3c
5. d	3.4.0	3d
Section Four		
1. b	4.1.0	4a
2. c	4.1.2	4a
3. d	4.2.0	4b
4. a	4.3.0	4c
5. b	4.4.0	4d
6. a	4.5.0	4e
Section Five		
1. c	5.1.1	5a
2. b	5.2.1	5b
3. d	5.3.1	5c
4. a	5.3.4	5c
5. b	5.4.1	5d

*Calculations for these answers are provided on the following page(s).

Section Review Calculations

SECTION 1.0.0

Question 6

$$\text{molarity} = \frac{\text{moles of solute}}{\text{liters of solution}}$$

$$\text{molarity} = \frac{5.0 \text{ mol}}{3\text{L}} = 1.7\text{M}$$

The molarity is **1.7 M**.

SECTION 2.0.0

Question 1

Calculate the volume of the block of wood by multiplying the lengths of its sides:

Volume = 6 cm × 6 cm × 6 cm = 216 cm³

Use the mass and volume to calculate the density:

$$\text{Density} = \frac{\text{Mass}}{\text{Volume}}$$

$$\text{Density} = \frac{160 \text{ g}}{216 \text{ cm}^3} = 0.74 \text{ g/cm}^3$$

Its density is **0.74 g/cm³**

Question 2

Put the information you have into the specific gravity equation, and solve for the missing value:

$$\text{Specific gravity} = \frac{\text{mass of specific volume of substance}}{\text{mass of the same volume of water}}$$

$$0.92 = \frac{\text{mass of a gallon of crude oil}}{8.3 \text{ lbs}}$$

0.92 × 8.3 lbs = mass of a gallon of crude oil = 7.6 lbs

The mass of one gallon of crude oil at room temperature is **7.6 lbs**.

NCCER CURRICULA — USER UPDATE

NCCER makes every effort to keep its textbooks up-to-date and free of technical errors. We appreciate your help in this process. If you find an error, a typographical mistake, or an inaccuracy in NCCER's curricula, please fill out this form (or a photocopy), or complete the online form at **www.nccer.org/olf**. Be sure to include the exact module ID number, page number, a detailed description, and your recommended correction. Your input will be brought to the attention of the Authoring Team. Thank you for your assistance.

Instructors – If you have an idea for improving this textbook, or have found that additional materials were necessary to teach this module effectively, please let us know so that we may present your suggestions to the Authoring Team.

NCCER Product Development and Revision
13614 Progress Blvd., Alachua, FL 32615

Email: curriculum@nccer.org
Online: www.nccer.org/olf

❏ Trainee Guide   ❏ Lesson Plans   ❏ Exam   ❏ PowerPoints   Other _____

Craft / Level: _____ Copyright Date: _____

Module ID Number / Title: _____

Section Number(s): _____

Description: _____

_____

_____

_____

_____

Recommended Correction: _____

_____

_____

_____

_____

Your Name: _____

Address: _____

_____

Email: _____ Phone: _____

Glossary

5V logic: A logic device that operates on +5 VDC. It has logic levels in which a logic-0 is close to 0 V and a logic-1 is close to 5 V.

Acid: A substance that produces hydrogen ions (H^+) in solution.

Address: A unique identifier that enables a PLC program to refer to specific modules and wiring terminal positions.

Algorithm: In the context of computer programming, this refers to code that performs a specific task. In analog control systems, it refers to symbols for components that perform various analog functions, such as PIDs and summers.

Amplitude: The strength of a signal, often represented by the height of the plot in a graph.

Analog-to-digital converter (ADC): A circuit that converts an analog voltage or current signal into a numeric equivalent.

AND gate: A digital gate (see *gate*) whose output is logic-1 only when both of its inputs are logic-1.

Atom: The smallest particle with a unique chemical identity.

Atomic mass: The average mass of the naturally occurring isotopes of a given element. Expressed in unified atomic mass units.

Atomic number: The number of protons in a particular element. This value is unique to each element.

Backbone: A high-speed network channel designed to connect multiple networks together, usually across a facility, campus, or geographically distant area. Backbones often use fiber optic technology.

Base: A substance that produces hydroxide ions (OH^-) in solution.

BCD number: A decimal digit represented by a four-bit binary number in the range of 0000 to 1001; stands for binary coded decimal.

Binary number: A number made up of digits that are either 0 or 1; also known as base 2.

Bit: A single logic-0 or logic-1 value. The term is a contraction of the words binary digit.

Blacklist: A list of network addresses that may not cross a firewall.

Bond: The process by which elements join together to form compounds.

Buffer gate: A digital gate with an output that can deliver more current or a different voltage than a normal gate. Logically, it doesn't change the signal in any way, so the output equals the input.

Buffer solution: A solution with a specific pH value that strongly resists changes in pH and so maintains its value reliably.

Bump test: A method of testing the response of a control system by suddenly changing the controller setpoint to produce a process disturbance.

Calibration: The process of adjusting an instrument so its output signal or level of energy is representative of the level of the non-adjustable input energy at any given point. Calibration involves test instruments traceable to a national standards body.

Cavitation: An undesirable pulsation in a fluid stream caused by flow changes that create pockets of low-pressure fluid mixed in with normal pressure. Cavitation can damage equipment and cause a process to oscillate.

Chemical reaction: A process that occurs when substances combine or break apart at the chemical level.

Chromatography: A family of analytical laboratory techniques in which substances are separated for individual identification.

Clock: A signal that oscillates between logic-0 and logic-1 at a particular frequency. It is used to synchronize digital circuits with each other.

Cloud-based storage: Remote file storage accessed through the Internet and provided by a service company that maintains one or more servers in various locations.

Colorimeter: A device that measures the quantity of specific suspended particles in a solution by measuring how particular wavelengths of light are absorbed when they pass through the solution.

Combination logic: A digital circuit made from a collection of gates.

Commissioning: This step involves documenting and testing an instrument channel using live process. Checks are usually incremental, beginning with manual control and progressing to full automatic.

Compound: A chemical combination of two or more elements.

Conductivity: The ability of a solution to carry an electric current due to the presence of ions. Measured in siemens per centimeter (S/cm.

Continuity tester: A basic piece of test equipment that verifies an electrical current can flow through a circuit.

Control code: A character that when received causes a device to modify its behavior in some way.

Counter: A digital circuit that stores a value that gets increased or decreased in response to an external signal.

Covalent bond: A chemical bond in which atoms share some of their valence shell electrons.

Critical frequency: The frequency at which a control system can become unstable if its gain is greater than 1.

Current loop trim: An adjustment made at the third stage of a smart instrument, which adjusts the relationship between the calculated output signal and the actual analog output signal.

Dampened oscillation method: A closed loop tuning method in which calculations are based on the period of the process response curve when its oscillations have been adjusted to die out.

Data table: The section of a PLC's memory map that contains data used by the control program.

Deadtime: The delay interval between the point when the controlled process variable changes and the point when the controller begins to respond to the change.

Decoder: A digital circuit that outputs a unique pattern in response to each input combination.

Density: A physical property of matter defined as mass per unit volume.

Derivative: A control method in which the controller's output depends on the rate of change of the process variable or error.

Differentiator: A control circuit or mechanism that outputs a signal based on the derivative of the input signal.

Digital-to-analog converter (DAC): A circuit that converts a number into an equivalent analog voltage or current signal.

Dissociate: Occurs when ionic compounds dissolve and separate into ions.

Drop: A controller, server, workstation, or other DCS component connected to a network.

Dry leg: A connection from the process that contains a dry, non-condensing gas.

Eavesdropping: Using an unauthorized electronic device to monitor wired or wireless network traffic in order to steal information.

Edge-triggered: A digital circuit that performs an operation when it detects a change in logic level.

Electrolytes: Substances that dissociate/ionize into ions when dissolved in water. Electrolytes make solutions electrically conductive.

Electrons: The negatively charged parts of an atom that orbit around the nucleus.

Electron shells: The path in which electrons orbit around the atom's nucleus.

Element: A substance composed of just one type of atom.

Elevated zero: A zero value that's above the true zero point.

Emissivity: The ratio of the quantity of electromagnetic radiation that an object emits to the amount that a theoretical perfect emitter would emit under the same conditions.

Encoded: The process of representing one kind of information as another. For example, a character encoding system represents each character with an equivalent number.

Encryption: A security measure against data interception in which the data is turned into an unreadable form using cipher techniques prior to being transmitted. It is returned to its original form upon receipt.

Enterprise-level: Technology designed to provide high-level information services for company executives in order to facilitate large-scale planning and decision-making.

Ethernet: Refers to a family of wired networking hardware standards with speeds ranging up to gigabits per second. A major networking standard since the 1980s.

Factory characterization: Information describing a sensor's characteristics that has been generated by the manufacturer and stored within the device.

Fail safe: A system that puts itself into a safe or minimally hazardous condition in the event of a failure.

Fiber optic network: A networking technology in which data is transmitted through fiber optic cables as pulses of light.

Field device: A general term for a sensor or actuator that interacts directly with a process. Field devices are at the bottom of the DCS hierarchy.

Fieldbus: A general term for an industrial process network and protocols used to connect instruments, sensors, and controllers together using simple cabling systems.

Firewall: A network device that divides a network into two sections and prevents unauthorized accesses or protocols from crossing between the sections.

First-out logic: The first element (possibly of several) that caused a process trip.

Five-point calibration: A method of calibrating an instrument in which the output level is checked and set against a simulated input level at five percentages of span: 0 percent, 25 percent, 50 percent, 75 percent, 100 percent, 75 percent, 50 percent, 25 percent, and 0 percent, in that order.

Flash point: The lowest temperature at which a liquid gives off vapors that will mix with air and be able to ignite.

Flip-flop: A digital memory circuit capable of storing a single logical value (one bit).

Formula unit: The most fundamental unit of an ionic compound.

Fulcrum: The pivot point around which a lever turns.

Functional: Diagram of a digital logic system expressed as connected logic components.

Gate: A digital device whose output is the result of a logical operation performed on its inputs.

Gateway: A network device that translates from one type of physical layer and/or protocol to another. Gateways allow normally incompatible systems to communicate.

Ground loop: An undesirable current flow through a cable's shield caused by grounding it at each end. The difference of ground potentials causes the current. Ground loops result in noise and signal distortion.

Head pressure: The pressure that results from a liquid column of a certain height.

Hexadecimal number: A number made up of the digits 0–9 and A–F; also known as base 16.

Hierarchical: Arranged in layers, with higher layers being more sophisticated and performing more supervisory functions than lower ones.

Historian: A program that keeps a record of process data for later analysis and to reveal trends. Historians can run on dedicated computers or on a server along with other programs.

Human-machine interface (HMI): Any device that allows a human to control and monitor the status of a computer or PLC.

Hydrocarbon: A chemical compound composed of hydrogen and carbon atoms.

Hydrometer: A device used to measure the specific gravity of a solution.

Inches of water: A measurement of pressure (1" H_2O = 0.03612628 psi, or 249 Pa).

Input/output (I/O): The component of a computer or PLC that allows it to interact with the outside world.

Instrument asset management system (IAMS): A software application that combines a database and communications system to keep detailed records about instrumentation. An IAMS guides calibrations and maintenance, keeps inventory, and provides alerts to instrument personnel.

Integral: A control method in which the controller's output depends on the amount and duration of the error signal.

Integrating process: A process that changes without limit as long as the controller is outputting a value above the process equilibrium level.

Integrator: A control circuit or mechanism that outputs a signal based on the integral of the input signal.

Interlock: A mechanical, electrical, or software safety feature that prevents certain events from occurring unless other necessary conditions are first met.

Interrupt: An event that causes a PLC processor to stop whatever it's doing and perform a specific task, usually one that's time-critical.

Ion: An atom, or group of atoms, that has a positive or negative electrical charge.

Ionic bond: A chemical bond formed by the transfer of electrons from one atom to another.

Ionize: To convert into an ion or ions; ionization occurs when covalent compounds break down into ions in solution.

Isolated: Connected in such a way that a control signal does not directly interact at the electrical level with a device.

Isotopes: Forms of the same element but with different numbers of neutrons in the nucleus.

Lag: The delay between an input change and a corresponding output response.

Language: In the context of computers and PLCs, a language is a set of instructions and an associated "grammar" that governs how they are used to develop programs that control the processor.

Level-triggered: A digital circuit that performs an operation when it detects a specific logic level.

Logic family: A collection of digital circuit components that are compatible with each other.

Logic level: The voltages that represent logic-0 and logic-1 for a particular digital logic family.

Look-up table: A table of input and output values stored within an instrument. The device compares its input signal with the input side of the look-up table until it finds a match. It then outputs the corresponding output value.

Loop checking: The process of performing a series of check on a completed instrument chain and verifying that each device is installed correctly and has continuity to other devices.

Malware: A class of malicious software that enters a system surreptitiously and that can cause damage, steal information, or spread to other systems. Malware includes viruses, spyware, Trojans, and other similar programs.

Memory map: A diagram of the way in which a PLC's memory is divided up and assigned to different tasks.

Metallic bond: A chemical bond in which the valence electrons of many atoms move freely between their atoms, forming a kind of electron sea.

Modules: Components in a PLC system that provide particular features, such as I/O or communications.

Molarity (M): The concentration of a solution equal to the number of moles of solute divided by the number of liters of solution.

Mole (mol): A quantity equal to 6.02×10^{23} particles (atoms, formula units, molecules, or ions).

Molecules: The most fundamental units of a covalent compound.

Motion encoder: An electromechanical device that converts a rotary or linear motion into a number representing the current position.

Multifunction loop calibrator: A test instrument that provides an energy source and measuring capabilities in the same equipment housing. Multifunction calibrators often perform many different types of calibrations.

NAND gate: An AND gate with an inverted output.

Nanosecond (ns): One billionth of a second ($1 \text{ ns} = 1/1,000,000,000 \text{ s}$).

Negative-going: A logic signal that's changing from logic-1 to logic-0.

Network switch: A kind of network hub to which many devices connect. Switches perform intelligent traffic management, channeling data between their ports based on hardware addresses.

Neutrons: Uncharged atomic particles that are similar in size to protons and are present in the nucleus of an atom.

Non-integrating process: A process that reaches a new equilibrium level in response to a change in controller output but does not keep changing once it reaches the new level. Also known as a self-regulating process.

Nonvolatile memory: Memory that does not require electrical power to maintain its contents. Flash memory is a typical example.

NOR gate: An OR gate with an inverted output.

NOT gate: A digital gate that outputs the logical opposite of its input (also known as an inverter).

Nucleus: The central part of an atom. The nucleus contains the protons and neutrons.

OR gate: A digital gate whose output is logic-1 if either or both of its inputs is logic-1.

Oscillating process: A process that continues reacting to a disturbance rather than returning to a stable condition in a reasonable period of time.

Oscillations: Repetitive changes in signal amplitude.

Over-the-hump time: The time that it takes a waveform to reach its peak amplitude and start moving downwards.

Overshoot: The tendency for the controlled variable to exceed the setpoint.

Oxidation-reduction potential (ORP): The ability of a solution to act as an oxidizer or reducing agent. Determined by measuring voltage polarity across a pair of electrodes.

Partitioning: Dividing process I/O across multiple drops or multiple I/O cards in order to prevent a failure from causing a process shutdown.

Parts per million (ppm): The concentration of solute in a solution, expressed as the parts of solute per million parts of solution.

Patch: A fix for defects discovered in software after its release. Many patches are security-oriented.

Periodic table: A grid arrangement in which the elements are ordered in a way that shows their similarities and repeating properties.

Permissive: A condition in a control system that must be met before an action can occur. For example, a group of series-wired sensor switches would be permissives in an electrical control system.

pH: A measurement scale that expresses how acidic or basic a solution is based on its concentration of hydrogen (H^+) ions; values typically range from 0 to 14.

Poise (P): A unit of dynamic viscosity. It is often used with the prefix *centi-*; one centipoise is one one-hundredth of a poise. It is derived from French and pronounced *pwaz*.

Positive-going: A logic signal that's changing from logic-0 to logic-1.

Propagation delay: A delay caused by the circuits inside a digital device. It is the time interval between the moment when the inputs change and the moment when the output responds.

Protocol: A set of rules for communications between two devices. Using a common protocol enables devices to communicate, even if they are made by different manufacturers.

Protons: Positively charged atomic particles that are similar in size to neutrons and are present in the nucleus of an atom.

Proving: A step involving the testing of an instrument channel using a simulated process. Control behavior is assessed and devices are calibrated.

RAID array: Stands for Redundant Array of Independent Disks; a group of hard drives acting as a unit in order to improve performance and/or provide error correction and redundancy.

Range: The minimum and maximum values that an instrument can measure.

Reaction rate method: An open loop tuning method in which calculations are based on the process reaction rate.

Real time: A computer system that responds with minimal and predictable delays to events as they happen. Critical control systems must be real time.

Real-time clock: A battery-backed timekeeping circuit embedded in a PLC that maintains the current time and date.

Reciprocal: The value by which you can multiply a given number in order to obtain a result of 1. The reciprocal of a number is usually written as a fraction, and is equal to is 1 over that number. For example, the reciprocal of 3 is ⅓.

Reformed: A maintenance step performed on electrolytic capacitors that have been stored too long and have dried out. A gradually increasing voltage is applied over a period of time to restore the electrolyte to normal condition.

Register: A digital memory circuit made from flip-flops and capable of storing a single word (see *word*) of information.

Repeats per minute (rpm): A measure of how fast a signal is being integrated.

Router: A network traffic manager that connects multiple networks together. Routers examine data at the protocol levels to decide where it should be forwarded.

Ruggedized: Designed to work reliably in a potentially harsh environment, such as one with unusual temperatures, electrical noise, or physical risk.

Salt: An ionic compound composed of at least one metallic and one nonmetallic element. Salts are formed from a neutralization reaction between an acid and a base.

Saturate: A condition in which a control element moves to its maximum position in response to an excessive control signal value.

Scan: A PLC processor cycle in which inputs are read, program logic is executed, and outputs are updated.

Selling: The final step of loop checking, which involves turning the commissioned loop over to the client or to another contractor.

Sensor trim: An adjustment made at the first stage of a smart instrument, which establishes the relationship between the input signal and the lower and upper limits of the calculated process variable.

Serial network: A broad umbrella term for various low-speed communications systems in which data travels one bit at a time between devices over a simple cabling system. Examples include RS-232, RS-422, and RS-485.

Server: A networked computer that provides information or services to other "client" machines.

Simulator: A piece of software that mimics the operation of a different computer system. Simulators allow software development and testing without the need to have the actual hardware present.

Solute: A substance that is dissolved in another substance, forming a solution.

Solvent: The substance in which another substance is dissolved to form a solution.

Span: The difference between an instrument's upper and lower measurement limits.

Spanning: Setting an instrument's output so that it is at its maximum value when the input process value is at the maximum value of the specified operating range.

Specific gravity: A physical property of matter defined as the ratio of a substance's mass to the mass of an equal volume of a reference substance, usually water or air.

Stability: The condition in which a control system responds to disturbances by returning quickly to the setpoint with oscillations that either remain constant or die out.

Steady state: The condition in which the controlled variable is essentially at the setpoint and is not changing significantly.

Stiction: The tendency for a valve stem to stick due to static (non-moving) friction and not respond smoothly to its actuator. Stiction tends to cause a response that jumps between positions rather than moving smoothly.

Stroke: To operate a valve through its entire range to confirm that it is working correctly.

Supervisory Control and Data Acquisition (SCADA): An industrial control system that emphasizes data collection and display. Human operators make and execute high-level control decisions over the process through the system.

Suppressed zero: A zero value that's below the true zero value.

Third-party solutions: A product developed by an outside company to solve a problem or meet a need associated with a particular manufacturer's technology.

Three-point calibration: A method of calibrating an instrument in which the output level is checked and set against a simulated input level at three different percentages of span: 0 percent, 50 percent, 100 percent, 50 percent, and 0 percent, in that order.

Time constant: A measurement of time that describes the response behavior of a particular control system. A single time constant is the time required for the controlled variable to change by 63.2 percent of the amount that it must change to reach steady state.

Time constant method: An open loop tuning method in which calculations are based on the process response curve's time constant.

Timing diagram: A diagram that shows the behavior of a digital circuit's signals with respect to time.

Total hydrocarbon content (THC): A measurement of all hydrocarbon molecules present in a sample.

Transfer function: A mathematical equation that describes an input-to-output relationship.

Transmission Control Protocol/Internet Protocol (TCP/IP): A family of communications protocols used to connect many different types of networked computers together. It is the protocol basis of the global Internet.

Truth table: A table that lists a digital circuit's output values for every possible input combination.

Tuning: Adjustments made to a loop controller to produce a desired response behavior to process disturbances.

Turbidity: The cloudiness of a solution containing suspended particles.

Ultimate period method: A closed loop tuning method in which calculations are based on the oscillation period that occurs when the controller is set to its ultimate gain or ultimate proportional band.

Unified atomic mass unit (u): A measurement unit of atomic mass based on $\frac{1}{12}$ the mass of a carbon-12 atom. One u equals about 1.660×10^{-27} kg.

Universal serial bus (USB): USB is a communications system built into many computers, PLCs, and portable electronic devices to allow simple, high-speed cabled connections.

Valence shell: The outermost electron shell of an atom. Chemical reactions involve the valence shell.

Viscosity: The internal friction of a fluid that opposes flow.

Volatile memory: Memory that requires uninterrupted electrical power to maintain its contents. Also known as RAM.

Vulnerability: A flaw in a piece of commercial software that can be exploited as a means of gaining access to a system.

Wet leg: A connection from the process that's intentionally filled with a reference fluid to prevent gas condensation from affecting the measurement.

Whitelist: A list of network addresses that may cross a firewall.

Wi-Fi: Refers to a particular family of wireless networking hardware and software standards designed for both general and special-purpose communications.

Word: A group of bits acting together as a single unit. Words are usually (but not always) multiples of four bits.

XNOR gate: An XOR gate with an inverted output.

XOR gate: A digital gate whose output is logic-1 if exactly one of its inputs is logic-1.

Zero and span calibration: A basic calibration procedure in which an instrument's zero and span points are adjusted to produce the proper relationship between the input and output minimum and maximum points.

Zeroing: Setting an instrument's output so that it is at its minimum value when the input process value is at the minimum value of the specified operating range.

Index

A
Abel flash point analysis system, (12409):21–22
Absorption spectrometry
   IR, (12409):42
   UV, (12409):42–44
Acid, (12409):1, 9, 49
Actuator field devices, (12407):7
Actuators, DCS, (12407):1
ADC. *See* Analog-to-digital converter (ADC)
Adding circuit, binary, (12406):22, 23
ADD instruction, (12406):25
Addition
   binary, (12406):22
   principles of, (12406):22
Address, (12406):13, 20, 34
Air bubble measurement of specific gravity, (12409):13
Alarm setpoints, (12407):24
Alarm systems
   DCS, (12407):23–24
   detailed point displays, (12407):24–25
   non-DCS, (12407):24
Algorithm, (12407):6, 7, 9, 36
Allen-Bradley address, (12406):20
American Standard Code for Information Interchange (ASCII), (12406):9–10
Amplitude, (12405):1, 2, 24
AMS Suite® by Emerson, (12410):14
Analog control graphics, DCS, (12407):21–22
Analog control systems, (12407):2–3
Analog instrument calibration, (12402):12–13
Analog instrument loops, calibrating, (12410):12
Analog I/O modules, PLCs, (12406):17
Analog logic diagram, (12407):9, 11
Analog signal
   data from, storing, (12406):15
   defined, (12402):12
Analog-to-digital converter (ADC), (12406):13, 17, 34
Analyzers
   magneto-dynamic oxygen, (12409):32
   paramagnetic oxygen, (12409):32
   pH, (12409):25–26
   process, (12409):12
   ratio turbidity, (12409):19
   reflection turbidity, (12409):18–19
   transmission turbidity, (12409):17–18
   UV absorption, (12409):44
   zirconium oxide oxygen, (12409):32
AND gate, (12406):1, 2–4, 29
Anti-malware software, (12407):32
Apps for mobile OWs, (12407):2
Arithmetic, principles of addition, (12406):22
Arithmetic circuits, (12406):22
Arithmetic instructions, LD programming, (12406):25–26
ASCII. *See* American Standard Code for Information Interchange (ASCII)
ASTM
   D56-70, flash point testing, (12409):22
   D93-71, flash point testing, (12409):22
Atom, (12409):1, 2, 49
Atomic mass, (12409):1, 2, 7–8, 49
Atomic number, (12409):1, 2, 49
Auto-ignition flash point analysis system, (12409):21–22
Avalanche effect, (12409):44

B
Backbone, (12407):6, 18, 36
Base, (12409):1, 9, 49
BCD. *See* Binary coded decimal (BCD) counters; Binary coded decimal (BCD) numbers; Binary coded decimal (BCD) storage registers
Binary adding circuit, (12406):22, 23
Binary coded decimal (BCD) counters, (12406):20
Binary coded decimal (BCD) numbers, (12406):11, 16, 18, 19, 29
Binary coded decimal (BCD) storage registers, (12406):15
Binary codes
   ASCII, (12406):9–10
   BCD (binary coded decimal), (12406):11
   gray code, (12406):11
Binary counters, (12406):20, 21
Binary numbers, (12406):7–9, 19–20, 22, 29
Binary storage registers, (12406):15
Bit, (12406):11, 29
Bit distribute (BTD) transfer instruction, (12406):26
Black light, (12409):42
Blacklist, (12407):27, 32, 36
Bonds
   atomic, (12409):1, 4, 49
   covalent, (12409):4
   ionic, (12409):4, 7
   metallic, (12409):7
   PLCs, (12406):14
BTD. *See* Bit distribute (BTD) transfer instruction
Bubblers, (12410):6–7
Buffer gate, (12406):1, 9, 29
Buffer solution, (12409):21, 24, 49
Building networks, (12407):2, 13
Bump test, (12405):1, 3, 24

C
Calibration
   DCS equipment, (12407):28
   defined, (12402):1, 33
   documentation, (12402):3, 4
   DP transmitters in liquid level applications, (12402):13–17
   input signals or energies, (12402):2
   instruments requiring, (12402):1
   output signals or energies, (12402):2
   process analyzer, (12409):12
   terminology, (12402):2–3
   traceable, (12402):2
   tuning vs., (12402):6
   UV absorption analyzers, (12409):44
   valid, (12402):2
Calibration process
   automatic, (12402):6
   electro-mechanical temperature transmitter, (12402):12
   five-point, (12402):3, 5–6
   HART® devices, (12402):21–24
   pneumatic differential pressure transmitters, (12402):12
   Rosemount™ 1151, (12402):17

Calibration process (*continued*)
   temperature transmitters, (12402):18
   three-point, (12402):5
   transducers, (12402):25
Calibration test equipment. *See also* Instruments
   analog
      differential pressure transmitters, (12402):13
      multifunction calibrators, (12402):13
   calibration of, (12402):1–2, (12410):12
   Crystal® HPC40, (12402):8–9
   Fluke® 719, (12402):8–9
   Fluke® 725, (12402):13
   Foxboro® 13A, (12402):10, 11
   GE Druck® DPI 620-IS, (12402):8–9
   loop calibrators
      Fluke® 725 loop calibrator, (12402):13
      multifunction loop calibrator, (12402):1, 33
   pneumatic
      differential pressure transmitters, (12402):9–12
      multifunction calibrators, (12402):8–9
   Rosemount™ 1151, (12402):16–17
   Siemens Series 50, (12402):10
   spanning, (12402):1, 3, 16–17, 33, (12410):9, 12, 31
   temperature transmitters, (12402):12
      with analog outputs, (12402):18
   Wallace and Tiernan® Wally Box®, (12402):8–9
   WIKA® Instrument, LP, (12402):8–9
   zeroing, (12402):1, 3, 16–17, 33, (12410):9, 12, 31
Capillary viscometer, (12409):16–17
Carbon dioxide
   monitoring, (12409):33–34
   properties, (12409):33–34
Carbon monoxide, measuring, (12409):33
Cavitation, (12410):19, 26, 31
Chemical composition analysis, (12409):38–40
Chemical property analysis, (12409):21–22
Chemical reactions
   atomic mass and, (12409):7–8
   bonding, (12409):4, 7
   defined, (12409):1, 4, 49
Chemistry, defined, (12409):2
Chromatography, (12409):12, 49
Clean rooms, (12409):37
Cleveland flash point analysis system, (12409):21–22
Clock, (12406):11, 13, 29
Clocked logic memory circuits
   clock signals, (12406):13, 14
   D flip-flop, (12406):15–16
   JK flip-flop, (12406):16
   RS flip-flop, (12406):13
   timing diagram, (12406):13, 14
Closed-cup flash point analysis, (12409):21
Closed loop tuning
   controller, function of the, (12405):7
   dampened oscillation method, (12405):18–19
   ultimate period method, (12405):16–18
Cloud-based storage, (12407):6, 11, 36
Colorimeter, (12409):12, 17–18, 49
Combination logic, (12406):1, 6, 29
Combination logic circuits, (12406):6–9
Commissioning a loop
   defined, (12410):1, 31
   documentation, (12410):16
   process, (12410):16–18
Communication
   I/O modules, PLCs, (12406):19
   minicomputers, (12407):3
   PLCs to external devices, (12406):22

Communication protocol
   DCS, (12407):2, 15
   fieldbus networks, (12407):2
   layers, (12407):15
   serial network, (12406):19
Communication protocols
   DeviceNet™, (12406):19
   DF1, (12406):19
   FOUNDATION™ Fieldbus, (12402):20, (12407):16–17
   HART®, (12402):20, 28–29, (12406):19, (12407):16, (12410):12
   Internet, (12407):15
   Modbus®, (12406):19, (12407):16
   PROFIBUS® (Process Bus), (12406):19, (12407):16
   TCP/IP, (12407):1, 3, 15, 37
Complete response equation, (12405):8
Compound, (12409):1, 4, 49
Concentration, (12409):8
Conductivity
   defined, (12409):21, 49
   measuring, (12409):26–27
   temperature effects, (12409):26–27
Conductivity cell, (12409):26
Construction Specification Institute (CSI) format, (12410):16
Continuity tester, (12410):1, 5, 31
Control code, (12406):7, 9, 34
Controllers. *See also* Programmable logic controllers (PLCs)
   closed loop tuning, (12405):7
   DCS
      applications, (12407):8–11
      building networks, (12407):2
      fieldbus networks, (12407):2
      function, (12407):1–2
      I/O, (12407):8–9
      operating systems, (12407):8
      PLCs compared, (12407):7–8
      redundancy, (12407):9
   pH, (12409):25–26
   programmable, (12406):1
   valves, smart, (12402):28–29
Control room plan
   control panel layout and wiring drawings, (12410):16
   section and detail drawings, (12410):16
   single-line diagram drawings, (12410):16
Control station graphics, DCS, (12407):21–22
Control valve positioners
   calibrating, (12402):27–28
   direct-acting, (12402):27–28
   electro-pneumatic, (12402):27
   function of, (12402):25
   pneumatic, (12402):25–26
   smart, (12402):28–29
Coulometric oxygen sensor, (12409):33
Count down (CTD) output instruction, (12406):24
Counters
   BCD, (12406):20
   binary, (12406):20, 21
   defined, (12406):19, 29
   discrete particle, (12409):38
   instructions, LD programming, (12406):23–25
   instructions, storing, (12406):15
   I/O modules, PLCs, (12406):18
   up/down, (12406):20
Count up (CTU) output instruction, (12406):24
Covalent bond, (12409):1, 4–5, 49
Covalent compound, (12409):8
Critical frequency, (12405):1, 2, 24
Critical processes, (12407):4

Crystal Engineering®, (12410):5, 6
Crystal® HPC40, (12402):8–9
CSI. *See* Construction Specification Institute (CSI) format
CTD. *See* Count down (CTD) output instruction
CTU. *See* Count up (CTU) output instruction
Current loop trim, (12402):23, (12410):9, 13, 31

D

D. *See* Data (D) flip-flop
DAC. *See* Digital-to-analog (DAC) converter
Dampened oscillation method, (12405):7, 18–19, 24
Damping, (12402):23
Data (D) flip-flop, (12406):15–16
Database servers, DCS, (12407):9, 12
Data collection, trends, (12407):25
Data comparison instructions, LD programming, (12406):26
Data table, PLCs, (12406):13, 15, 34
Data transfer instructions, LD programming, (12406):26
DCS. *See* Distributed control system (DCS)
Deadtime, (12405):1, 3–4, 24
Debugging PLCs, (12406):19
Decimal addition, (12406):22
Decimal numbers, (12406):7–8, 19
Decoders, (12406):22–26, 29
DeltaV FOUNDATION fieldbus, (12407):16–17
DeltaV™, Emerson Process Management, (12407):16
Densitometer, (12409):13
Density, (12409):12, 49
Derivative, (12405):1, 3, 24
Derivative time equation, (12405):10–11
Detailed point displays, (12407):24–25
DeviceNet™, (12406):19
DF1, (12406):19
Differential pressure (DP) transmitter
   analog, (12402):13
   head pressure measurements, (12402):13
   level measurements
      closed, pressurized vessel installations, (12402):15–16
      open vessel installations, (12402):13–15
   loop checking, (12410):3
   pneumatic, (12402):9–12
Differentiator, (12405):7, 10, 24
Digital circuits
   clocked logic memory circuits
      clock signals, (12406):13, 14
      D flip-flop, (12406):15–16
      JK flip-flop, (12406):16
      RS flip-flop, (12406):13
      timing diagram, (12406):13, 14
   combination logic, (12406):6–9
   counter, (12406):19
   gates
      buffer gate, (12406):1, 9, 29
      defined, (12406):1, 2, 29
      AND gate, (12406):1, 2–4, 29
      logical operators, (12406):4
      multiple, combination logic circuits, (12406):6–9
      NAND gate, (12406):1, 5, 29
      NOR gate, (12406):1, 5–6, 29
      NOT gate, (12406):1, 5, 29
      OR gate, (12406):1, 4, 29
      schematic symbols, (12406):2, 3
      truth table, (12406):1, 2, 29
      XNOR gate, (12406):1, 6, 29
      XOR gate, (12406):1, 4–5, 29
   schematic symbols, (12406):2
Digital logic diagram, (12407):9, 10
Digital-to-analog (DAC) converter, (12406):13, 17, 34

Direct-acting positioners, (12402):26–28
Discrete control graphics, DCS, (12407):21–22
Discrete I/O modules, PLCs, (12406):16
Discrete particle counters (DPCs), (12409):38
Displacement measurement of specific gravity, (12409):13–14
Dissociate, (12409):1, 9, 49
Distributed control system (DCS)
   actuators, (12407):1
   alarm systems, (12407):23–24
   communication protocols, (12407):15
   communications, (12407):2
   compared with
      PLCs, (12407):4, 7–8, 30
      safety instrumented systems, (12407):4–5
      SCADA, (12407):4
   controllers
      applications, (12407):8–9
      building networks, (12407):2
      failures, (12407):30
      fieldbus networks, (12407):2
      function, (12407):1–2
      I/O, (12407):8–9
      operating systems, (12407):8
      PLCs compared, (12407):7–8
      redundancy, (12407):9
   field device, (12407):6, 7, 36
   hierarchy, (12407):1, 13
   historically, (12407):1, 2–3, 9, 25
   maintenance
      calibration, (12407):28
      history files, (12407):27, 28
      importance of, (12407):27
      preventive, scheduling, (12407):27
      production downtime, (12407):27
      on-site, (12407):28
   network connections, (12407):2
   networks
      building networks, (12407):2, 13
      failures, (12407):30–31
      fieldbus, (12407):2, 13, 15–16, 17
      higher-level, (12407):17–18
   power supply repairs, (12407):30
   process, monitoring and controlling, (12407):2
   protocols, (12407):15
   security, (12407):31–32
   sensors, (12407):1
   servers
      database, (12407):9, 12
      enterprise-level, (12407):13
      file server, (12407):9
      function, (12407):2, 9, 11
      historian, (12407):12–13, 25
      operating systems, (12407):13
      software, (12407):11–12
      web server, (12407):9
Distributed control system (DCS), repairing
   expertise, acquiring
      equipment knowledge, (12407):31
      tool skills, (12407):31
   failures
      components, (12407):30
      controllers, (12407):30
      field devices, (12407):29
      networks, (12407):30–31
      power supply, (12407):30
      troubleshooting, (12407):12, 29
   operator approvals and, (12407):28–29

Distributed control system (DCS), repairing (continued)
    production downtime and, (12407):28
    on-site, (12407):28–29
    spare in-kind, (12407):28
Distributed control system (DCS) workstations
    engineering, (12407):13
    function, (12407):13
    operating systems, (12407):3, 13
    operator (HMIs)
        components, (12407):20
        function, (12407):2, 13
        location, (12407):2
        pre-DCS, (12407):19–20
        types of, (12407):2, 3
        updating, (12407):20
    operator (HMIs), graphics
        alarming system, (12407):23–24
        analog control, (12407):21–22
        detailed point displays, (12407):24–25
        discrete control, (12407):21–22
        historical trends, (12407):25
        informational screens, (12407):23
        minimizing, (12407):20
        navigation, (12407):23
        process, (12407):20–21
        trends, (12407):25
    smart technology, (12407):2, 13
DIV instruction, (12406):26
DN. See Timer done (DN) output
DP. See Differential pressure (DP) transmitter
DPCs. See Discrete particle counters (DPCs)
Drop, (12407):1, 2, 36
Dry leg, (12402):8, 15, 33

E

Eavesdropping, (12407):27, 31, 32, 36
Edge-triggered, (12406):11, 13, 29
Electrochemical sensors, (12409):35
    high-temperature, (12409):31–32
Electrolytes, (12409):1, 9, 49
Electrons, (12409):1, 2, 49
Electron shells, (12409):1, 2, 49
Electro-pneumatic valve positioners, (12402):27
Elements
    atomic mass, (12409):2
    atomic number, (12409):2
    compounds, (12409):4
    defined, (12409):1, 2, 49
    isotope variations, (12409):2
    Periodic Table of the, (12409):1, 2–6, 50
    symbols, (12409):2
Elevated zero, (12402):8, 10, 17, 33
Emerson AMS Suite®, (12410):14
Emerson Process Management DeltaV™, (12407):16
Emissivity, (12409):37, 40, 49
Encoded, (12406):7, 34
Encoders
    data from, storing, (12406):15
    motion encoder, (12406):7, 11, 34
    PLC I/O module, (12406):18
    position, (12406):18
Encryption, (12407):6, 19, 36
Energy balance equation, (12405):7–8
Engineering (EWs) workstations, DCS, (12407):13
Enterprise-level, (12407):6, 13, 36
Enterprise-level servers, DCS, (12407):13
EQU. See Equal-to (EQU) comparison instruction
Equal-to (EQU) comparison instruction, (12406):26
Ethernet, (12406):13, 19, 34

Ethernet networks, (12407):2, 3, 15, 17–19
EWs. See Engineering (EWs) workstations, DCS
Extech®, (12410):5

F

Factory characterization, (12402):20, 33
Fail safe, (12407):1, 5, 36
Falling ball viscometer, (12409):15–16
FBD. See Function block diagram (FBD) programming
Federal Standard 209E, cleanliness, (12409):37
Fiber optic networks, (12407):1, 2, 15, 17–18, 36
FID. See Flame ionization detector (FID)
Fieldbus, (12407):1, 36
Fieldbus networks
    DCS, (12407):2, 13, 15–16, 17
    security, (12407):32
Field devices, (12407):6, 7, 36
File server, DCS, (12407):9
Fire point, (12409):21
Firewall, (12407):6, 19, 36
First-out logic, (12407):6, 23, 36
Fisher® 3582i Series electro-pneumatic positioner, (12402):26, 28
Fisher® 3582 Series pneumatic positioner, (12402):26, 27
Fisher® FIELDVUE™ DVC 6000 Series digital valve controller, (12402):28–29
5V logic, (12406):1, 2, 29
Five-point calibration, (12402):1, 33
Five-point calibration process
    generally, (12402):3, 5–6
    temperature transmitters, (12402):18
Flame detectors, (12409):44
Flame ionization detector (FID), (12409):35–36
Flapper, (12402):9
Flash point, (12409):21, 49
Flash point determination, (12409):21–22
Flip-flops
    clocked
        D flip-flop, (12406):15–16
        JK flip-flop, (12406):16
        RS flip-flop, (12406):13
    connected into registers, (12406):17
    defined, (12406):11, 29
    NAND RS flip-flop, (12406):12–13
    RS flip-flop, (12406):11–12
Flowcharts, troubleshooting, (12410):26–27
Flow transmitter, pneumatic DP transmitter as, (12402):10–11
Fluke® 719, (12402):8–9
Fluke® 724, (12402):18
Fluke® 725 loop calibrator, (12402):13
Fluke® 725 multifunction process calibrator, (12410):10–11
Fluke® 754, (12402):22, (12410):12–13
Fluke® 789 ProcessMeter™, (12410):5
Fluke® test and calibration meters, (12410):5, 6
Formula unit, (12409):1, 4, 49
FOUNDATION™ Fieldbus, (12402):20, (12407):16–17
Foxboro®, (12407):16
Foxboro® 13A, (12402):10, 11
Fulcrum, (12402):8, 9, 33
Functional, (12407):6, 9, 36
Function block diagram (FBD) programming, (12406):21
Furnace design, (12406):7–8

G

Galvanic oxygen sensor, (12409):33
Gas analysis
    carbon dioxide, (12409):33–34

carbon monoxide, (12409):33
hydrogen sulfide, (12409):34–35
oxygen, (12409):31–33
total hydrocarbons, (12409):35–36
Gas chromatography, (12409):38–40
Gates
   AND gate, (12406):1, 2–4, 29
   buffer gate, (12406):1, 9, 29
   defined, (12406):1, 2, 29
   logical operators, (12406):4
   NAND gate, (12406):1, 5, 29
   NOR gate, (12406):1, 5–6, 29
   NOT gate, (12406):1, 5, 29
   OR gate, (12406):1, 4, 29
   schematic symbols, (12406):2, 3
   truth table, (12406):1, 2, 29
   XNOR gate, (12406):1, 6, 29
   XOR gate, (12406):1, 4–5, 29
Gateway, (12407):6, 19, 36
GE Druck®, (12410):5, 6
GE Druck® DPI 620-IS, (12402):8–9
Global warming, (12409):34
Graphics, operator workstations (DCS)
   alarming system, (12407):23–24
   analog control, (12407):21–22
   detailed point displays, (12407):24–25
   discrete control, (12407):21–22
   historical trends, (12407):25
   informational screens, (12407):23
   minimizing, (12407):20
   navigation, (12407):23
   process, (12407):20–21
   trends, (12407):25
Graphics symbols, (12407):21
Gray code, (12406):11, 15, 18
Greater-than (GRT) comparison instruction, (12406):26
Greenhouse gases, (12409):34, 35
Ground loop, (12410):1, 3, 31
Grounds, PLCs, (12406):14
GRT. *See* Greater-than (GRT) comparison instruction

H

Hardware installation detail drawings, (12410):16
HART Communication Foundation, (12402):20
HART® (Highway Addressable Remote Transducer) protocol, (12402):20, 28–29, (12406):19, (12407):16, (12410):12
HART® (Highway Addressable Remote Transducer) smart devices
   asset management communication, (12410):12–13
   calibrating, (12402):21–24, (12410):12–13
   transmitter interface modules, PLCs, (12406):19
   with electrical continuity testers, (12410):5
Head pressure, (12402):8, 13, 33
Hexadecimal numbers, (12406):9, 19, 20, 29
Hierarchical, (12407):1, 36
High-temperature electrochemical sensors, (12409):31–32
Historian
   DCS, (12407):12–13, 25
   defined, (12407):6, 36
Historical trends, (12407):25–26
History files, DCS maintenance, (12407):27, 28
HMI. *See* Human-machine interfaces (HMIs)
Holding registers, (12406):15
Human-machine interfaces (HMIs). *See also* Operator workstations (OWs), DCS
   defined, (12406):1, 2, 34
   SCADA, (12407):4
   smart, (12406):5

Hydrocarbon, (12409):12, 16, 49
Hydrogen sulfide, measuring, (12409):34–35
Hydrometer, (12409):12, 13, 49

I

IAMS. *See* Instrument asset management system (IAMS)
IEC
   61508, SIS, (12407):4
   61511, SIS, (12407):4
IL. *See* Instruction list (IL) programming
Inches of water, (12402):1, 2, 33
Inductive conductivity instruments, (12409):27
Industrial control, (12407):1
Industrial Ethernet, (12407):19
Informational graphics screens, DCS, (12407):23
Infrared imaging, (12409):40–42
Infrared radiation, measuring, (12409):40–42
Input/output (I/O), (12406):1, 34
Input/output (I/O) modules
   DCS controllers, (12407):8–9
   PLCs
      address schemes, (12406):19–20
      analog, (12406):17
      communication, (12406):19
      data table function, (12406):15
      discrete, (12406):16
      encoder/counter, (12406):18
      module addressing, (12406):19–20
      numerical data interface, (12406):17
      smart transmitter interface, (12406):18
      specialized, (12406):17–19
      stepper motor control, (12406):18
      temperature monitoring/control, (12406):18
Input registers, (12406):15
Inside the box, (12406):22
Instruction list (IL) programming, (12406):22
Instrument asset management system (IAMS), (12410):9, 14, 31
Instrument index, (12410):16
Instrument location and conduit plan, (12410):16
Instruments. *See also* Calibration test equipment
   look-up table, (12402):20, 21, 33
   operating range limit, (12402):2
   range, (12402):1, 2, 33
   requiring calibration, (12402):1
   span, (12402):1, 2–3, 33
   spanning, (12402):1, 3, 33, (12410):9, 12, 31
   zeroing, (12402):1, 3, 33, (12410):9, 12, 31
Integral, (12405):1, 2, 24
Integral time equation, (12405):10
Integrating process, (12405):1, 5–6, 24
Integrator, (12405):7, 10, 24
Interlock, (12410):9, 11, 31
Internet, (12407):3, 15, 32
Interrupt, (12406):13, 34
Inverter, (12406):see NOT gate
I/O. *See* Input/output (I/O)
Ion, (12409):1, 8–9, 49
Ionic bond, (12409):1, 4, 49
Ionic compound, (12409):8
Ionization, (12409):8
Ionize, (12409):1, 9, 49
Isolated, (12406):1, 3, 34
Isotopes, (12409):1, 2, 49

J

Jackson turbidimeter, (12409):17
JK flip-flop, clocked, (12406):15

JMP. *See* Jump (JMP) output instruction
JSR. *See* Jump to subroutine (JSR) output instruction
Jump (JMP) output instruction, (12406):27
Jump to subroutine (JSR) output instruction, (12406):27

K
Key switch, PLC, (12406):14

L
Ladder diagram (LD) programming
   about, (12406):21
   instruction categories
      arithmetic, (12406):25–26
      data comparison, (12406):26
      data transfer, (12406):26
      program control, (12406):26–27
      relay, (12406):22–23
      timer and counter, (12406):23–25
   symbolic instruction set, (12406):22
Lag, (12405):1, 4, 24
Language, (12406):21, 34
Languages, programming
   development tools, (12406):22
   function block diagram (FBD), (12406):21
   instruction list (IL), (12406):22
   ladder diagram (LD), (12406):21–27
   sequential function chart (SFC), (12406):21–22
   structured text (ST), (12406):22
LBL output instruction, (12406):27
LCD displays/lights, PLC, (12406):14, 15
LD. *See* Ladder diagram (LD) programming
LES. *See* Less-than (LES) comparison instruction
Less-than (LES) comparison instruction, (12406):26
Level measurements
   closed pressurized vessel installations, (12402):15–16
   open vessel installations, (12402):13–15
Level-triggered, (12406):11, 13, 29
Linus® operating system, (12407):13
Loggers, (12407):13
Logical operators, (12406):4
Logic family, (12406):1, 29
Logic level, (12406):1, 2, 29
Logic window, (12406):1–2
Look-up table, (12402):20, 21, 33
Loop checking
   cable shields, (12410):3
   conduit, (12410):3
   control room components, (12410):3
   defined, (12410):1, 31
   drain wires, (12410):3
   equipment
      bubblers, (12410):6–7
      continuity testers, (12410):5
      hand pump for pneumatic testing, (12410):6
      network testing, (12410):6
      telephone receivers, (12410):5–6
      tone testers, (12410):6
   field transmitter, (12410):2–3
   loop sheets, (12410):4
   mechanical inspection, purpose, (12410):1–2
   primary element, (12410):2
   repairs, (12410):3
   signal noise, (12410):3
   steps in, (12410):1
   tag numbers, (12410):4
   tubing, (12410):3
   wiring
      electrical continuity test, (12410):5–6
      mechanical inspection, (12410):3

Loop continuity tests
   cables
      fiber optic cables, (12410):6
      network cables, (12410):6
   electrical continuity, (12410):5–6
   equipment, (12410):5–6
   pneumatic pressure testing, (12410):6–7
   purpose, (12410):4
Loop current adjustments, (12402):23–24
Loop diagrams, (12410):16, 17
Loop troubleshooting
   documentation, (12410):24
   fundamental steps
      example, (12410):21–24
      identify the loop components, (12410):20
      identify the problem, (12410):19–20
      repair the loop, (12410):24
      understand the loop, (12410):20–21
   normal-abnormal state compare, (12410):20–21
   recurrent problems, (12410):20
Loop troubleshooting tools
   flowcharts, (12410):26–27
   listen to the operator, (12410):19, 24
   loop diagram, (12410):24–25
   maintenance record, (12410):20, 24
   panel graphic, (12410):24
Loop tuning, (12405):see tuning

M
M. *See* Molarity (M)
Magneto-dynamic oxygen analyzer, (12409):32
Malware, (12407):27, 32, 36
Masked move (MVM) transfer instruction, (12406):26
Master control relay (MCR) output instruction, (12406):27
Matter, defined, (12409):2
MCR. *See* Master control relay (MCR) output instruction
Memory, PLCs
   allocation, (12406):14–15
   data (I/O) table, (12406):13, 15, 34
   diagnostics, (12406):15
   executive program, (12406):15
   nonvolatile, (12406):13, 14, 34
   processor work area, (12406):15
   scratch pad, (12406):15
   storage registers, (12406):15
   user programs, (12406):15
   volatile, (12406):13, 14, 34
Memory devices, digital
   clocked logic
      clock generator, (12406):13
      clock signals, (12406):13, 14
      D flip-flop, (12406):15–16
      edge-triggered, (12406):13, 14
      JK flip-flop, (12406):16
      level-triggered, (12406):13, 14
      RS flip-flop, (12406):13
      timing diagram, (12406):13, 14
   flip-flops, (12406):11
      NAND RS flip-flop, (12406):12–13
      RS flip-flop, (12406):11–12
   registers, (12406):11, 17, 29
   words, (12406):11, 16–17, 29
Memory failure, (12407):30
Memory map, (12406):13, 34
Metallic bond, (12409):1, 7, 49
Metalloids, (12409):2
Metals, (12409):2
Methane, (12409):35
Microprocessor, (12406):14, (12407):3

Microsoft® operating systems, (12407):3, 13
Modbus®, (12406):19, (12407):16
Module addressing, (12406):19–20
Modules, (12406):1, 34
Mol. *See* Mole (mol)
Molarity (M), (12409):1, 8, 49
Mole (mol), (12409):1, 7, 49
Molecule, (12409):1, 7, 49
Motion encoder, (12406):7, 11, 34
MOV. *See* Move (MOV) transfer instruction
Move (MOV) transfer instruction, (12406):26
MUL instruction, (12406):25–26
Multifunction loop calibrator, (12402):1, 33
MVM. *See* Masked move (MVM) transfer instruction

N

NAND gate, (12406):1, 5, 29
NAND RS flip-flop, (12406):12–13
Nanosecond (ns), (12406):1, 8, 29
National Electrical Code® (NEC®), (12406):27
National Fire Protection Association (NFPA) Standards
   85, *Boiler and Combustion System Hazards Code*, (12407):4
   86, furnace safeguards, (12406):7–8
National Institute of Standards and Technology (NIST), (12402):1, (12407):32
Natural gas, (12409):34
Navigation, graphic displays, (12407):23
Negative-going, (12406):1, 9, 29
Networks
   basics, (12407):15
   DCS
      building networks, (12407):2, 13
      failures, (12407):30–31
      fieldbus, (12407):2, 13, 15–16, 17
      higher-level, (12407):17–18
   encryption, (12407):6, 19, 36
   Ethernet, (12407):2, 3, 15, 17–19
   failures, (12407):30–31
   fiber optic, (12407):15, 17–18
   historically, (12407):3
   industrial Ethernet, (12407):19
   OSI Model, (12407):15
   protocol layers, (12407):15
   RS-422, (12407):15
   RS-485, (12407):15
   satellite, (12407):15
   wireless, (12406):13, 19, 34, (12407):2, 15, 18, 19
Network security
   access control, (12407):32
   anti-malware software, (12407):32
   encryption, (12407):19, 32, 36
   firewall, (12407):19, 32, 36
   Internet access, (12407):32
   passwords, (12407):32
Network switch, (12407):6, 18, 36
Network traffic management devices
   routers, (12407):6, 18–19, 37
   switches, (12407):6, 18, 36
Neutralization, (12409):10
Neutrons, (12409):1, 2, 49
Newtonian fluid, (12409):16
NFPA. *See* National Fire Protection Association (NFPA) Standards
NIST. *See* National Institute of Standards and Technology (NIST)
Non-integrating process, (12405):1, 4–5, 24
Nonmetals, (12409):2
Nonvolatile memory, (12406):13, 14, 34, (12407):30
NOR gate, (12406):1, 5–6, 29
NOR RS flip-flop, clocked, (12406):13, 15
NOT gate, (12406):1, 5, 29
ns. *See* Nanosecond (ns)
Nuclear detectors, (12409):14
Nucleus, (12409):1, 2, 49
Nuisance alarms, (12407):24
Numbers
   binary, (12406):7–9, 19–20, 22, 29
   binary coded decimal (BCD), (12406):11, 16, 18, 19, 29
   conversions
      binary-decimal/decimal-binary, (12406):7–8
      hexadecimal-binary/binary-hexadecimal, (12406):9
   digital, (12406):19–20, 29
   hexadecimal, (12406):19, 20, 29
Numerical data I/O, PLCs, (12406):17

O

Omega®, (12410):5
Open-cup flash point analysis, (12409):21
Open loop tuning
   basis of, (12405):7
   procedure overview, (12405):11–13
   reaction rate method, (12405):15–16
   time constant method, (12405):13–15
Open vessel installations, level measurements, (12402):13–15
Operating range limit, (12402):2
Operator workstations (OWs), DCS. *See also* Human-machine interfaces (HMIs)
   components, (12407):20
   function, (12407):2, 13
   graphics
      alarming system, (12407):23–24
      analog control, (12407):21–22
      detailed point displays, (12407):24–25
      discrete control, (12407):21–22
      historical trends, (12407):25
      informational screens, (12407):23
      minimizing, (12407):20
      navigation, (12407):23
      process, (12407):20–21
      trends, (12407):25
   location, (12407):2
   types of, (12407):2, 3
   updating, (12407):20
Optical microscopy, (12409):37–38
Optocouplers, (12406):3
OR gate, (12406):1, 4, 29
ORP. *See* Oxidation-reduction potential (ORP)
Oscillating process, (12410):19, 24, 31
Oscillating process, loop troubleshooting
   checking instruments in the field, (12410):25–26
   check the transmitter, (12410):26
   gather information, (12410):24–25
   identify possible causes, (12410):25
   locate the problem, (12410):26
   using a flowchart, (12410):26–27
   verify the problem, (12410):24
Oscillations, (12405):1, 2, 24
OSHA 1910.106(a), flash point defined, (12409):22
OSI Model networks, (12407):15
Output registers, (12406):15
Overshoot, (12405):1, 24
Over-the-hump time, (12405):7, 21, 24
OWs. *See* Operator workstations (OWs), DCS
Oxidation, (12409):27
Oxidation-reduction potential (ORP), (12409):21, 27–29, 49
Oxidation-reduction potential (ORP) probes, (12409):27–29

Oxidation-reduction potential (ORP) reference electrode, (12409):27–28
Oxygen, measuring, (12409):31–33

P

P. *See* Poise (P); Proportional (P) control
P&ID. *See* Piping and instrumentation diagram (P&ID)
Paramagnetic oxygen analyzer, (12409):32
Particulates, measuring, (12409):37–38
Partitioning, (12407):6, 8, 36
Parts per million (ppm), (12409):1, 8, 49
Passwords, (12407):32
Patch, (12407):6, 12, 36
Pensky-Martens flash point analysis system, (12409):21–22
Periodic Table of the Elements, (12409):1, 2–6, 50
Permissive, (12407):6, 23, 36
Permissive/trip graphics, (12407):23
Personal computers
   for DCS, (12407):2, 3
   operating systems, (12407):8–9
   PLCs compared, (12406):14, 15
Personnel protection indicators, (12409):34
pH
   defined, (12409):1, 9, 50
   measuring
      analyzer/controller, (12409):25–26
      electrodes, (12409):23–25
      indicator scale, (12409):23
      instruments, (12409):23
      probes, (12409):23–25
      reference electrode, (12409):24
pH scale, (12409):9, 10, 23
PI. *See* Proportional+integral (PI) control
PID. *See* Proportional+integral+derivative (PID) control; Proportional+integral+derivative (PID) loop equation
Piping and instrumentation diagram (P&ID), (12407):20–21, (12410):16
PLCs. *See* Programmable logic controllers (PLCs)
Pneumatic calibration, (12402):8–12
Pneumatic calibration equipment
   differential pressure transmitters, (12402):9–12
   multifunction calibrators, (12402):8–9
Pneumatic location and tubing plan, (12410):16
Pneumatic valve positioners, (12402):25–27
POH scale, (12409):9, 10
Poise (P), (12409):12, 14, 50
Position encoders, (12406):18
Positioners
   calibrating, (12402):27–28
   direct-acting, (12402):27–28
   electro-pneumatic, (12402):27
   function of, (12402):25
   pneumatic, (12402):25–26
   smart, (12402):28–29
Positive-going, (12406):1, 9, 29
Power distribution panel schedule drawing, (12410):16
Power supply
   digital technology, (12406):1–2
   failures, DCS, (12407):30
   PLCs, (12406):13–14
ppm. *See* Parts per million (ppm)
Problem-resolving mechanisms, PLCs vs. personal computers, (12406):14
Process analyzer, (12409):12
Process control, importance of, (12407):1
Process gain equation, (12405):8–9
Process graphics, DCS, (12407):20–21

Processor module, PLCs
   components
      key switch, (12406):14
      LCD displays/lights, (12406):14, 15
      memory, (12406):14–15
      microprocessor, (12406):14
      real-time clock, (12406):13, 14, 34
      USB/serial port, (12406):14, 19
   scans, (12406):13, 15, 34
PROFIBUS® (Process Bus), (12406):19, (12407):16
PROFIBUS® DP, (12407):16
PROFIBUS® PA, (12407):16
Program control instructions, LD programming, (12406):26–27
Programmable controllers, (12406):1
Programmable logic controllers (PLCs)
   applications, (12406):14
   architecture, (12406):1
   binary codes
      ASCII, (12406):9–10
      BCD (binary coded decimal), (12406):11
      gray code, (12406):11
   compared with
      DCS controllers, (12407):4, 7–8, 30
      hardwired systems, (12406):2–5
      personal computers, (12406):14, 15
   defined, (12406):1
   dynamic system checks, (12406):30
   examples
      large-capacity, heavy-duty, (12406):2
      medium-capacity, (12406):2
      micro, (12406):1
   hardware
      by size of PLC, (12406):2
      components, (12406):2
      equipment racks, (12406):13
      network connections, (12406):14, 19
      power supplies and grounds, (12406):13–14
   historically, (12407):4
   installation, (12406):28–29
   I/O modules
      address schemes, (12406):19–20
      analog, (12406):17
      communication, (12406):19
      data table function, (12406):15
      discrete, (12406):16
      encoder/counter, (12406):18
      LED status indicators, (12406):16
      numerical data interface, (12406):17
      smart transmitter interface, (12406):18
      specialized, (12406):17–19
      stepper motor control, (12406):18
      switching technologies, (12406):16
      temperature monitoring/control, (12406):18
      wiring, (12406):29–30
   networks, (12407):4
   number systems
      binary, (12406):7–8, 11
      hexadecimal, (12406):9
   optocouplers, (12406):3
   problem-resolving mechanisms, (12406):14
   programming computers, attaching, (12406):19
   purchasing considerations, (12406):14
   reconfiguring, (12406):4–5
Programmable logic controllers (PLCs), processor module
   components
      Ethernet ports, (12406):19
      key switch, (12406):14

LCD displays/lights, (12406):14
  memory, (12406):14–15
  microprocessor, (12406):14
  real-time clock, (12406):13, 14, 34
  USB/serial port, (12406):14, 19
 function, (12406):34
 scans, (12406):13, 15, 34
Programmable logic controllers (PLCs), programming
  guidelines, (12406):28
 languages
  development tools, (12406):22
  function block diagram (FBD), (12406):21
  instruction list (IL), (12406):22
  ladder diagram (LD), (12406):21–27
  sequential function chart (SFC), (12406):21–22
  structured text (ST), (12406):22
Programming. *See* Ladder diagram (LD) programming;
  Programmable logic controllers (PLCs), programming;
  Software
 sequential function chart (SFC) programming, (12406):21–22
 testing during proving, (12410):11
Propagation delay, (12406):1, 8, 29
Proportional (P) control, (12405):2
Proportional+integral (PI) control, (12405):2
Proportional+integral+derivative (PID) control, (12405):2
Proportional+integral+derivative (PID) loop equation, (12405):11
Proportional band equation, (12405):9–10
Protocol, defined, (12407):1, 2, 36. *See also* Communication protocol
Protons, (12409):1, 2, 50
Proving
 alarms, testing, (12410):11
 asset management systems, loop integration, (12410):14
 calibrating a loop
  analog instrument loops, (12410):12
  loop tuning vs., (12410):11
  process, (12410):12
  smart instrument loops, (12410):12–14
 defined, (12410):1, 31
 interlock testing, (12410):11
 programming, testing the, (12410):11
 purpose, (12410):9
 simulation in, reasons for, (12410):9
 test equipment, (12410):9–11

R

RAID. *See* Redundant Array of Independent Disks (RAID) array
Range, (12402):1, 2, 33
Range changes, (12402):23
Rapid flash point analysis system, (12409):21–22
Ratio turbidity analyzer, (12409):19
Reaction rate method, (12405):7, 15–16, 24
Real time, (12407):6, 8, 37
Real-time clock, (12406):13, 14, 34
Reciprocal, (12405):1, 2, 24
Redundancy, DCS controllers, (12407):9
Redundant Array of Independent Disks (RAID) array, (12407):6, 11, 37
Reflection turbidity analyzer, (12409):18–19
Reformed, (12407):27, 30, 37
Register, (12406):11, 17, 29
Relay instructions, LD programming, (12406):22–23
Repeats per minute (rpm), (12405):7, 10, 24
Reset-set (RS) flip-flop, (12406):11–12
 clocked, (12406):13, 15

Resistance temperature detectors (RTDs), (12402):18
Response, (12405):1
RET. *See* Return (RET) output instruction
Retentive time on (RTO) output instruction, (12406):24
Return (RET) output instruction, (12406):27
Rosemount, Inc., (12402):20
Rosemount™, (12406):19
Rosemount™ 1151, (12402):16–17
Rosemount™ 1151DP, (12402):13
Rosemount™ 3051, (12410):12–13
Rotating disc viscometer, (12409):15–16
Router, (12407):6, 18–19, 37
rpm. *See* Repeats per minute (rpm)
RS. *See* Reset-set (RS) flip-flop
RS-422 networks, (12407):15
RS-485 networks, (12407):15
RTDs. *See* Resistance temperature detectors (RTDs)
RTO. *See* Retentive time on (RTO) output instruction
Ruggedized, (12406):1, 34

S

Safety instrumented systems (SIS), (12407):4
Salt, (12409):1, 10, 50
SAMA. *See* Scientific Apparatus Manufacturers Association (SAMA)
Satellite networks, (12407):15
Saturate, (12405):7, 10, 24
SBR. *See* Subroutine (SBR) output instruction
SCADA. *See* Supervisory Control and Data Acquisition (SCADA)
Scans, (12406):13, 15, 34
Schematic symbols, (12406):2, 3
Schneider Electric™ Triconex® system, (12407):4
Scientific Apparatus Manufacturers Association (SAMA), (12407):20
Selling, (12410):1, 16, 31
Semiconductor sensors, (12409):34–35
Sensors
 DCS, (12407):1
 factory characterization, (12402):20, 33
Sensor trim, (12410):9, 13, 31
Sequential function chart (SFC) programming, (12406):21–22
Serial-in, parallel-out (SIPO) registers, (12406):17
Serial network, (12406):13, 19, 34
Serial ports, (12406):14, 19
Servers
 DCS
  database, (12407):9, 12
  enterprise-level, (12407):13
  file server, (12407):9
  function, (12407):2, 9, 11
  historian, (12407):12–13, 25
  operating systems, (12407):13
  software, (12407):11–12
  web server, (12407):9
 defined, (12407):1, 3, 37
 security, (12407):32
SFC. *See* Sequential function chart (SFC) programming
Shields, PLCs, (12406):14, 17
Shift registers, (12406):17
Siemens, (12407):16
Siemens Series 50, (12402):10
Simulator, (12406):21, 22, 34
Single point of failure, (12407):9
Sinking, (12406):16
SIPO. *See* Serial-in, parallel-out (SIPO) registers
SIS. *See* Safety instrumented systems (SIS)
Smartphones, mobile OWs, (12407):2, 13

INSTRUMENTATION LEVEL FOUR INDEX   I.9

Smart technology
  asset management, (12410):14
  field devices, DCS, (12407):7
  HMIs, (12406):5
  mobile OWs, (12407):2, 13
  positioners, (12402):28–29
  transmitters
    calibrating HART® devices, (12402):21–24, (12410):12–13
    co-existing with existing technology, (12407):16
    communication protocols, (12402):20
    focus on speed and reliability, (12407):16
    I/O modules, PLCs, (12406):18
    loop checking, (12410):3
Software
  anti-malware, (12407):32
  malware, (12407):27, 32, 36
  operating systems, DCS, (12407):3, 8, 13
  pogramming languages
    development tools, (12406):22
    function block diagram (FBD), (12406):21
    instruction list (IL), (12406):22
    ladder diagram (LD), (12406):21–27
    sequential function chart (SFC), (12406):21–22
    structured text (ST), (12406):22
  programming languages
    development tools, (12406):22
    function block diagram (FBD), (12406):21
    instruction list (IL), (12406):22
    ladder diagram (LD), (12406):21–27
    sequential function chart (SFC), (12406):21–22
    structured text (ST), (12406):22
Software servers, DCS, (12407):11–12
Solute, (12409):1, 8, 50
Solvent, (12409):1, 8, 50
Sourcing, (12406):16
Span
  defined, (12402):1, 2–3, 33
Spanning
  defined, (12402):1, 3, 33
  HART® devices, (12402):23
  Rosemount™ 1151, (12402):17
  temperature transmitters, (12402):18
  zero and span calibration, (12410):9, 12, 31
Span standard, (12409):27
Specification sheets, (12410):16
Specific gravity
  defined, (12409):12, 50
  equation, (12409):12–13
  measuring
    air bubble method, (12409):13
    densitometer, (12409):13
    displacement method, (12409):13–14
    hydrometer, (12409):13
    nuclear detectors, (12409):14
Spectrometry, IR, (12409):42
Spectrophotometer, UV, (12409):42
ST. See Structured text (ST) programming
Stability, (12405):1, 24
Steady state, (12405):1, 4, 24
Stepper motor control, (12406):18
Stiction, (12405):7, 21, 24
Stroke, (12410):16, 17, 31
Structured text (ST) programming, (12406):22
SUB instruction, (12406):25
Subroutine (SBR) output instruction, (12406):27
Sulfur dioxide gas, (12409):34

Supervisory Control and Data Acquisition (SCADA), (12407):1, 3–4, 37
Suppressed zero, (12402):8, 13, 17, 33
Symbols
  for analog control, (12407):9
  digital logic, (12407):9
  for elements, (12409):2
  on P&ID drawings, (12407):20–21

T

Tablets, mobile OWs, (12407):2, 13
TAG flash point analysis system, (12409):21–22
TCP/IP. See Transmission Control Protocol/Internet Protocol (TCP/IP)
Technologies, digital
  logical levels, (12406):1–2
  logic families, (12406):1
  power supplies, (12406):1–2
Temperature monitoring/control, (12406):18
Temperature transmitters
  electro-mechanical, (12402):12
  with analog outputs, (12402):18
  with resistance temperature detectors, (12402):18
  with thermocouples, (12402):18
THC. See Total hydrocarbon content (THC)
Thermistor, (12409):27
Thermocouples, (12402):2, 18, (12406):18
Thermography, (12409):40
Third-party solutions, (12407):1, 2, 37
Three-point calibration, (12402):1, 3, 33
Three-point calibration process
  generally, (12402):5
  temperature transmitters, (12402):18
Thumbwheels, (12406):16
Time constant, (12405):1, 4, 24
Time constant equation, (12405):8
Time constant method, (12405):7, 13, 24
Timer done (DN) output, (12406):23–25
Timer instructions, LD programming, (12406):23–25
Timer off delay (TOF) output instruction, (12406):24
Timer on delay (TON) output instruction, (12406):23–24
Timing diagram, (12406):11, 13, 29
TOF. See Timer off delay (TOF) output instruction
TON. See Timer on delay (TON) output instruction
Tone tester, (12410):6
Total hydrocarbon content (THC), (12409):31, 35, 50
Transducers, (12402):25
Transfer function, (12402):20, 21, 33
Transmission Control Protocol/Internet Protocol (TCP/IP), (12407):1, 3, 15, 37
Transmission turbidity analyzer, (12409):17–18
Transportation Department (U.S.), transported substances flash point, (12409):21
Trend displays, (12407):25–26
Triconex® system, (12407):4
Truth table, (12406):1, 2, 29
Tuning
  defined, (12405):1, 24, (12410):11
  graphical models, (12405):3–4
  process types
    integrating, (12405):5–6
    non-integrating, (12405):4–5
  proportional control, (12405):2–3
  reasons for, (12405):1
  stability, factors that influence, (12405):1–2
  terms and definitions, (12405):3–4
Tuning equations
  complete response, (12405):8

derivative time, (12405):10–11
energy balance, (12405):7–8
integral time, (12405):10
PID loop, (12405):11
process gain, (12405):8–9
proportional band, (12405):9–10
time constant, (12405):8
Tuning methods
  closed loop
    controller, function of the, (12405):7
    dampened oscillation method, (12405):18–19
    ultimate period method, (12405):16–18
  open loop
    basis of, (12405):7
    procedure overview, (12405):11–13
    reaction rate method, (12405):15–16
    time constant method, (12405):13–15
  visual loop
    apparent instability, (12405):19–20
    incremental changes, (12405):19
    sluggish response, (12405):19–21
Turbidity
  defined, (12409):12, 50
  factors of, (12409):16
  measuring, (12409):17–19

U

u. *See* Unified atomic mass unit (u)
Ultimate period method, (12405):7, 16–18, 24
Ultraviolet (UV) radiation, measuring, (12409):42–44
Unified atomic mass unit (u), (12409):1, 2, 50
Universal serial bus (USB), (12406):1, 34
Universal serial bus (USB) port, (12406):14, 19
Unix®-based operating systems, (12407):13
Up/down counters, (12406):20
USB. *See* Universal serial bus (USB)
UV. *See* Ultraviolet (UV) radiation, measuring
UV absorption analyzers, (12409):44

V

Valence shell, (12409):1, 4, 50
Valve controllers, smart, (12402):28–29
Viscometers, (12409):15–16
Viscosity
  defined, (12409):12, 14, 50
  measuring, (12409):15–17
  oil, (12409):14
  temperature effects, (12409):14–15
Visual loop tuning
  apparent instability, (12405):19–20
  incremental changes, (12405):19
  sluggish response, (12405):19–21
Volatile memory, (12406):13, 14, 34
Volcanoes, (12409):34
Vulnerability, (12407):27, 32, 37

W

Wallace and Tiernan® Wally Box®, (12402):8–9
Web server, (12407):9
Wet leg, (12402):8, 15, 33
Whitelist, (12407):27, 32, 37
Wi-Fi
  corporate, (12407):18
  defined, (12406):13, 34
  function, building networks, (12407):2
  PLC connections, (12406):19
  protocol layers, (12407):15
  security, (12407):19, 32

WIKA®, (12410):6
WIKA® Instrument, LP, (12402):8–9
WirelessHART®, (12407):32
Wireless networks. *See* Wi-Fi
Words, (12406):11, 15, 16–17, 29
Workstations, DCS
  engineering (EWs), (12407):13
  function, (12407):13
  operating systems, (12407):13
  operator (OWs)
    components, (12407):20
    function, (12407):2, 13
    location, (12407):2
    pre-DCS, (12407):19–20
    types of, (12407):2, 3
    updating, (12407):20
  operator (OWs), graphics
    alarming system, (12407):23–24
    analog control, (12407):21–22
    detailed point displays, (12407):24–25
    discrete control, (12407):21–22
    historical trends, (12407):25
    informational screens, (12407):23
    minimizing, (12407):20
    navigation, (12407):23
    process, (12407):20–21
    trends, (12407):25
  smart technology, (12407):2, 13
Workstation security, (12407):32

X

XNOR gate, (12406):1, 6, 29
XOR gate, (12406):1, 4–5, 29

Y

Yokogawa, (12407):16

Z

ZCL. *See* Zone control (ZCL) output instruction
Zero, suppressed, (12402):8, 13, 17, 33
Zero and span calibration, (12410):9, 12, 31
Zeroing
  defined, (12402):1, 3, 33
  HART® devices, (12402):23
  Rosemount™ 1151, (12402):17
  temperature transmitters, (12402):18
Zero standard measurement, (12409):27
Zirconium oxide oxygen analyzer, (12409):32
Zone control (ZCL) output instruction, (12406):27